Handbook of Energy

Volume I

Handbook of Energy
Volume I

Edited by **George Thomson**

NY RESEARCH PRESS
New York

Published by NY Research Press,
23 West, 55th Street, Suite 816,
New York, NY 10019, USA
www.nyresearchpress.com

Handbook of Energy: Volume I
Edited by George Thomson

International Standard Book Number: 978-1-63238-242-9 (Hardback)

Printed in China.

Contents

Preface

Energy, at its simplest, can be defined as the ability or the capacity of a system to do work. It is an elementary entity that is linked to almost all living and non-living organisms in some form or the other. In moving or changing and growing, each and every one of the systems that we encounter everyday is absorbed in doing work, and uses energy to do such work. Every living organism does work, and needs energy intake in some form or the other. Man also creates machines that does work for him, and that extract energy from fuels. Energy is thus one of the building blocks of a civilized society. There are various types of energy, each with its own set of characteristics and behaviours as well as general and specific applications in the real world such as Thermal energy, Mechanical energy, solar energy, Thermal energy, Chemical energy, Electrical energy, Electromagnetic energy, and Nuclear energy. As of now Humanity needs potent, sustainable and dependable energy sources. It is already that time when civilization needs to research and develop alternative energy sources as our current rate of fossil fuel usage is soon going to lead to an energy crisis this century. In order to survive and move beyond this energy crisis many researchers in energy industry are concocting new ways to extract energy from renewable sources. Awareness and government pressure is growing. Such awareness is also forcing researchers to look beyond fossil fuels. Thus there is a need for skilled researchers in this rapidly developing field.

This book is an attempt to compile and collate all available research on energy under one umbrella. I am grateful to those who put their hard work, effort and expertise into these researches as well as those who were supportive in this endeavour.

Editor

1

Analyzing the Impact of Ambient Temperature Indicators on Transformer Life in Different Regions of Chinese Mainland

Cui-fen Bai,[1] Wen-Sheng Gao,[1] and Tong Liu[2]

[1] Department of Electrical Engineering, Tsinghua University, Beijing 100084, China
[2] Electric Power Research Institute, CSG, Guangzhou, Guangdong 510080, China

Correspondence should be addressed to Cui-fen Bai; cuifenbai@gmail.com

Academic Editors: A. Klimenko and A. Piacentino

Regression analysis is applied to quantitatively analyze the impact of different ambient temperature characteristics on the transformer life at different locations of Chinese mainland. 200 typical locations in Chinese mainland are selected for the study. They are specially divided into six regions so that the subsequent analysis can be done in a regional context. For each region, the local historical ambient temperature and load data are provided as inputs variables of the life consumption model in IEEE Std. C57.91-1995 to estimate the transformer life at every location. Five ambient temperature indicators related to the transformer life are involved into the partial least squares regression to describe their impact on the transformer life. According to a contribution measurement criterion of partial least squares regression, three indicators are conclusively found to be the most important factors influencing the transformer life, and an explicit expression is provided to describe the relationship between the indicators and the transformer life for every region. The analysis result is applicable to the area where the temperature characteristics are similar to Chinese mainland, and the expressions obtained can be applied to the other locations that are not included in this paper if these three indicators are known.

1. Introduction

The life of a transformer is defined as the life of the paper insulation [1]. The paper insulation deteriorates with several factors, primarily including temperature, moisture content, and oxygen content. In all these factors, temperature has a major impact because the contribution of moisture or oxygen to the insulation deterioration can be minimized with modern oil preservation systems [2].

The temperature of a transformer, often considered as hot spot temperature (HST), is primarily controlled by ambient temperature and load, where ambient temperature is the dominated factor. On one hand, some types of load are directly affected by ambient temperature, such as the cooling load in summer: the higher the ambient temperature is, the greater the load will be. On the other hand, the load capability of a transformer is generally governed by ambient temperature. Transformers have to operate under the prescribed limit of HST, and ambient temperature is an uncontrollable factor in the influential factors of HST (ambient temperature and load); therefore, if the ambient temperature is high, the load capability of a transformer is always low, and vice versa. From these two aspects, it can be seen that ambient temperature is an important factor in estimating the transformer life.

There are some literatures that studied the impact of an increase of ambient temperature on transformer life. Reference [3] used a standardized engineering approach according to IEEE Standard C57.91 [4] to preliminarily assess the impact of increased ambient temperature on transformer loss of life at five locations in the USA. For an assumed 4°C rise in ambient temperature during 1900–2100, the predicted loss of life in the interval from 1990 to 2045 rises approximately 32% at a relatively warm location and 8% at a relatively cold location. Reference [5] applied an improved model of top oil temperature rise differing from the IEEE C57.91 model to assess the impact of ambient temperature change on transformer life, based on seven transformers and four loading test beds specified in IEEE C57.115. With the assumption of temperature rise of 3.5°C in 100 years, a probabilistic model for ambient temperature rise was performed to conclude that a reduction in the life of a transformer was about 3–6 years for the case studied, and there was a marked difference in the mean life of a transformer for several different loading conditions.

These literatures focus on the impact of ambient temperature rise on transformer life, and the impact of different temperature characteristics on transformer life at different locations is not included. Actually, the ambient temperature characteristics of one location have a great impact on the local transformer life. For example, the transformer life at a warmer area is shorter than that at a colder area. Furthermore, there are many indicators portraying temperature characteristics in meteorology, and the key issue related to transformer life prediction and power system operation is which indicators are most important for the transformer life.

This paper focuses on quantitatively analyzing the impact of different ambient temperature characteristics on transformer life at different locations of Chinese mainland and attempts to find the most important temperature indicators for transformer life estimation based on regression analysis. Chinese mainland is selected for study due to its vast territory and diverse climates. In practical situations, difference in latitude, longitude, or altitude results in complex temperature characteristics; different temperature characteristics cause different values of transformer life. The life consumption model in IEEE Std. C57.91-1995 [2] is employed to estimate different values of transformer life at 200 typical locations of Chinese mainland. These locations are specially divided into six regions. For each region, the local historical temperature and load data are provided as inputs variables of the life consumption model to estimate the transformer life at every location. Then, the partial least squares regression (PLSR) method is applied to construct the regression between the transformer life and five temperature indicators. Finally, based on a criterion to measure the contribution of temperature indicators in PLSR, three indicators are considered the most important factors and involved in the regression analysis for every region. The relationship between the transformer life and these three temperature indicators is formulated with a simple and acceptable equation for every region, and the equations can be used for life estimation at the locations that are not included in this paper.

2. Transformer Life Estimation at Different Locations

This section presents the calculation process of the transformer life at different locations of Chinese mainland based on the life consumption model in IEEE Std. C57.91-1995.

2.1. Introductions to IEEE Life Consumption Model. The extensively employed life consumption model in IEEE Std. C57.91-1995 [2] is applied in this paper. In the model, the transformer life is established as a function of hot spot temperature (noted as θ_H) which is computed by ambient temperature and load. Then, θ_H is provided as an input variable to acquire the thermal aging acceleration factor F_{AA} [2]:

$$F_{AA} = \exp\left(\frac{15000}{383} - \frac{15000}{\theta_H + 273}\right) \quad (1)$$

when θ_H is at the reference temperature 110°C F_{AA} is equal to 1. Based on this definition, the loss of life in a given time period is proposed as [2]

$$F_{EQA} = \frac{\sum_{n=1}^{N} F_{AA_n} \Delta t_n}{\sum_{n=1}^{N} \Delta t_n}, \quad (2)$$

$$\%\ \text{Loss of life} = \frac{F_{EQA} \times t \times 100}{\text{Normal insulation life}}, \quad (3)$$

where F_{EQA} is the equivalent aging factor in the total time period, F_{AA_n} is the aging acceleration factor during the time interval Δt_n, N is the total number of time intervals, and "Normal insulation life" is the proposed value of transformer life at the reference temperature 110°C.

According to the model, ambient temperature and load are two important root causes to the life consumption. The configuration of these two variables will be proposed subsequently to calculate the transformer life. In addition, to simplify the calculation process of the transformer life, ambient temperature and load are assumed to be annually cyclic so that only a sample annual temperature curve and a sample annual load curve on an hourly basis are needed to calculate the transformer life. With this assumption, the life value obtained for one location may not be very accurate, but it has little effect on the analysis result because the aim of this paper is to compare the life values at different locations, and the hypothesis and predigestion in calculation process for all locations are the same.

2.2. Hourly Ambient Temperature Curve. Ambient temperature is an input variable of the IEEE model. We obtain historical temperature records from the total of 200 meteorological stations for this research [6]. These locations have a variety of temperature and load characteristics so that they can provide a relatively comprehensive description of different climate and load types in Chinese mainland. Figure 1 illustrates the 200 locations on the map, and the important cities (capital cities, municipalities directly under the central government, or capitals of autonomous region) are labeled specially. The locations cover the latitude range approximately from 19°N to 51°N, longitude range from 76°E to 132°E, and historical temperature records are available from 1951 to 2011 for most meteorological stations. The temperature data includes daily maximum temperature ($T_{\max,d,y}$, $d = 1, 2, \ldots, 365$, $y = 1, 2, \ldots, 61$) and minimum temperature ($T_{\min,d,y}$).

To obtain a sample annual temperature curve on an hourly basis for every location, mean daily maximum temperature ($\overline{T_{\max,d}}$) and mean daily minimum temperature ($\overline{T_{\min,d}}$) are calculated by averaging the data set in 61 years, according to

$$\overline{T_{\max,d}} = \frac{1}{61}\sum_{y=1}^{61} T_{\max,d,y} \quad (d = 1, 2, \ldots, 365),$$
$$\overline{T_{\min,d}} = \frac{1}{61}\sum_{y=1}^{61} T_{\min,d,y} \quad (d = 1, 2, \ldots, 365). \quad (4)$$

Then mean daily maximum temperature and mean daily minimum temperature in 365 days of the sample year are

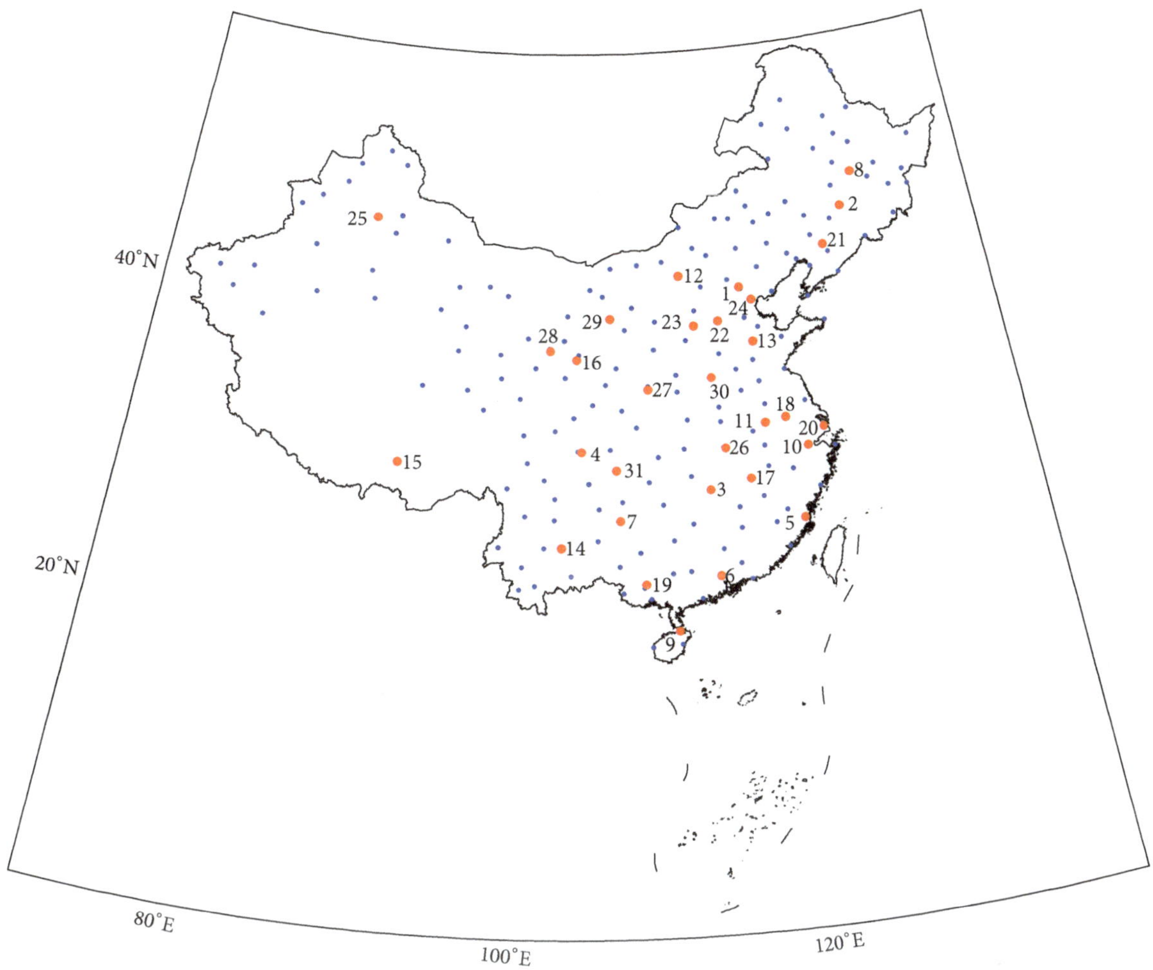

FIGURE 1: 200 meteorological stations in Chinese mainland. The stations in important cities are marked in red.

acquired. They are used to obtain the hourly temperature curve, according to [7]

$$T_{h,d} = \overline{T_{\max,d}} - \alpha_h \left(\overline{T_{\max,d}} - \overline{T_{\min,d}}\right) \tag{5}$$
$$(h = 1, 2, \ldots, 24, \; d = 1, 2, \ldots, 365),$$

where $T_{h,d}$ is the temperature at hour h in the day d and α_h is the temperature ratio shown in Figure 2 [7]. Hence, the hourly temperature curves in the sample year for the 200 locations are formed.

2.3. Hourly Load Curve. Load is the other input variable of the IEEE model. We set the hourly load curve for every location according to the statistics data [8, 9]. The load repeats both daily and annually, described by hourly and monthly load variation curves. For hourly load variation curves, 200 locations are hypothesized to have the same shape as Figure 3 because the actual difference is not significant.

However, the monthly load variation curves are significantly different from location to location due to many influential factors. Based on the load statistics [8, 9], locations are divided into six regions as Figure 4 illustrates, including

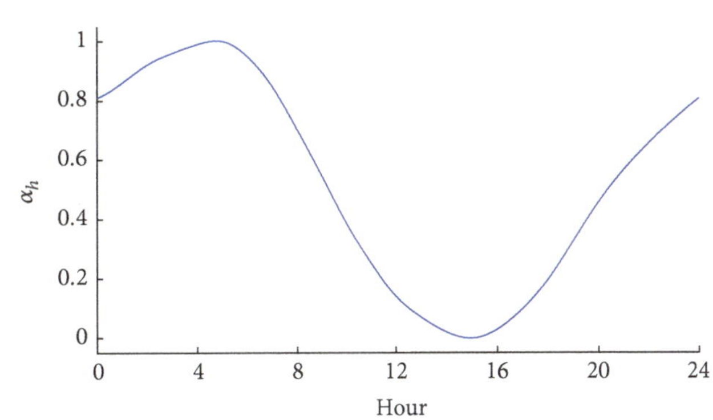

FIGURE 2: Temperature ratio, a parameter used to obtain the hourly temperature [7].

Dongbei, Huabei, Huadong, Huazhong, Xibei, and Nanfang, and the number of locations included in every region is given in Table 1. The locations in one region have similar load characteristics, and the difference between two regions becomes large. The regional average monthly load variation curves are given in Figure 5 according to [8, 9]. It shows

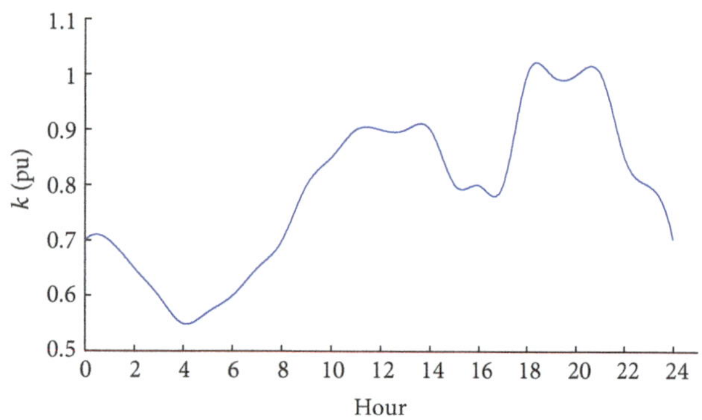

FIGURE 3: Hourly load variation in a day.

FIGURE 4: 200 locations are divided into six regions.

that every region generally has a summer load peak and a winter load peak in one year, but some have larger winter load (Figure 5(a)), and some have larger summer load (Figure 5(b)). Accordingly, the monthly load variation curves can be roughly classified into two categories.

The curves in Figures 3 and 5 are used conjunctionally to form the hourly load curve in the sample year, according to

$$k_{h,m} = \frac{k_m}{\overline{k_h}} \cdot k_h, \quad h = 1, 2, \ldots, 24; \; m = 1, 2, \ldots, 12, \tag{6}$$

TABLE 1: The number of locations in every region.

Region	The number of locations
Dongbei	35
Huabei	34
Huadong	19
Huazhong	34
Xibei	49
Nanfang	29

TABLE 2: The parameters of the sample transformer.

Parameter	Value
$\Delta\theta_{TO,R}$	45°C
$\Delta\theta_{H,R}$	40°C
R	1.0
τ_{TO}	3.5 h
m_w	0.8
n_o	1.0

where $k_{h,m}$ is the load factor in hour h of the month m, k_m is the average monthly load factor in Figure 5, k_h is the hourly load variation in Figure 3, and $\overline{k_h}$ is the average load factor in Figure 3.

As a result, all locations in one region have the same hourly load curve. Such treatment aims to do the analysis in Section 3.

2.4. Local Transformer Life Estimation. The local hourly ambient temperature and load curves in the sample year are inputted into the IEEE life estimation model. The parameters of a sample transformer used are shown in Table 2, most of which are extracted from [10]. The "Normal insulation life" in (3) is set to be 180000 h according to [2]. The calculation results of the transformer life values at all locations are shown in Figure 6. The range for the value of life is about 10.6–149.3 years. It can be seen that the distribution of transformer life has a certain geographical feature, which will be discussed in details in the next section.

3. Analyzing the Impact of Ambient Temperature on Transformer Life

Section 2.4 obtained the transformer life values at 200 locations and concluded that the transformer life has a geographical feature. In this section, we will discuss the reason for it by analyzing the temperature and load characteristics. Then, the relationship between ambient temperature indicators and transformer life will be quantified by regression analysis.

3.1. The Temperature and Load Characteristics in Chinese Mainland. Figure 6 shows that the distribution of transformer life has a certain geographical feature, which can be inferred from the geographical feature of temperature and load. After the hourly temperature curve in the sample year is got in Section 2.2, many indicators describing the temperature characteristics can be acquired, such as

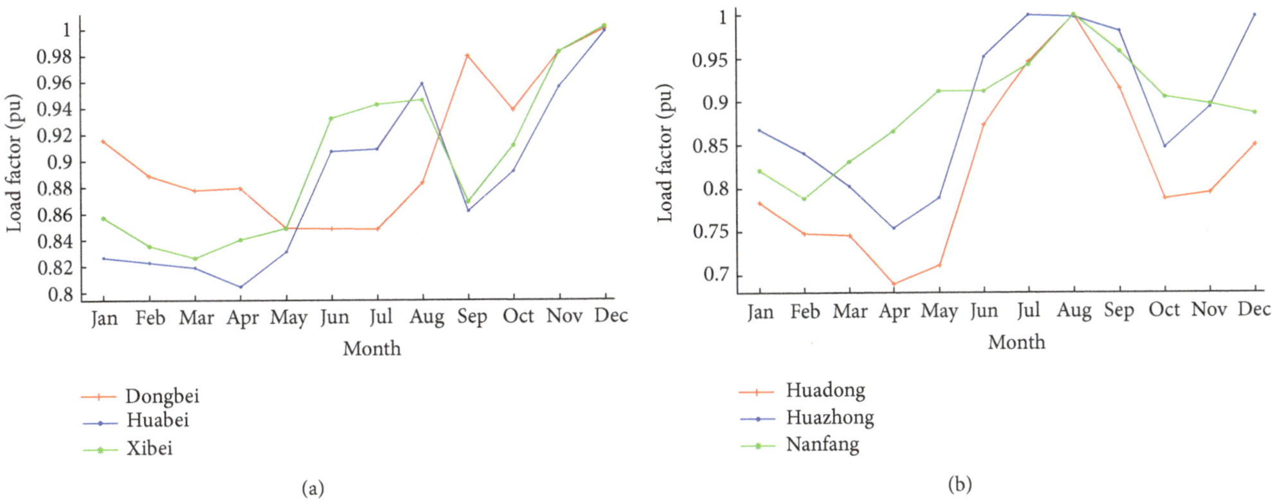

FIGURE 5: Monthly load variation curves of the six regions [8, 9].

annual maximum (minimum) temperature and annual mean temperature (AMT). Figure 7 illustrates the annual mean temperature and annual temperature range (ATR) at the 200 locations to give a general description of the temperature characteristics in Chinese mainland. AMT shows the annual mean temperature level at one location, and ATR is the mean temperature difference between the warmest and coldest months during one year, demonstrating the temperature amplitude variation mostly from summer and winter. ATR can be obtained according to

$$\mathrm{ATR} = \max_{m}\left(\frac{1}{n}\sum_{d=1}^{n}\left(\frac{1}{24}\sum_{h=1}^{24}T_{h,d,m}\right)\right) - \min_{m}\left(\frac{1}{n}\sum_{d=1}^{n}\left(\frac{1}{24}\sum_{h=1}^{24}T_{h,d,m}\right)\right), \tag{7}$$

where $T_{h,d,m}$ is the temperature in hour h of the day d of the month m and n is the number of days in the month m. Figure 7(a) shows that the locations can be approximately divided into north and south, and it is warmer in south than in north on annual mean temperature level basis. Figure 7(b) illustrates that the temperature difference from large to small is in the northern area, central area, and southern area. Generally, the temperature levels in the warmest month are similar all across the country, and the differences of ATR primarily come from the coldest month. Therefore, the northern area has the lowest temperature level in the coldest month, and the climate of the southern area is more moderate correspondingly.

For the load characteristics, it can be found from Figures 4 and 5 that North China has larger winter load, and South China has larger summer load. The load has characteristics similar to the temperature. One reason is that the temperature is a primary influential factor of the load. For the regions in Figure 5(a), the cold weather of winter causes a larger heating load in winter compared to the cooling load in summer. This phenomenon is more pronounced in Dongbei compared to the other two in Figure 5(a) due to the significant larger winter load resulting from its extreme cold weather in winter. For the regions in Figure 5(b), their indistinct seasons and mild temperature in all year result in larger cooling load in summer than the heating load in winter.

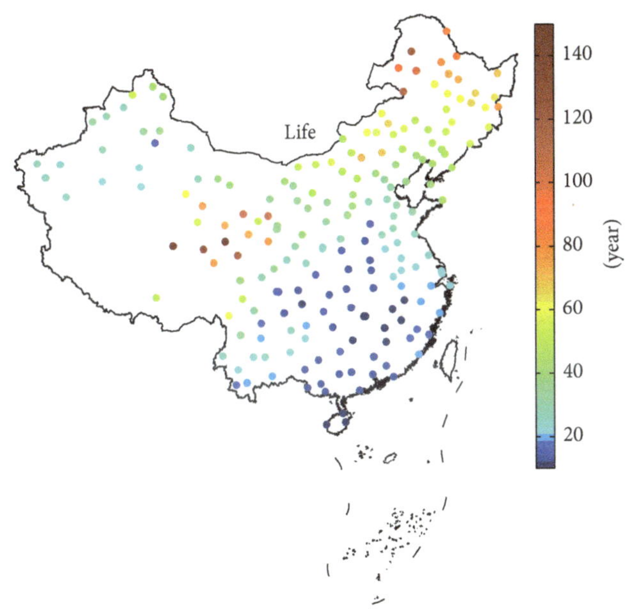

FIGURE 6: Calculation results of the transformer life values at 200 locations.

The discussion above demonstrates the preliminary characteristics of temperature and load distribution in Chinese mainland. Different types of ambient temperature and load will result in different values of life; therefore, the distribution of transformer life also has its characteristics at these locations. In addition, the ambient temperature characteristics largely affect the characteristics of the load. Therefore, the ambient temperature is an extremely important factor influencing the life of a transformer in normal operation.

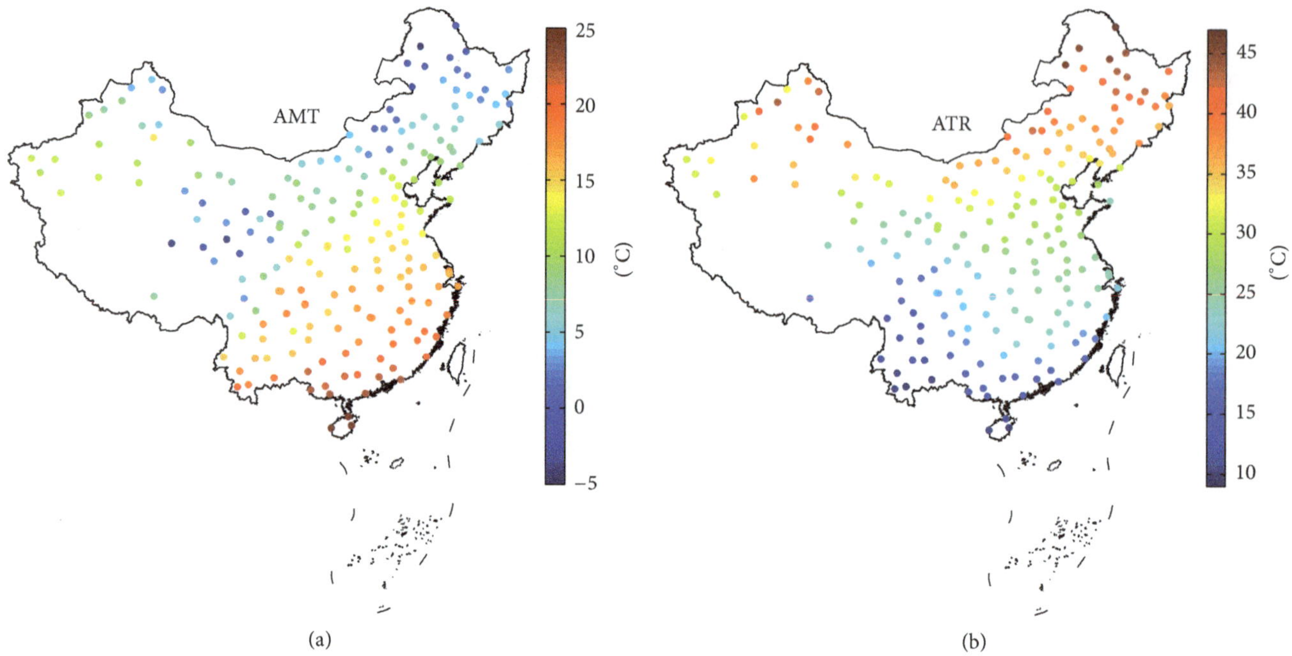

FIGURE 7: (a) Annual mean temperature at 200 locations; (b) annual temperature range at 200 locations.

3.2. The Impact of Ambient Temperature on Transformer Life. From Figures 6 and 7, we can find that there are a rough positive correlation between the transformer life and AMT and a rough negative correlation between it with ATR. It is worth mentioning that because all locations in one region have the same hourly load curve in the sample year as discussed in Section 2.3, the differences in transformer life for the locations in one region are considered primarily from the different temperature characteristics. The treatment is a starting point for the analysis of the relationship between temperature characteristics and transformer life. Consequently, the similarities between the life distribution and the temperature can be found, and their relationship can be investigated and analyzed in a regional context.

3.2.1. The Selection of Regression Variables. The regression technique [11], a widely used relationship assessment method, is performed to assess the relationship between transformer life and ambient temperature and find out the significantly influential temperature indicators. The temperature indicators involved in the regression analysis are picked out firstly. AMT is chosen as a representative temperature indicator which reflects the overall temperature level of one location. Generally, it affects the transformer life strongly because the life is consumed greatly when the ambient temperature is high according to the Arrhenius theory [12] (when the ambient temperature is high, the transformer is generally subjected to a heavier load meanwhile which also increases the life consumption). Based on this theory, another two indicators, the mean temperature of the warmest day (MTWD) and the mean temperature of the warmest month (MTWM), are included. Additionally, diurnal temperature range (DTR) and ATR are involved to take account into the temperature range. DTR is defined as the temperature difference between the maximum and the minimum temperatures during the day, which describes the temperature amplitude variation in one day. These five temperature indicators can provide a relatively comprehensive description of the temperature characteristics of one location, and they are noted as x_1(AMT), x_2(MTWD), x_3(MTWM), x_4(DTR), and x_5(ATR).

The dependent variable involved in the analysis is the transformer life. An indicator-life scatterplot can preliminarily describe the relationship between transformer life and temperature indicators. For example, Figure 8 illustrates the relationship for two indicators in Huazhong, respectively, where it is approximately expressed in an exponential form. The similar conclusion can also be drawn for the other five regions. Actually, this fact is associated with the exponential form of (1). The natural logarithm of the variable "transformer life" is taken before the regression analysis so that the multiple linear regression (MLR) can be conducted conveniently. The variable "transformer life" after the natural logarithm, noted as y, is considered the dependent variable of the regression analysis.

3.2.2. Partial Least Squares Regression. Correlation coefficient analysis is applied to check the correlation of these variables. Table 3 shows the result for Huadong as an example. It can be seen that there are strong correlation between y and every x_i $(i = 1, 2, \ldots, 5)$, which confirms the important influence of temperature indicators on the transformer life.

In addition, there is also a close correlation between every x_i, which identifies the multiple correlations among the independent variables. This multicollinearity problem must be paid attention in order to find the most crucial temperature indicators influencing the transformer life and analyze the relationship between the indicators and the transformer life.

TABLE 3: Correlation coefficient matrix in Huadong.

	y	x_1	x_2	x_3	x_4	x_5
y	1	−0.9633	−0.9897	−0.9919	0.8493	−0.6177
x_1		1	0.9363	0.9351	−0.8738	0.3976
x_2			1	0.9986	−0.8724	0.6866
x_3				1	−0.8617	0.6946
x_4					1	−0.437
x_5						1

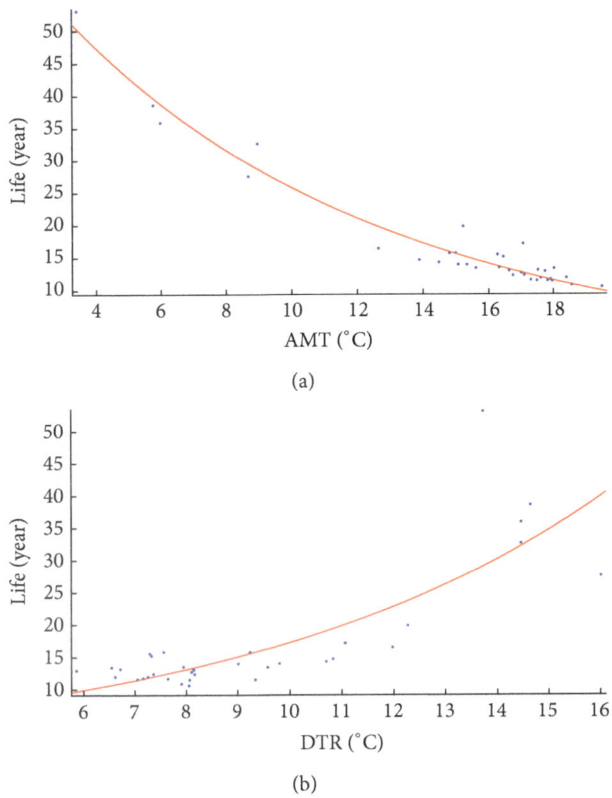

FIGURE 8: The relationship between transformer life and temperature indicators in Huazhong. (a) Life versus annual mean temperature. (b) Life versus diurnal temperature range.

Under this situation, the classical MLR is hard to employ because it will obtain an unstable or impractical solution [13], and PLSR is a good alternative because it is more robust [13–16].

PLSR combines the functions of principal component analysis and MLR, aiming at analyzing a set of dependent variables from a set of independent variables [15]. It constructs a new set of orthogonal factors extracting from the independent variables, often called latent variables or components, to predict the dependent variables. The latent variables capture most of the information in the independent variables and have the best predict power; therefore, the regression can be constructed by using fewer components than the numbers of independent variables, and the dimensionality of the regression problem is reduced [14, 15]. In addition, because finding the most influential variables on the dependent variables interests many analysts, the variable importance in the projection (VIP) scores obtained by the PLSR has been paid an increasing attention as an important measure of each independent variable [17, 18]. VIP is a criterion to measure the contribution of independent variables to the regression in PLSR, and its analytical definition can be found in [17]. VIP will be applied to measure the importance of each temperature indicator.

3.2.3. The Calculation Result and Discussion. Based on the PLSR, one can construct an estimation formula of transformer life:

$$\hat{y} = \beta_0 + \beta_1 x_1 + \beta_2 x_2 + \beta_3 x_3 + \beta_4 x_4 + \beta_5 x_5, \qquad (8)$$

where β_i $(i = 0, 1, \ldots, 5)$ are the regression coefficients. The red markers in Figure 9 demonstrate that with these five indicators, the predicted value of the transformer life from (8) (horizontal axis) and the actual value of life (vertical axis) are in good agreement, which verifies the correctness of the regression result.

Figure 10 illustrates the VIP values of the five variables in these six regions. It can be seen that the variables have different important levels in different regions. According to their relevance in explaining y, $\mathbf{X}$ could be classified as follows: VIP > 1.0 (highly influential), 0.8 < VIP < 1.0 (moderately influential), and VIP < 0.8 (less influential) [18]. It can be found that x_1 and x_3 have great influence on the transformer life in every region. In other words, the annual mean temperature and mean temperature of the warmest month are important for the transformer life. For AMT, it is not surprising because transformer life has a strongly positive correlation with it based on the aforementioned discussion. The importance of MTWM implies that the loss of life in the warmest month accounts for a large percentage in the whole year. Accordingly, controlling the load level in the warmest month becomes especially significant in order to postpone the lifetime of the transformer in warm regions.

It can be also seen from Figure 10 that only parts of temperature indicators are strongly related to the transformer life in every region, some indicators are apparently less relevant than the others, and the VIP values of some are smaller than 0.8. Additionally, there are some similarities that x_1, x_2, and x_3 are highly influential in most regions except for Huadong. The difference between Huadong and the other regions may be attributed to the less sample data in Huadong compared to the other regions as shown in Table 1. This result implies that AMT, MTWD, and MTWM of one location affect the life of

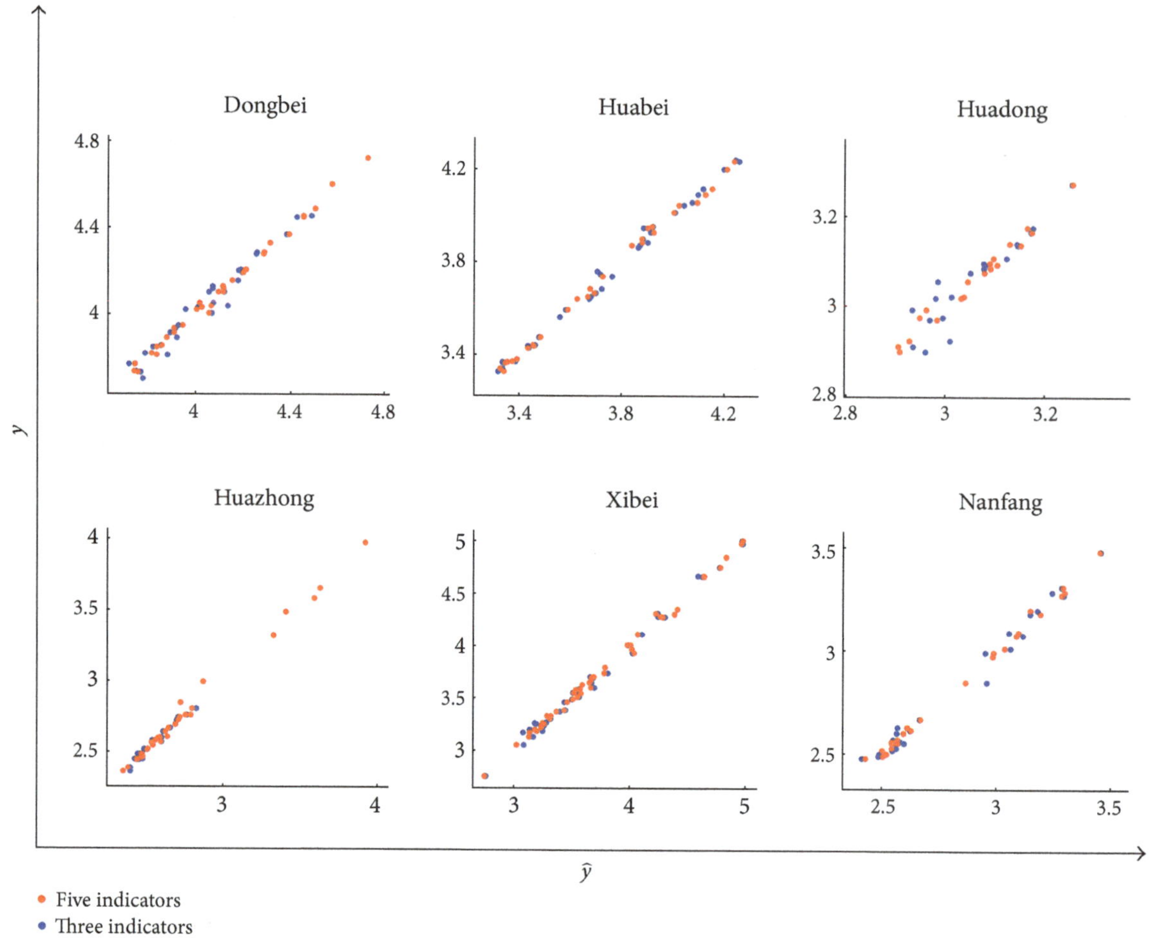

FIGURE 9: The real value of transformer life (y) versus the predicted one ($\hat{y}$).

TABLE 4: The regression coefficients.

Region	$\hat{y} = \beta_0 + \beta_1 x_1 + \beta_2 x_2 + \beta_3 x_3$			
	β_0	β_1	β_2	β_3
Dongbei	6.6077	−0.0238	−0.0535	−0.0517
Huabei	5.8204	−0.0225	−0.0388	−0.0393
Huadong	5.9147	−0.0197	−0.0366	−0.0520
Huazhong	4.6873	−0.0197	−0.0300	−0.0303
Xibei	5.3413	−0.0251	−0.0312	−0.0312
Nanfang	4.9882	−0.0301	−0.0311	−0.0306

transformer strongly. In contrast, the temperature difference, either day or year, is less important in the life estimation. Considering the convenience of practical application, we attempt to construct the PLSR with only x_1, x_2, and x_3. The regression coefficients are listed in Table 4, and Figure 9 also gives the regression result with three indicators. It can be seen from Figure 9 that the difference between the result with five indicators and with three indicators is small, and an acceptable accurate level still can be achieved with only x_1, x_2, and x_3. In this case, the values of VIP of these three temperature indicators in every region are almost equal except that the VIP of x_2 is a little smaller than x_1, x_3 in Huadong.

Conclusively, AMT, MTWD, and MTWM are considered the most important indicators influencing the life of transformers in all regions of Chinese mainland. The conclusion is meaningful due to the large span in latitude and longitude and a wide variety of climatic conditions in Chinese mainland, and it is applicable to the area where the temperature characteristics are similar to Chinese mainland. In addition, a unified and common regression model is constructed with these three variables involved for all these six regions in

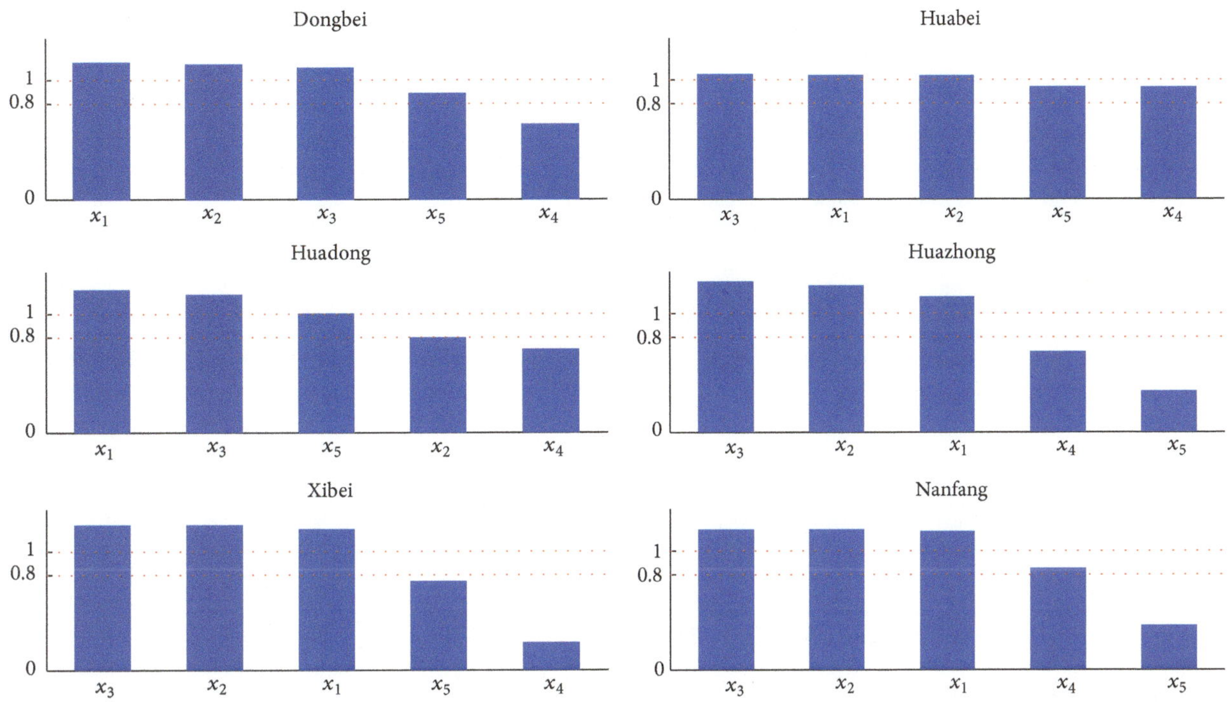

Figure 10: Variable importance in the projection scores of five variables in six regions: x_1(AMT), x_2(MTWD), x_3(MTWM), x_4(DTR), and x_5(ATR).

Chinese mainland. It can be applied to the other locations that are not included in this paper if these three variables are known.

4. Conclusion

Ambient temperature is an important factor in estimating transformer life. 200 locations divided into six regions in Chinese mainland are selected to study the impact of various temperature characteristics on transformer life. According to the historical records of ambient temperature and the statistics data of load curves, preliminary characteristics of temperature and load distribution in Chinese mainland are discussed. Different types of ambient temperature and load will result in different values of life. Based on the IEEE life model, the calculated result shows that the transformer life in the 200 locations ranges from 10.6 to 149.3 years.

Additionally, the assumption of the load configuration causes the difference in transformer life at different locations in one region to be considered primarily from the different temperature characteristics at different locations. Consequently, the effect of temperature indicators can be investigated and analyzed in a regional context.

To quantitatively analyze the impact of different ambient temperature characteristics on transformer life, PLSR is performed with five independent variables (annual mean temperature, mean temperature of the warmest day, mean temperature of the warmest month, diurnal temperature range, and annual temperature range) for every region. These five indicators can provide a comprehensive description of the temperature characteristics and give a well prediction of the transformer life at one location. Furthermore, considering the convenience of practical application, we use VIP as a contribution criterion to assess the importance of the independent variables, and annual mean temperature, mean temperature of the warmest day, and mean temperature of the warmest month are considered the most important indicators influencing the life of transformers for all regions. The conclusion is applicable to the area where the temperature characteristics are similar to Chinese mainland, and the regression model with these three variables can be applied to the other locations that are not included in this paper.

Acknowledgment

This work is supported by the National High-tech R&D Program (863 Program) of China (Grant no. 2011AA05A120).

References

[1] M. Wang, A. J. Vandermaar, and K. D. Srivastava, "Review of condition assessment of power transformers in service," *IEEE Electrical Insulation Magazine*, vol. 18, no. 6, pp. 12–25, 2002.

[2] IEEE Guide for Loading Mineral-Oil-Immersed Transformers IEEE Std C57.91-1995. 1996.

[3] J. W. Stahlhut, G. T. Heydt, and N. J. Selover, "A preliminary assessment of the impact of ambient temperature rise on distribution transformer loss of life," *IEEE Transactions on Power Delivery*, vol. 23, no. 4, pp. 2000–2007, 2008.

[4] IEEE Guide for Loading Mineral-Oil-Immersed Overhead and Pad-Mounted Distribution Transformers Rated 500 kVA and Less With 65 Degrees C or 55 Degrees C Average Winding Rise ANSI/IEEE Std C57.91-1981. 1981.

[5] B. R. Sathyanarayana, G. T. Heydt, and M. L. Dyer, "Distribution transformer life assessment with ambient temperature rise projections," *Electric Power Components and Systems*, vol. 37, no. 9, pp. 1005–1013, 2009.

[6] "China Meteorological Data Sharing Service System," 2012, http://www.cma.gov.cn/2011qxfw/2011qsjgx/.

[7] P.-L. Chen, S.-W. Cao, and J.-X. Guo, *Air Conditioning Load Calculation Theory and Method*, Tongji University Press, Shanghai, China, 1987.

[8] W. Chen, F. Zhou, X.-Y. Han, and B.-G. Shan, "Analysis on load characteristic of state grid," *Electric Power Technologic Economics*, pp. 25–29, 2008.

[9] X.-M. Jin, B.-Y. Wu, and J. Mo, "The interconnection schemes and relative key technology for China Southern Power Grid and Vietnam Nation Grid," *Southern Power System Technology Research*, vol. 2, pp. 26–39, 2006.

[10] C.-Y. Lee, H.-C. Chang, and C.-C. Liu, "Emergency dispatch strategy considering remaining lives of transformers," *IEEE Transactions on Power Systems*, vol. 22, no. 4, pp. 2066–2073, 2007.

[11] D. G. Kleinbaum and L. L. Kupper, *Applied Regression Analysis and Other Multivariable Methods*, Brooks/Cole, Calif, USA, 4th edition.

[12] T. W. Dakin, "Electrical insulation deterioration treated as a chemical rate phenomenon," *Transactions of the American Institute of Electrical Engineers*, vol. 67, pp. 113–122, 1948.

[13] P. Geladi and B. R. Kowalski, "Partial least-squares regression: a tutorial," *Analytica Chimica Acta*, vol. 185, pp. 1–17, 1986.

[14] P. H. Garthwaite, "An interpretation of partial least squares," *Journal of the American Statistical Association*, vol. 89, pp. 122–127, 1994.

[15] H. Abdi, "Partial least squares regression and projection on latent structure regression (PLS Regression)," *Wiley Interdisciplinary Reviews*, vol. 2, no. 1, pp. 97–106, 2010.

[16] H.-W. Wang, Z.-B. Wu, and J. Meng, *Partial Least-Squares Regression-Linear and Nonlinear Methods*, National Defence Industry Press, Beijing, China, 2006.

[17] I.-G. Chong and C.-H. Jun, "Performance of some variable selection methods when multicollinearity is present," *Chemometrics and Intelligent Laboratory Systems*, vol. 78, no. 1-2, pp. 103–112, 2005.

[18] M. Olah, C. Bologa, and T. I. Oprea, "An automated PLS search for biologically relevant QSAR descriptors," *Journal of Computer-Aided Molecular Design*, vol. 18, no. 7-9, pp. 437–449, 2004.

2

Relationships among Energy Price Shocks, Stock Market, and the Macroeconomy: Evidence from China

Rong-Gang Cong[1,2] **and Shaochuan Shen**[3]

[1] *School of Economics and Management, North China Electric Power University, Beijing 102206, China*
[2] *Centre for Environmental and Climate Research (CEC), Lund University, 223 62 Lund, Sweden*
[3] *State Key Laboratory Breeding Base of Green Chemistry Synthesis Technology, College of Chemical Engineering and Materials Science, Zhejiang University of Technology, Hangzhou 310014, China*

Correspondence should be addressed to Rong-Gang Cong; ronggang.cong@cec.lu.se

Academic Editors: K. J. Chua and A. Piacentino

This paper investigates the interactive relationships among China energy price shocks, stock market, and the macroeconomy using multivariate vector autoregression. The results indicate that there is a long cointegration among them. A 1% rise in the energy price index can depress the stock market index by 0.54% and the industrial value-adding growth by 0.037%. Energy price shocks also cause inflation and have a 5-month lag effect on stock market, which may result in the stock market "underreacting." The energy price can explain stock market fluctuations better than the interest rate over a longer time period. Consequently, investors should pay greater attention to the long-term effect of energy on the stock market.

1. Introduction

Since 2002, the global crude oil price has risen rapidly. In July 2008, price of both WTI crude oil and Brent crude oil went beyond $140 per barrel. This results in fluctuation in the international market, causing the domestic oil price to increase, which in turn pushes the energy price up for coal and electricity [1]. As a result, the nationwide retail commodity price index also raises to some extent. All of these factors have a profound effect on China's macroeconomy. In the meantime, owing to appreciation of the China currency (RMB), and in anticipation of expansion in the Chinese economy during the Olympics Games, the Chinese stock market has steadily increased since 2006, which, as a consequence, has attracted investors world wide. As a result, observing whether the shocks in energy price are transmitted to the stock market will receive considerable attention from investors. It might be argued that this relationship reflects the efficiency of the Chinese stock market.

Extant research includes analyses of relationships between energy price, a country's macroeconomics, and its stock market. Research on the relationship between energy price and the macroeconomy is mainly focused on the relationship between crude oil price and GDP/GNP and the channels through which crude oil price influences the macroeconomy. In the former, the majority of scholars have found a reverse relationship between oil price and GDP/GNP which meant that rising oil price would depress GDP/GNP [2–4]. In the mid-1980s, however, the linear reverse relationship became not so evident. Various scholars studied the asymmetric relationship between oil price and macroeconomic activities [5–7]. Lardic and Mignon [8] found that while standard cointegration is rejected, there is evidence for asymmetric cointegration between oil prices and GDP in the major European countries. In recent years, scholars measured the elasticity of GNP to oil price [9]. In the latter, it is generally assumed that oil price influenced the macroeconomy at the demand and supply levels. As to the demand level, on the one hand, the rise of oil price can cause inflation [10] and suppress consumption [11, 12] and investment [13, 14]. If the Central Bank cannot increase the currency supply, the interest rate may tend to rise, which may influence the aggregate demand [15, 16]. On the other hand, the rise in oil price may transfer wealth from oil-importing countries to those exporting. This may deteriorate the purchasing power of oil-importing countries and affect their international trade [17]. As regards to the supply level, as oil is a major material for production, price increases will raise production costs and cause industries to scale-down or transfer to low energy-intensive industries [18]. This, in-turn, may affect production and increase unemployment [14, 19]. Meanwhile, the rise of production costs may result in cost-driven inflation. If the cost increase is transferred to industries downstream, their production cost will increase. So any rise in oil price, influences various industries in different ways. Fan et al. [20] researched the impact of rising international crude oil price on China's GDP, investments, consumptions, import, export, and so on.

As shown previously, scholars have extensively researched the relationship between oil price shocks and a country's macroeconomy. But there is relatively little research on the relationship between oil price shocks and the financial markets—the exception being those that focused on developed countries. For example, Jones and Kaul [21], based on a standard cash flows/dividends valuation model, researched the stock market of US, Canada, Japan, and England. They found that the change of oil price had a decisive effect on the four countries' real stock returns. Sadorsky [22] identified that oil price, and its volatility, played an important part in explaining the real stock returns. The movement of oil price explained more than interest rates for the forecasting variances. Papapetrou [23] researched the Greek stock market. Likewise, he also found that oil price played an important component in explaining stock price movements, and positive oil price shocks suppressed real stock returns. Ewing and Thompson [24] also examined the cyclical relationships among industrial production, consumer prices, unemployment, and stock prices using time series filtering methods.

Several scholars showed that oil price and its shocks influenced various industries differently [25]. A common held view is that shocks are beneficial for oil companies upstream, yet has an adverse effect on companies downstream and other industries. For example, Huang's research [26], based on correlative coefficient method and a VAR model, used the S & P 500 index, twelve industries' stock price indices, and three oil company stock prices. He found that crude oil future returns had significant abilities to explain oil companies' stock returns, which could be seen as their lead index, but had little effect on the total market. Faff and Brailsford [27] used an enlarged market model to research several industries returns in the Australian stock market. They found that oil price had an effect on stock prices. They also found that the oil and gas industry and diverse resources industry had positive sensitivities, while papermaking, packing, and transportation industry had negative sensitivities. Using Johansen cointegration test, Hammoudeh et al. [28] found that one-month to four-month WTI oil future price shocks explained oil extracting, refining, and marketing companies' stock price movement. Sadorsky [29] took Canadian companies as an example. Using the stock market index, energy price, interest rates, and exchange rates as explanatory variables, he found that the rise of the stock market index and oil price had a positive effect on oil companies' returns, while the rise of interest rates and exchange rates had a negative effect. Lanza et al. [30] used VAR/VECM models to research the relationships among six large oil companies, various stock markets, and the spread of crude oil future and spot price. They found that the greater the spread, the higher the oil companies' stock prices.

In summary, it can be stated that there are relationships among oil price, macroeconomy, and the stock market, which have been tested in several developed countries. Whether these relationships exist in China is the focus of this paper. As the Chinese economic dependency on energy increases, any rise in energy price has a significant effect on the macroeconomy. In essence, the stock market is a virtual economy. To know whether it is affected by energy price, this paper has focused on the Chinese Shanghai stock market using a VAR model. Based on impulse response functions and forecasting variance decomposition, it analyzes the interactive responses among several economic variables and energy price.

The paper is structured as follows. We begin with a brief introduction of the VAR model and then describe the data in Section 2. Section 3 presents the empirical results. In Section 4, we discuss the results within the context of the Chinese situation. Next, we conclude with a summary and propose suggestions in Section 5. Finally, possible future work is presented.

2. Methodology and Data

2.1. Methodology of VAR Model. We have selected to use a vector autoregression (VAR) method [31]. Since VAR model requires all variables in the system to be stationary, a unit root test is initially completed. Here we choose ADF (augmented Dickey-Fuller test) method. Take oil price series as an example. The fundamental principle of ADF test is

$$\Delta \mathrm{OIL}_t = \gamma \mathrm{OIL}_{t-1} + \sum_{i=1}^{p} \beta_i \Delta \mathrm{OIL}_{t-i} + u_t,$$

$$t = 1, 2, \ldots, T \text{ no constant or linear time trends},$$

$$\Delta \mathrm{OIL}_t = \gamma \mathrm{OIL}_{t-1} + a + \sum_{i=1}^{p} \beta_i \Delta \mathrm{OIL}_{t-i} + u_t, \quad (1)$$

$$t = 1, 2, \ldots, T \text{ constant},$$

$$\Delta \mathrm{OIL}_t = \gamma \mathrm{OIL}_{t-1} + a + \delta t + \sum_{i=1}^{p} \beta_i \Delta \mathrm{OIL}_{t-i} + u_t,$$

$$t = 1, 2, \ldots, T \text{ constant and linear time trends}.$$

The letter "a" is a constant. δt is a linear trend function. Δ represents differentiation. $u_t \sim$ i.i.d. $N(0, \sigma^2)$ which form is chosen depends on a series graph. p is the optimum lag order, which depends on AIC (Akaike Information Criterion). The original hypothesis and the alternative hypothesis can be written as

$$H_0 : \gamma = 0,$$
$$H_1 : \gamma < 0. \tag{2}$$

The original hypothesis for oil price time series has a unit root which is a nonstationary series. The alternative hypothesis is the oil price time series that does not have unit roots, which is a stationary series. As to nonstationary time series, another stationary test is needed. If the test shows that the first-order differentiation is stationary, the series is named as I (1); otherwise the second differentiation should be tested.

IP_t, OIL_t, r_t, and STOCK_t stand for industry production, oil price, nominal interest rates, and stock market index, respectively. $y_t = (\mathrm{IP}_t, \mathrm{OIL}_t, R_t, \text{ and } \mathrm{STOCK}_t)'$. The VAR model can be constructed as follows:

$$y_t = A_1 y_{t-1} + A_2 y_{t-2} + \cdots + A_p y_{t-p} + \varepsilon_t \quad t = 1, 2, \ldots, T, \tag{3}$$

where p is lag orders, which is determined by AIC and SC information criterion. T is the size of the sample. $A_1, A_2, \ldots, A_p$ and $\mathbf{B}$ are parameter matrices. ε_t are random disturbances which can correlate in the same time but cannot correlate with their lag variables and the variables on the right of the functions.

If IP_t, OIL_t, R_t, and STOCK_t can be tested as the same number of the unit roots, which can be assumed as I (1), we can use JJ method to test whether there are cointegration among the series. In other words, whether there are long-term stable relationships among nonstationary variables.

Based on VAR model, we can also use impulse response functions and forecasting variance decomposition to explain the model established. Impulse response functions can be used to test the effect of a standard variance shock on the endogenous variables and their future values. Its fundamental rationale is as follows.

If L is defined as a lag operator, $Lx_t = x_{t-1}$, it can be derived from (3) as follows:

$$\begin{aligned} \left(I_k - A_1 L - \cdots - A_p L^p\right) y_t &= \varepsilon_t, \\ y_t &= \left(I_k - A_1 L - \cdots - A_p L^p\right)^{-1} \varepsilon_t, \\ &= \left(I_k + C_1 L + C_2 L^2 + \cdots\right) \varepsilon_t. \end{aligned} \tag{4}$$

The first variable of y_t (industry production) can be written as

$$\mathrm{IP}_t = y_{1t} = \sum_{j=1}^{4} \left(c_{1j}^{(0)} \varepsilon_{jt} + c_{1j}^{(1)} \varepsilon_{jt-1} + \cdots\right) \quad t = 1, 2, \ldots, T, \tag{5}$$

where $c_{1j}^{(q)}$ is the first row and first line element of C_q, which can be represented as $c_{1j}^{(q)} = \partial \mathrm{IP}_{t+q} / \partial \varepsilon_{jt}$. It shows at period IP_{t+q}'s response to a shock of y_{jt} (IP_t, OIL_t, r_t, or STOCK_t) at the condition that other variables keep constant. We refer to it as the impulse-response function, which is similar to the shock multiplier effect analysis. As there are correlations in the same period among various functions' random variances in VAR model, it needs to construct an orthogonal matrix to transform shocks from correlation in the same period to noncorrelation. There are many methods to accomplish this goal. Here we used general impulse method [32] which does not depend on the variables' orders in VAR model.

Variance decomposition decomposes the forecasting variances by various variables shocks. We can use it to estimate the importance of various structural shocks. Its fundamental logic is as follows: take industrial production as an example. It can be known from (5) that what is in the brackets is the total ε_j's effect on IP_t up to now. If it can be assumed that there are not series correlations among ε_j, the variances are

$$\begin{aligned} &E\left[\left(c_{1j}^{(0)} \varepsilon_{jt} + c_{1j}^{(1)} \varepsilon_{jt-1} + c_{1j}^{(2)} \varepsilon_{jt-2} + \cdots\right)^2\right] \\ &\quad = \sum_{q=0}^{\infty} \left(c_{1j}^{(q)}\right)^2 \sigma_{jj} \quad j = 1, 2, \ldots, 4. \end{aligned} \tag{6}$$

Additionally, if there are no correlations among disturbances in the same period the variance of IP_t is the sum of four variances above mentioned as follows:

$$\operatorname{var}(\mathrm{IP}_t) = \sum_{j=1}^{4} \left\{ \sum_{q=0}^{\infty} \left(c_{1j}^{(q)}\right)^2 \sigma_{jj} \right\} \quad t = 1, 2, \ldots, T. \tag{7}$$

Because the variance of IP_t can be decomposed into four irrelative effect, we can define the criterion as follows to measure the contribution of various disturbance to the variance of IP_t:

$$\begin{aligned} \mathrm{RVC}_{j \to 1}(\infty) &= \frac{\sum_{q=0}^{\infty} \left(c_{1j}^{(q)}\right)^2 \sigma_{jj}}{\operatorname{var}(\mathrm{IP}_t)} \\ &= \frac{\sum_{q=0}^{\infty} \left(c_{1j}^{(q)}\right)^2 \sigma_{jj}}{\sum_{j=1}^{4} \sum_{q=0}^{\infty} \left(c_{1j}^{(q)}\right)^2 \sigma_{jj}} \quad j = 1, 2, \ldots, 4. \end{aligned} \tag{8}$$

The relative variance contribution (RVC) measures the effect of the jth variable on IP_t. This paper decomposes each endogenous variable's shock to four parts related to every function's disturbance, which can be used to know the relative importance of various shocks to endogenous variables in the model. Comparing the importance varying with time, we also estimate the lag of the variable's effect.

2.2. Data Sources. In this paper, we do not choose oil price data. Drawing upon lessons from Papapetrou [23], we choose domestic energy price index. Different from commodities price index for fuel that he chose in his paper, we choose purchase price index for fuel power. Since Chinese oil prices are not in accordance with international markets, we do not choose oil spot price or future price in international markets. Interestingly, though, since June 2006, the home refined oil price started to adjust in accordance with the movement

TABLE 1: The test result of unit root.

Variables	ADF test	Test type (c, t, p)	Critical value
IP	−2.003340	$(c, t, 2)$	−3.160198*
dIP	−1.973988	(0, 0, 11)	−3.165046*
d^2IP	−4.952033	(0, 0, 10)	−2.598416***
OIL	−1.808898	$(c, t, 1)$	−3.159780*
dOIL	−5.270185	(0, 0, 0)	−2.593824***
R	−1.922089	$(c, 0, 0)$	−2.585861*
dR	−2.995834	(0, 0, 11)	−2.598416***
STOCK	−1.045074	(0, 0, 0)	−2.585861*
dSTOCK	−7.767847	(0, 0, 0)	−2.593824***

Note. [1]The meanings of various variables in the table are as follows: IP is industry value added discounted by price index, OIL is discounted purchase price index for fuel power, R stands for real interest rates, and STOCK is monthly price index of Shanghai stock market.
[2]c, t, and p in test type stand for constant, trend, and lag orders, respectively.
[3]At three remarkable levels, when ADF value is greater than critical value, corresponding series has unit root.
[4]***, **, and * stand for 1%, 5%, and 10% critical levels, respectively.
[5]d stands for the first differential of the variables, and d^2 stands for the second differential of the variables.

of oil price in international market. Based on Jiao et al.'s [33] research, the domestic oil price responded 78.4% to the change of crude oil cost, and it had a four-month lag. Some of the international oil price shocks were absorbed by oil companies. The purchase energy price index can accurately reflect the domestic oil price in a timely manner.

Considering that the energy price has a close relationship to industry, we chose industry value adding as a proxy for the macroeconomy. As for the stock market, we selected the monthly Chinese Shanghai stock market index and considered the bonus and stock dividend. According to traditional financial theory, it is generally assumed that the stock price is equal to the discount of stock's future cash. So interest rates are important to stock price. Following Sadorsky [22] and Papapetrou [23], we introduce interest rates to dynamic analysis. Interest rates can be divided into nominal interest rates and real interest rates. This paper chooses the latter. We have selected one year to mature a lump deposit for total withdrawal as a representative and eliminate the interest tax. With exception of the interest rate data sourced from the website of Bank of China, all the data is from the National Bureau of Statistic of China. With due consideration of availability of information, we selected monthly data between January 2000 and December 2010. The industry value-added data and the energy price data are both adjusted by Census X12 method and discounted by consumer price index. All the variables are in logarithmic form. As interest rates are in percentages, we define the logarithm of it as $\log(1 + r/100)$.

3. Results and Analysis

3.1. The Stationary Test of Data. In Table 1, ADF test shows that industrial value added has two unit roots, and other variables have a unit root.

3.2. The Determination of Lag Orders. Through testing on 1 lag order to 5 lag orders, AIC and SC can be obtained from Table 2.

TABLE 2: The choose of lag orders in the model.

Lag	AIC	SC	Log L
0	−11.09981	−10.9762	420.2427
1	−20.11605	−19.49806*	774.3520
2	−20.36989	−19.2575	799.8710
3	−20.37243*	−18.7656	815.9662
4	−20.16755	−18.0664	824.2831
5	−20.30166	−17.7061	845.3124

Note: * above stand for the minimum values in the columns of AIC and SC.

Through test iterations, we find when the largest lag order equals to three, the AIC reaches the minimum. SC reaches the minimum when the lag order equals to one. So LR test is chosen as tradeoff. The original hypothesis test is the largest lag order equals to one. The testing statistic is as follows:

$$\mathrm{LR} = -2 * (l_1 - l_3) = -2 * (774.352 - 815.9662) = 83.2284, \quad (9)$$

where l_1 and l_3 are the whole log likelihood function values when p equals to one and three, respectively. In the original hypothesis, the statistic has gradually conformed to χ^2 distribution whose degrees of freedom are thirty-two. The accompanied probability is $1.9 * 10^{-6}$. So we chose a three lag order VAR model, because the original hypothesis is rejected.

3.3. The Estimation and Test of the Model. To give the variables the same integration order, we choose the first lag order to the third lag order values of OIL, R, STOCK, and dIP as an alternative of IP as endogenous variables and ordinary least squares method to estimate the model. The results are as follows:

$$\begin{pmatrix} d\mathrm{IP}_t \\ \mathrm{OIL}_t \\ R_t \\ \mathrm{STOCK}_t \end{pmatrix} = \begin{pmatrix} -0.92^* & -0.175 & 0.73 & 0.12^* \\ -0.056 & 1.27^* & -0.223 & 0.0048 \\ -0.11^* & 0.118^* & 1.17^* & 0.017 \\ 0.36 & -0.26 & -1.48 & 1.056 \end{pmatrix} \begin{pmatrix} d\mathrm{IP}_{t-1} \\ \mathrm{OIL}_{t-1} \\ R_{t-1} \\ \mathrm{STOCK}_{t-1} \end{pmatrix} + \begin{pmatrix} -0.69^* & 0.19 & 1.75 & 0.006 \\ -0.02 & -0.35^* & -0.163 & 0.008 \\ -0.079^* & -0.14 & -0.067 & -0.005 \\ -0.03 & 0.23 & 1.82 & -0.09 \end{pmatrix} \begin{pmatrix} d\mathrm{IP}_{t-2} \\ \mathrm{OIL}_{t-2} \\ R_{t-2} \\ \mathrm{STOCK}_{t-2} \end{pmatrix} + \begin{pmatrix} -0.15 & -0.122 & -2.8^* & -0.2^* \\ -0.02 & 0.04 & 0.29 & -0.045 \\ -0.01 & 0.017 & -0.19 & -0.015 \\ 0.28 & 0.18 & 0.47 & 0.03 \end{pmatrix} \begin{pmatrix} d\mathrm{IP}_{t-3} \\ \mathrm{OIL}_{t-3} \\ R_{t-3} \\ \mathrm{STOCK}_{t-3} \end{pmatrix} + \begin{pmatrix} 1.05^* \\ 0.43^* \\ 0.06 \\ -0.67 \end{pmatrix}, \quad (10)$$

where * stands for coefficients that are significant.

TABLE 3: The series cointegration test result.

The original hypothesis	Eigenvalue	Trace statistic (P value)	λ-max statistic (P value)
No cointegration vector	0.606545	96.20150 (0.0000)*	73.69024 (0.0000)*
More than one cointegration vector	0.154912	22.51127 (0.2709)	13.29684 (0.4253)
More than two cointegration vector	0.089195	9.214422 (0.3459)	7.380688 (0.4451)
More than three cointegration vector	0.022945	1.833735 (0.1757)	1.833735 (0.1757)

Note: * above means that the original hypothesis is significantly rejected.

TABLE 4: The model test results.

Stability test	The coefficients of all roots are less than one So the model estimated is stable
Autocorrelation LM test	LM = 18.88 (0.2749) No series autocorrelation
Heteroskedasticity test	χ^2 (240) = 231.2206 (0.6462) No heteroskedasticity
Jarque-Bera normal test	JB (8) = 7.11161 (0.5246) The residual series conform to normal distribution

As four series in the model are I (1) series, the condition for cointegration is satisfied. The lag order is determined to 2. Johansen cointegration test result shows that there is a cointegration relationship among four series above which demonstrates that there is a prolonged stationary relationship among macroeconomy, energy price, and the stock market (Table 3).

The cointegration can be expressed as

$$d\mathrm{IP} = -0.037^{*}\mathrm{OIL} - 0.14R - 0.02^{*}\mathrm{STOCK} + 0.36. \quad (11)$$

We can see from the previous equation that in the longer term, the 1% rise in energy price will have an adverse effect of 0.037% on industrial value adding and an adverse effect of 0.54% on the stock market.

Testing using VAR model obtains the following results (Table 4).

Generally, it is difficult to analyze the dynamic relationships among variables based on the estimated parameters in the VAR model. Consequently, we use impulse-response function and variance decomposition to analyze the interactive relationships among them.

3.4. Impulse-Response Function and Variance Decomposition. Figure 1 shows the impulse-response function curves simulated by analytic method, based on VAR model.

We consider the response of four variables to one S.D. (standard deviation) innovation of energy price.

For one standard deviation innovation of energy price (1.1% of energy price), interest rate apparently respond. In 3 months, it rises to the peak (0.2%). In 5 months, the effect keeps stable, which represents the level of interest rate with a 1% stable increase.

One standard deviation innovation of energy price to the growth rate of industrial value adding has the greatest adverse effect (0.2%) in 2 months. Also, there is a shock following. But it has a negative convergence trend in the long term.

As for one standard deviation innovation of energy price, the stock price responds adversely to its peak in 3 months (1%). But in 6 months, the adverse effect disappears.

Next, we consider the effect of four variables' shocks on the energy price index (Figure 2).

One standard deviation shock for interest rate shows that energy price is affected most adversely in four months and remains so for a long time. The residual variances are still half of its peak in a year.

Energy price has little response to one standard deviation shock of industrial value adding.

The stock price rises 5%. In three months the energy price has a downward trend. Twenty months later, the adverse effect has reached its minimum value (2%).

We decompose the variance to stock market price based on VAR model and analytic method (above mentioned). The results are as follows.

In Table 5, the first and seventh columns are the periods that are set to a maximum of twenty due to limited space. The data in the S.E. (standard error) column are the forecasting variance of STOCK in various periods, which are caused by the change of the present or future value. dIP, OIL, R, and STOCK are the contribution of the innovations to forecasting variance, respectively, which sum to 100. Figure 3 is the variance decomposition where the periods are prolonged to thirty.

It can be seen from Figure 3 that, at the first month, the stock price can only explain 94% of its forecasting variance. The energy price can explain 0.14%, the industrial value-added can explain 0.4%, and the change rate of interest rate can only explain 5.03%. In fifteen months, the importance of energy price exceeds interest rates. At the twentieth month, the interest rates and energy price can explain 7.7% and 9.56%, respectively, while the industrial value added can only explain 0.24%.

4. Discussion

4.1. There Is a Cointegration among Stock Market, Macroeconomy and Energy Price. In this paper, we find that there is a long-term stable correlation among the stock market, macroeconomy and energy price. Stock market is often said to be the barometer of one's macroeconomy, which means that they should have a close relationship. But it is a controversial issue in China. Chinese scholars generally have two views. One is that the stock market has a weak positive correlation with the macroeconomy. The other perspective is that the correlation

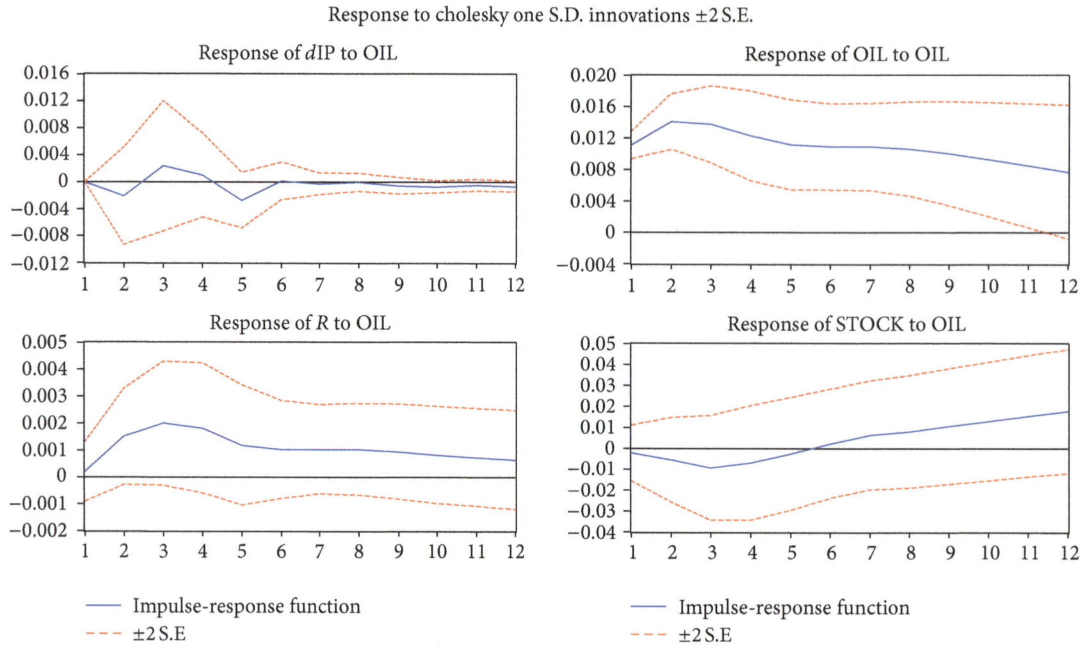

Figure 1: Generalized impulse responses to one S.D. shock for energy price changes. Note that the horizontal axis is the period. The vertical axis is the explanation level of dependent variables to independent variables. In the model, we fix the periods at 12 months.

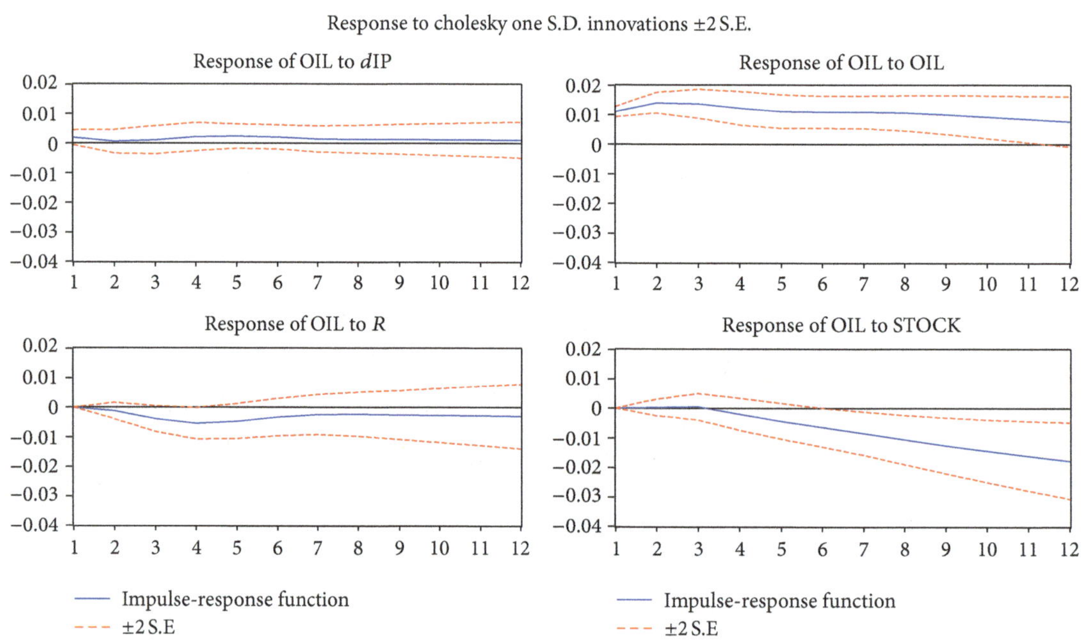

Figure 2: Generalized impulse responses of energy price to one S.D. shock for other variables changes. Note that the horizontal axis is the period. The vertical axis is the explanation level of dependent variables to independent variables. In the model, we fix the periods at 12 months.

is not significant or is negative. Scholars holding this latter view believe that the Chinese stock market aims at financing but not investing. The stock market can not evaluate the value of companies correctly. Therefore, the stock market is not the barometer of the Chinese macroeconomy. Based on a capital asset pricing model, the real value of stock equals to the discount of its future cash flows. As a result, the value of stock relates to future economy and also is affected by discount rates (interest rate). The rising energy price affects the Chinese macroeconomy and pushes up the inflation rate. The cointegration obtained in this paper demonstrates that energy price has a significant effect on the stock market, which is an extension of existing literature. In equation (11), we find that in the long-term, energy price has a negative

TABLE 5: Generalized variance decomposition of STOCK.

Periods	S.E.	*d*IP	OIL	*R*	STOCK	Periods	S.E.	*d*IP	OIL	*R*	STOCK
1	0.029264	0.395151	0.141588	5.030712	94.43255	11	0.043981	0.342164	2.011849	3.523414	94.12257
2	0.040527	0.430799	0.470383	8.225033	90.87379	12	0.044001	0.318452	2.603733	3.671406	93.40641
3	0.041955	0.289069	1.067355	8.672081	89.97149	13	0.044018	0.29907	3.283352	3.936467	92.48111
4	0.043575	0.848855	1.11992	7.35657	90.67465	14	0.044035	0.283251	4.046391	4.29525	91.37511
5	0.043702	0.688591	0.943073	6.128618	92.23972	15	0.044049	0.270742	4.875873	4.736972	90.11641
6	0.043732	0.573093	0.803602	5.102978	93.52033	16	0.044062	0.26073	5.760365	5.249124	88.72978
7	0.043834	0.502843	0.839256	4.414149	94.24375	17	0.044074	0.253142	6.685862	5.814719	87.24628
8	0.043892	0.446533	0.952725	3.926427	94.67431	18	0.044085	0.247541	7.636951	6.422031	85.69348
9	0.043942	0.407653	1.182682	3.620365	94.7893	19	0.044096	0.243609	8.599254	7.054811	84.10233
10	0.043956	0.371269	1.533481	3.489268	94.60598	20	0.044105	0.241156	9.556565	7.699988	82.50229

Cholesky ordering: *d*IP, OIL, *R*, and STOCK.

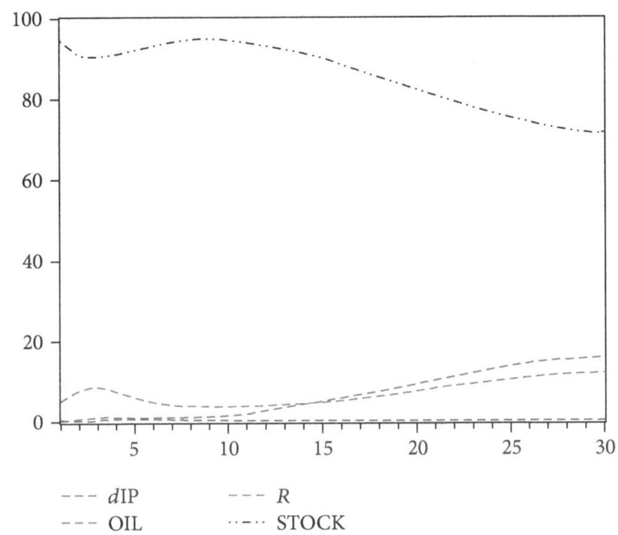

FIGURE 3: Variance decomposition of STOCK.

relationship with industrial value adding and the stock market. The rise in energy price affects the macroeconomy, on the one hand, and depresses the stock market on the other hand.

4.2. The Positive Shock of Energy Price Can Push up the Real Interest Rate, Affect Industry Productions, and Depress the Stock Market Price. On the basis of realizing the long-term relationship among them, we want to know the response of stock prices and the macroeconomy to the short-term shock of the energy price. Based on impulse response function, we have seen that the positive shock of energy price can push-up the real interest rate, affect industry, and depress the stock market.

The shock of energy price can affect real interest rates to some extent. In China, since 2002, the international crude oil price has risen from \$20 per barrel to nearly \$100 per barrel. In turn, this drives up oil related energy products and total prices. Therefore, China has the risk of cost-push inflation. The rise of international crude oil price will increase the domestic price of the refined oil and related chemical engineering products. The rise of the refined oil price can further increase the price of transportation, industrial production, and residential gas. Moreover, the rise in price of related chemical engineering products will increase the cost of plastic, rubber, chemical fiber, and other inputs. Finally, the total social price level rises. Inflation makes the supply of currency relatively insufficient. The controllable real currency on the whole will decrease. Consequently, the rise of energy price will drive up the total interest rate level.

As a major input to industry, the rise in energy price can affect economic growth. In the Chinese industry, chemical engineering, metallurgy, and some other energy-intensive industries are seriously affected by the rise in oil price. However, following advances in technology and the upgrading of industrial infrastructure, the effect becomes less and less in the long term.

The stock market is not completely efficient with regard to recognizing energy price. There is a phenomenon of underreaction. As a collection of energy-input companies, the rise of energy price will increase their cost and decrease their profit. If the stock market is efficient, the rise in energy price will depress the stock market immediately. But in this paper, we find that the rise in energy price will depress the stock market during a five-month lag. As the energy price in the public domain, it means there is conservatism in most investors' decisions, as acknowledged by Barberis et al. [34]. In other words, investors find it difficult to renew their views and have an underreaction to the importance of the price of energy. As a limitation in cognitive abilities, they find it hard to evaluate the importance of energy price to stock value. This conforms to the hypothesis that "information diffuses gradually" suggested by Hong and Stein [35].

4.3. The Effect on Energy Price of the Industrial Demand: Increasing Interest Rates and the Stock Market. We have discussed earlier that the energy price will have a significant effect on industrial production, real interest rates, and the stock market. But we still want to know the effect of three shocks on energy price.

To control the inflation caused by the rise of energy price increasing the real interest rate is very efficient. It is because, on the one hand, high interest rates change people's controllable income by increasing savings and decreasing consumption on energy relating products, and on the other hand, high interest rates will cause the capital to depreciate and reduce people's wealth, which reduces their consumption. As to industries, high interest rates increase their capital cost and depresse their investment, which reduces their energy use for relating products. Since 2007, the People's Bank of China has increased the interest rates five times and decreased the interest rate tax. Currently, a sixth increase in interest rates is being considered. It can be forecasted that the People's Bank of China will still control inflation by increasing interest rates as price increase further.

The shock of industrial growth has a small, but long effect on energy price. So the demand from industry on energy is one of the reasons that cause a rise in the energy price. But the effect is slight.

The bull stock market can attract a good deal of money, which reduces the currency. The Chinese stock market has risen from 1000 points to 6000 points since 2005. It has attracted a large amount of money. So after the 2008 Olympics, or when the appreciated RMB has been realized, a lot of money will be taken back from stock market. Additionally, there is risk of inflation in the near future. Consequently, the rise in energy price will be severe.

4.4. Energy Price, over the Longer Term, Has Better Ability to Explain the Stock Market than Interest Rates. From the results of forecasting decomposition, we find that over the longer term, the greatest effect of three factors on stock price is energy price, interest rates, and industrial value adding. In the short term, the increase in interest rates is a *bad news* and can be perceived easily by all investors. So in the short term, the interest rates are more important than the other two factors. But in the long term, the rise in energy price will become gradually apparent to people and have an effect on people's anticipation. So it will be the more important factor in the longer term. As industries are only a part of stock market, the effects from them on the total market are limited. There is separation between the real economy and the virtual economy in China. Consequently, it has a poor explanation in explaining fluctuations in the stock market.

5. Conclusions and Future Work

Through the analysis of results and discussion, we can draw the following conclusions.

(1) There is a stable long-term relationship among energy price, industrial value adding, real interest rates, and stock price. In the longer term, the rise in energy price will drive up the real interest rates and have an adverse effect on industrial value adding and the stock market.

(2) The shocks in energy price have a lasting effect on real interest rates, which to some extent causes long-term inflation. The shocks in energy price have a notable effect on industrial value adding. But the effect tends to be weaker in the long term. In the short time, the shocks in energy price have an effect on the stock market.

(3) To control the rise in energy price, we should consider the macroeconomy and stock market, in addition to the international and domestic energy market. The rising industrial value adding increases the energy demand, which to some extent increases the energy price. But the rise of interest rates is favorable for controlling the energy price.

(4) The effect of energy on the stock market is hard to ignore. Through forecasting the decomposition of stock prices, we can identify that the stock price is chiefly affected by itself. What is a contradiction to the general view is that energy price and interest rates have greater effect on the stock market than industrial value adding. In the long term, energy price is even more important than interest rates. It indicates to investors that they should not only consider the interest rates but also consider the energy price for their investment.

This paper discusses the relationship between energy price and stock market at the macrolevel. The relationships between the shocks of energy price and the stock prices of various industries and oil companies remain to be researched. Meanwhile, opportunities for future research include the simulation and analysis of China economic transition and sustainable development at various levels of energy price, and the optimal design of subsidy policy for rising oil price [36].

Conflict of Interests

The authors of the paper do not have a direct financial relation with the commercial identity mentioned in this paper that might lead to a conflict of interests for any of the authors.

Acknowledgments

The authors gratefully acknowledge the good research environments provided by the Centre for Environmental and Climate Research (CEC), Lund University. The authors thank the anonymous referees for their helpful suggestions and corrections on the earlier draft of the paper which improved the contents. This research is funded by "Biodiversity and Ecosystem services in a Changing Climate (BECC) and SAPES." The funders had no role in study design, data collection and analysis, decision to publish, or preparation of the paper. The contents are the responsibility of the authors and do not necessarily reflect the views of Centre for Environmental and Climate Research (CEC), Lund University.

References

[1] K. Hanabusa, "The effect of 107th OPEC ordinary meeting on oil prices and economic performances in Japan," *Renewable and Sustainable Energy Reviews*, vol. 16, pp. 1666–1672, 2012.

[2] M. R. Darby, "The price of oil and world inflation and recession," *American Economic Review*, vol. 72, no. 4, pp. 738–751, 1982.

[3] J. D. Hamilton, "Oil and the macroeconomy since World War II," *Journal of Political Economy*, vol. 91, pp. 228–248, 1983.

[4] A. F. Darrat, O. W. Gilley, and D. J. Meyer, "US oil consumption, oil prices, and the macroeconomy," *Empirical Economics*, vol. 21, no. 3, pp. 317–334, 1996.

[5] M. K. Anton, "Oil and the macroeconomy when prices go up and down: an extension of Hamilton's results," *Journal of Political Economy*, vol. 97, pp. 740–744, 1989.

[6] S. P. A. Brown and M. K. Yücel, "Energy prices and aggregate economic activity: an interpretative survey," *Quarterly Review of Economics and Finance*, vol. 42, no. 2, pp. 193–208, 2002.

[7] M. P. Clements and H. M. Krolzig, "Can oil shocks explain asymmetries in the US business cycle?" *Empirical Economics*, vol. 27, no. 2, pp. 185–204, 2002.

[8] S. Lardic and V. Mignon, "The impact of oil prices on GDP in European countries: an empirical investigation based on asymmetric cointegration," *Energy Policy*, vol. 34, no. 18, pp. 3910–3915, 2006.

[9] A. Fayyad and K. Daly, "The impact of oil price shocks on stock market returns: comparing GCC countries with the UK and USA," *Emerging Markets Review*, vol. 12, no. 1, pp. 61–78, 2011.

[10] L. J. Álvarez, S. Hurtado, I. Sánchez, and C. Thomas, "The impact of oil price changes on Spanish and euro area consumer price inflation," *Economic Modelling*, vol. 28, no. 1-2, pp. 422–431, 2011.

[11] J. B. Taylor, "The lack of an empirical rationale for a revival of discretionary fiscal policy," *American Economic Review*, vol. 99, no. 2, pp. 550–555, 2009.

[12] J. Elder and A. Serletis, "Oil price uncertainty in Canada," *Energy Economics*, vol. 31, no. 6, pp. 852–856, 2009.

[13] S. van Wijnbergen, "Oil price shocks, unemployment, investment and the current account: an intertemporal disequilibrium analysis," *The Review of Economic Studies*, vol. 52, pp. 627–645, 1985.

[14] N. D. Uri and R. Boyd, "Economic impact of the energy price increase in Mexico," *Environmental and Resource Economics*, vol. 10, no. 1, pp. 101–107, 1997.

[15] J. L. Pierce, J. J. Enzler, D. I. Fand, and R. J. Gordon, "The effects of external inflationary shocks," *Brookings Papers on Economic Activity*, vol. 5, no. 1, pp. 13–61, 1974.

[16] K. A. Mork, O. Olsen, and H. T. Mysen, "Macroeconomic responses to oil price increases and decreases in seven OECD countries," *Energy Journal*, vol. 15, no. 4, pp. 19–35, 1994.

[17] J. D. Sachs, R. N. Cooper, and S. Fischer, "The current account and macroeconomic adjustment in the 1970s," *Brookings Papers on Economic Activity*, vol. 12, no. 1, pp. 201–282, 1981.

[18] M. Erturk, "Economic analysis of unconventional liquid fuel sources," *Renewable and Sustainable Energy Reviews*, vol. 15, no. 6, pp. 2766–2771, 2011.

[19] N. D. Uri, "Crude oil price volatility and unemployment in the United States," *Energy*, vol. 21, no. 1, pp. 29–38, 1996.

[20] Y. Fan, J. L. Jiao, Q. M. Liang, Z. Y. Han, and Y. M. Wei, "The impact of rising international crude oil price on China's economy: an empirical analysis with CGE model," *International Journal of Global Energy Issues*, vol. 27, no. 4, pp. 402–424, 2007.

[21] C. M. Jones and G. Kaul, "Oil and the stock markets," *Journal of Finance*, vol. 51, no. 2, pp. 463–491, 1996.

[22] P. Sadorsky, "Oil price shocks and stock market activity," *Energy Economics*, vol. 21, no. 5, pp. 449–469, 1999.

[23] E. Papapetrou, "Oil price shocks, stock market, economic activity and employment in Greece," *Energy Economics*, vol. 23, no. 5, pp. 511–532, 2001.

[24] B. T. Ewing and M. A. Thompson, "Dynamic cyclical comovements of oil prices with industrial production, consumer prices, unemployment, and stock prices," *Energy Policy*, vol. 35, no. 11, pp. 5535–5540, 2007.

[25] R. G. Cong, Y. M. Wei, J. L. Jiao, and Y. Fan, "Relationships between oil price shocks and stock market: an empirical analysis from China," *Energy Policy*, vol. 36, no. 9, pp. 3544–3553, 2008.

[26] R. D. Huang, R. W. Masulis, and H. R. Stoll, "Energy shocks and financial markets," *Journal of Futures Markets*, vol. 16, no. 1, pp. 1–27, 1996.

[27] R. W. Faff and T. J. Brailsford, "Oil price risk and the Australian stock market," *Journal of Energy Finance and Development*, vol. 4, pp. 69–87, 1999.

[28] S. Hammoudeh, S. Dibooglu, and E. Aleisa, "Relationships among U.S. oil prices and oil industry equity indices," *International Review of Economics and Finance*, vol. 13, no. 4, pp. 427–453, 2004.

[29] P. Sadorsky, "Risk factors in stock returns of Canadian oil and gas companies," *Energy Economics*, vol. 23, no. 1, pp. 17–28, 2001.

[30] A. Lanza, M. Manera, M. Grasso, and M. Giovannini, "Long-run models of oil stock prices," *Environmental Modelling and Software*, vol. 20, no. 11, pp. 1423–1430, 2005.

[31] C. A. Sims, "Macroeconomics and reality," *Econometrica*, vol. 48, pp. 1–48, 1980.

[32] H. H. Pesaran and Y. Shin, "Generalized impulse response analysis in linear multivariate models," *Economics Letters*, vol. 58, no. 1, pp. 17–29, 1998.

[33] J. L. Jiao, Y. Fan, and Y. M. Wei, "VECM based analysis of gasoline/diesel prices anti-symmetry," *Chinese Journal of Management Science*, vol. 14, pp. 97–102, 2006.

[34] N. Barberis, A. Shleifer, and R. Vishny, "A model of investor sentiment," *Journal of Financial Economics*, vol. 49, no. 3, pp. 307–343, 1998.

[35] H. Hong and J. C. Stein, "A unified theory of underreaction, momentum trading, and overreaction in asset markets," *Journal of Finance*, vol. 54, no. 6, pp. 2143–2184, 1999.

[36] R.-G. Cong and M. Brady, "How to design a targeted agricultural subsidy system: efficiency or equity?" *PLoS ONE*, vol. 7, Article ID e41225, 2012.

3

Advances in Energy Conservation of China Steel Industry

Wenqiang Sun, Jiuju Cai, and Zhu Ye

State Environmental Protection Key Laboratory of Eco-Industry, Institute of Thermal and Environmental Engineering, Northeastern University, Shenyang, Liaoning 118019, China

Correspondence should be addressed to Wenqiang Sun; neu20031542@163.com

Academic Editors: A. Greco, A. J. N. Khalifa, and A. Parlak

The course, technical progresses, and achievements of energy conservation of China steel industry (CSI) during 1980–2010 were summarized. Then, the paper adopted *e-p* method to analyze the variation law and influencing factors of energy consumptions of large- and medium-scale steel plants within different stages. It is pointed out that energy consumption per ton of crude steel has been almost one half lower in these thirty years, with 60% as direct energy conservation owing to the change of process energy consumption and 40% as indirect energy conservation attributed to the adjustment of production structure. Next, the latest research progress of some key common technologies in CSI was introduced. Also, the downtrend of energy consumption per ton of crude steel and the potential energy conservation for CSI during 2011–2025 were forecasted. Finally, it is indicated that the key topic of the next 15 years' research on the energy conservation of CSI is the synergistic operation of material flow and energy flow. It could be achieved by the comprehensive study on energy flow network optimization, such as production, allocation, utilization, recovery, reuse, and resource, according to the energy quantity, quality, and user demand following the first and second laws of thermodynamics.

1. Energy Conservation Process and Changes of Energy Consumption

Since the late 1970s, energy conservation has been put in an important position in China's steel industry (CSI). CSI has been making every attempt to energy conservation and emission reduction, and remarkable achievements have been made by saving energy and reducing the nonenergy material consumption. CSI has experienced great changes from 1980 to 2010. It not only broke the proposition of "steel industry is sunset industry" put forward by some scholars 30 years ago [1], but also supported the fast economy development in China. In the past 30 years, CSI made brilliant achievements and technical progress in energy conservation. It alleviated the tense energy supplements situation, improved the utilization efficiencies of resource and energy, and promoted the sustainable fast growth of steel industry [2, 3].

1.1. CSI Energy Conservation Process. The research subjects of energy conservation of CSI have five different levels, as shown in Figure 1. Practice of energy conservation in 30 years confirms that studying single equipment or equipment unit is not enough to reduce the energy consumption. What else need studying are the higher levels of energy conservation problems, such as production department composed by some single equipment, integrated steelworks composed by several production departments, and even steel industry composed by some integrated steelworks [4, 5]. Distinguished from traditional energy conservation like "single equipment and its units," the higher levels are referred as "systems energy conservation." Since 1978, CSI has gone through several important developing stages, such as single equipment energy conservation stage, systems energy conservation at production department level stage, and that at integrated steelworks level stage.

1.1.1. 1978–1980, Starting Period of CSI Energy Conservation. Before the oil crisis in 1973, "energy conservation" was still an unfamiliar concept in China [6]. There were not any energy statistics or energy consumption evaluation in metallurgical, chemical, and other industrial departments. CSI decided to carry out energy consumption survey and energy statistics collection in the national key steelworks and local important steelworks until 1978 when oil crisis happened again.

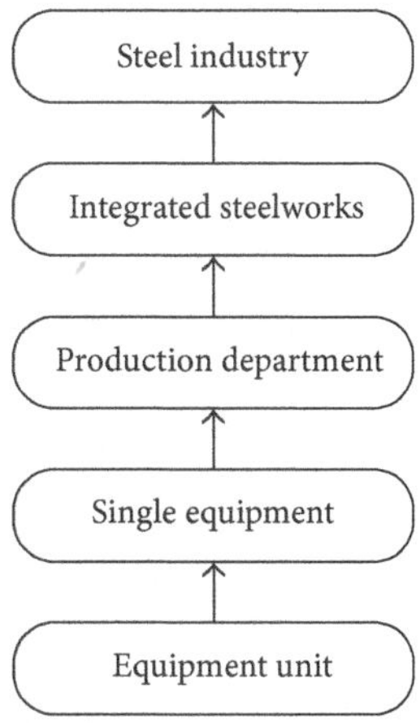

FIGURE 1: Five levels in energy conservation subjects.

Then CSI establishes the energy consumption indices system composed of energy consumption per ton of crude steel and process and equipment energy consumption and publishes the interim provisions of "energy balance and energy consumption indices calculation method for steel industry" and so on. After two years of starting period, the energy consumption and energy waste in CSI were lessened through forming energy conservation team, promoting energy conservation function, strengthening energy management, and blocking energy leaking (commonly known as "sweeping movable assets") and so forth. In 1980, CSI produced 37.12 Mt crude steel with a total consumption of 75.72 Mtce ("ce" is the abbreviation for "standard coal equivalent," hereafter; and 1 tce is equivalent to 29.31 MJ). Hence, the industry average comprehensive energy consumption per ton of crude steel is 2.039 tce/t. And the comprehensive energy consumption per ton of crude steel in large- and medium-sized steel enterprise is 1.646 tce/t; whilst the comparable energy consumption per ton of crude steel of those is 1.285 tce/t. The year 1980 is the first year of CSI energy conservation era. The energy consumption per ton of crude steel in large- and medium-sized steel enterprises in 1980 represents the initial energy consumption level of CSI. Since then, energy conservation of CSI enters a new development level every 10 years or so.

1.1.2. 1981–1990, Single Equipment and Production Department Energy Conservation Stage. For these 10 years, the main research subject of CSI energy conservation is single equipment, such as heating furnace, soaking pit, open hearth furnace, and hot blast stove. Later it extended to the production department, such as rolling process, steelmaking process, iron-making process, coke-making process, sintering process, and even auxiliary raw material plants like iron alloy, carbon products and refractory materials [7]. To promote and regulate the energy conservation of all kinds of equipment and production departments, CSI made 15 "calculation methods of heat balance test" for metallurgical furnaces and 19 "energy conservation rules" for production departments. And it also promoted some dozen energy conservation technologies, such as the transformation of heating furnace, thick bed layers sintering operation, waste heat utilization of flue gas from metallurgical furnaces, and liquid core rolling of large steel ingot. And it organized level upgrade and other activities for the purpose of carrying out energy conservation of single equipment and production department [8]. During these 10 years, energy consumption of CSI dropped clearly from 1.285 tce/t in 1980 to 1.017 tce/t in 1990. In the period, energy consumption per ton of crude steel of CSI declined by 268 tce/t with a dropping rate of 20.9%.

1.1.3. 1991–2000, Systems Energy Conservation Stage Focusing on Material Flow Optimization. During this period, CSI turned its eyes from single equipment and production department energy conservation to systems energy conservation [9, 10]. Meanwhile, CSI changed views of energy conservation to both energy and nonenergy materials conservation. Energy conservation in these 10 years is mainly contributed to the production structure adjustment and the process flow optimization [11], for example, eliminating backward production technology like die casting, blooming, cogging, cupola, open hearth furnace, and small electric stove and developing new technologies like continuous casting and rolling, hot delivery and hot charging, and changing backward process flow "multiheating" to "single heating." The main technical progress of energy conservation during 1991–2000 is the steel manufacturing process flow optimization specialized in reducing iron material consumption, that is, the reduction of ferrite flow [12, 13]. The adjustment of production structure actively pushed forward by CSI in the Ninth-Five-Year Plan period is focused on continuous casting. That improved the relationships between two production departments and made the steel manufacturing process large-sized, sequential, and automatic gradually. Also it created conditions to reduce the process energy consumption of each production department greatly. Therefore, energy consumption per ton of crude steel in some enterprises reached the world advanced level at that time. And most enterprises' energy consumption indices achieved a new level on the basis of the first 10 years, with a drop from 1.017 tce/t in 1990 to 0.781 tce/t in 2000. In the second 10 years, the energy consumption declined by 236 tce/t with a dropping rate of 18.4% compared with data in 1980.

1.1.4. 2001–2010, Systems Energy Conservation Stage Focusing on Energy Flow Optimization. The production structure adjustment and process flow optimization (namely, material flow optimization) in the Ninth-Five-Year Plan period laid a solid foundation for further energy conservation (i.e., energy flow optimization) of CSI. Energy flow and energy flow network are the main research subjects from 2001 to 2010 [14]. CSI paid its attention to developing and promoting the key common energy conservation technologies and studying topics of production, transformation, storage, allocation, utilization, recovery, reuse, and resource of every energy medium. Corresponding technologies [15] include high-temperature air combustion (HTAC) of fuel gas, blast furnace gas dry dust cleaning (BFG-BDC), Linz-Donawitz gas dry dust cleaning (LDG-BDC), coke dry quenching (CDQ), recovery of waste heat, blast furnace top gas recovery turbine unit (TRT),

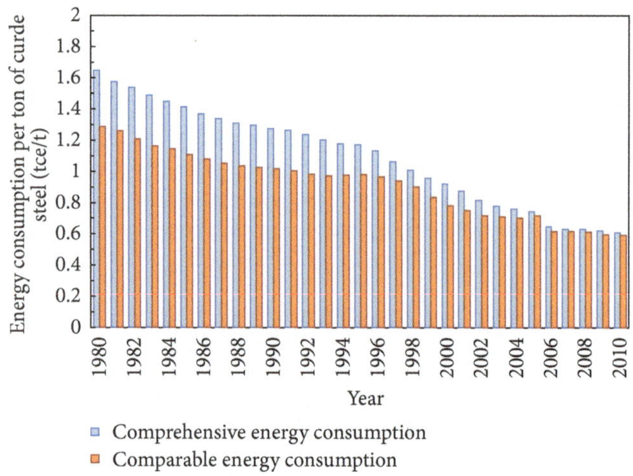

FIGURE 2: Changes in CSI energy consumption.

moisture removable blast, and coal moisture control (CMC). The application of these key common technologies, as well as the investment of large energy-saving equipment and energy management system (EMS), improved the integrated steelworks' equipment level and energy conversion function [16]. In these 10 years, process energy consumptions of sintering, iron-making, steel-making, rolling, and coke-making got a further reduction; and waste heat recovery and reuse rate were improved generally. Most enterprises had advanced world-level energy consumption. The energy consumption per ton of crude steel of CSI dropped from 0.781 tce/t in 2000 to 0.690 tce/t in 2010. In the third 10 years, energy consumption of CSI continuously declined by 91 tce/t with a dropping rate of 7.1% compared with data in 1980.

1.2. Changes of Energy Consumption per Ton of Crude Steel. Figure 2 shows the change of energy consumption per ton of crude steel from 1980 to 2010. There are two cylindrical curves in the figure. The upper one describes the change of comprehensive energy consumption from 1980 to 2010, with a drop from 1.646 tce/t in 1980 to 0.605 tce/t in 2010 [17–19], and the dropping rate is 63.24%. The lower one presents the change of comparable energy consumption in the 30 years, with a drop from 1.285 tce/t in 1980 to 0.590 tce/t in 2005, and the dropping rate is 54.08%. It can be found from Figure 2 that the energy consumption per ton of steel dropped by almost half from 1980 to 2010 [20–22].

1.3. Analysis of Energy-Saving Effect. Energy consumption per ton of crude steel is defined as the ratio of total energy consumption to steel production in statistical period [23]. But the definition is too simple to find its components and influence factors. To analyze the energy consumption change, energy-saving effect, and its related factors, it is necessary to rewrite the definition to further *e-p* analysis expression [24, 25]; namely,

$$E = \sum_{i=1}^{n} (e_i \cdot p_i), \quad (1)$$

where e_i is the energy consumption of each process in statistical period, tce/t-product; p_i is the ratio of process production to crude steel production in statistical period, which is referred to the steel ratio coefficient, t-product/t. It can be seen from (1) that there are two direct factors influencing energy consumption per ton of crude steel. One is steel ratio coefficient of each production department; the other is process energy consumption of each production department. It is clear that the variation of energy consumption per ton of crude steel in statistical period is equal to the energy consumptions difference between the start and the end of the statistical period [26]; namely,

$$\Delta E = \sum_i p_i'' e_i'' - \sum_i p_i' e_i' = \sum_i e_i'' (p_i'' - p_i') + \sum_i p_i' (e_i'' - e_i'), \quad (2)$$

where ΔE is the energy consumption variation in statistical period. The differences in two brackets represent the variation of steel ratio coefficient and process energy consumption of each process from the start to the end in statistical period, respectively. The first part of the right-hand side in (2) is energy conservation attained by the adjustment of steel ratio coefficients (i.e., structural adjustment) called indirect energy conservation. And the second part is energy conservation attained by the adjustment of process energy consumption called direct energy conservation. The total energy conservation is equal to the addition of direct and indirect energy conservation. Equation (2) is applied to analyze the influences of process energy consumption and steel ratio coefficient on energy consumption per ton of crude steel and calculate the contribution rate of the direct and indirect energy conservation in different periods.

Steel ratio coefficients of CSI in representative years during 1980–2010 are shown in Table 1, and process energy consumptions are shown in Table 2. Table 1 shows that steel ratio coefficients of just basic oxygen furnace (BOF) steelmaking and steel rolling have an ascendant trend because of the increase of BOF steel production and finished product rate of CSI. Steel ratio coefficients of BOF and steel rolling rose by 84.84% and 23.53%, respectively, while those of other production departments are falling like coking, sintering, iron-making, and electric arc furnace (EAF) steel-making, and the largest decline rate is 49.05% for coke-making. Table 2 presents that all process energy consumptions are reduced. BOF is falling with a fastest fall rate of 100.30% followed by 80.42% and 76.69% for EAF steel-making and steel rolling, respectively. Based on (2) and data in Tables 1 and 2, variation of energy consumption per ton of crude steel, direct energy conservation, indirect energy conservation, and energy conservation contribution rate of CSI in each Five-Year Plan period are calculated and listed in Table 3. It can be found in Table 3 that the energy conservation of CSI is 595 kgce/t in the 30 years, with 60% direct energy conservation of 356 kgce/t and 40% indirect energy conservation of 239 kgce/t.

TABLE 1: Steel ratio coefficients of CSI during 1980–2010 (t-product/t-crude steel).

Year	Coke-making	Sintering	Iron-making	Open hearth	BOF	EAF	Steel rolling
1980	0.793	1.727	1.082	0.320	0.488	0.192	0.765
1985	0.706	1.638	1.012	0.263	0.521	0.216	0.800
1990	0.632	1.602	0.990	0.198	0.589	0.213	0.811
1995	0.580	1.619	0.992	0.137	0.667	0.196	0.812
2000	0.402	1.517	0.922	0.008	0.877	0.115	0.890
2005	0.404	1.334	0.875	NA	0.883	0.117	0.920
2010	0.334	1.269	0.944	NA	0.902	0.098	0.945

NA is the abbreviation for "not available" due to the elimination of open hearth.

TABLE 2: Process energy consumptions of CSI during 1980–2010 (kgce/t-product).

Year	Coke-making	Sintering	Iron-making	Open hearth	BOF	EAF	Steel rolling
1980	217	104	556	200	67	378	266
1985	187	90	535	156	48	325	211
1990	185	79	526	123	38	303	187
1995	180	78	509	111	31	319	177
2000	159	70	465	140	29	274	119
2005	147	65	457	—[a]	36	201	89
2010	106	53	408	—	−0.2[b]	74[c]	62

[a]NA is the abbreviation for "not available" due to the elimination of open hearth, [b]negative value is due to the negative energy steel-making technology, [c]and the great decline of process energy consumption of EAF is because that China has adjusted the standard coal coefficient of electricity from 0.404 kgce/kWh to 0.1229 kgce/kWh since 2006.

2. Progress on Key Common Energy-Saving Technologies

Reviewing the 30 years of energy conservation process mentioned above, it may be observed that two aspects are promoting the energy-saving technologies. One is digestion and reinnovation of introduced foreign advanced technologies, and the other is the formation and development of domestic systems energy conservation technology. The cross-fusion of the two aspects makes a series of key common energy-saving technologies arise in all phases of CSI at the historic moment [27–29].

2.1. Waste Heat Recovery and Reusing Technology. Due to the continuous optimization of steel manufacturing process and reduction of process energy consumption, recovering and reusing of waste heat and sources from all production departments have drawn more and more attention from steelworks [30, 31]. Partly estimated, there will be 8.44 GJ waste heat when 1 ton of crude steel is produced, accounting for 37% of the energy consumption per ton of crude steel [32]. The waste heat may be carried by finished or semifinished product, molten slag, waste gas, cooling water, and so forth, as shown in Figure 3. If taking the chemical energy of LDG and blast furnace (BF) top gas waste pressure into account, the waste heat and energy will reach 9.58 GJ/t for the BF-BOF route. As shown in Table 4, the current waste heat recovery level is about 3.00 GJ/t, accounting for 31.3% of the total waste heat and energy.

The recovery and utilization of waste heat must be based on both of the first and second laws of thermodynamics. So attention should be paid not only to the amount of heat recovery, but also to the loss of effective energy in recovery process. Therefore, the cascade utilization principle was strongly advocated in recent years of CSI according to the quantity, quality (temperature), and user demand. By the end of 2010, the number of CDQ devices under operation and construction had been 159 [33]. That achieves the recovery of 0.5 t–0.6 t steam at about 3.9 MPa and 450°C, which can generate electricity of 95 kWh–105 kWh [34]. There were 597 sets of TRT, with an average and highest power generation of 32 and 56 kWh per ton of hot metal in 2010. And CSI has 30 CCPP under operation and construction and 66 sets of sintering machines equipped with waste heat recovery devices [35]. Selective recovery and cascade utilization of sintering waste heat [36], the most representative case, not only reduce the energy consumption and waste gas emissions but also ameliorate the sintering process and sinter quality. Obviously, that happens on the condition that hot air about of 200°C is directly used as the combustion air of ignition furnace or for drying materials or hot air sintering.

2.2. Raw Material and Fuel Pretreatment Technology. Coal and lean ore are the dominating energy and ore in China, which brings many difficulties to energy conservation and environment protection for CSI. It creates conditions to reduce process energy consumption and waste emissions of pretreating raw material and fuel including iron ore, coal for coking, crude gas, and so forth. Therefore, CSI has always adhered to the principle of fine materials and improved the grade of iron ore and charging ore continuously [37]. Now the grade of charging ore in large-scale BF is controlled at

TABLE 3: Variation and effect of energy consumption per ton of crude steel for CSI during 1980–2010.

Items		1980–1985	1985–1990	1990–1995	1995–2000	2000–2005	2005–2010
Variation	kgce/t	−179	−89	−37	−199	−67	−24
	%	30	15	6	33	11	4
Direct parts	kgce/t	−124	−57	−20	−103	−38	−14
	%	69	63	55	52	57	60
Indirect parts	kgce/t	−55	−33	−17	−96	−29	−10
	%	31	37	45	48	43	40

Data in column 2005–2010 excluding the influence of the change of standard coal coefficient of electricity.

TABLE 4: Data of waste heat and resource and its recovery situation of CSI.

Waste resource type	Technical parameter	Amount of waste resource (GJ/t)	Average recovery resource (GJ/t)	Recovery rate (%)
BF top gas waste pressure	0.15 MPa–0.25 MPa	0.30	0.25	83.3
LDG chemical energy	8360 kJ/m^3	0.84	0.58	69.0
Waste heat	High, moderate, and low temperature	8.44	2.17	25.7
Total		9.58	3.00	31.3

The steel ratio coefficient of iron-making process is set as 1.0.

about 57% [38]. In the 21st century, the new technologies and equipment for raw material and fuel pretreatment mainly include the hot air sintering technology with circular cooler waste gas as heat source [36], CMC technology with waste gas from coke oven as heat source [39], moisture removable blast and gas dehydration technology with recovered low-temperature stream as power source [40, 41], BFG-BDC [42, 43], and LDG-BDC [44]. These techniques reduce the moisture in charging coal, coke, gas and blast furnace blowing and also provide a space for the low-temperature waste heat.

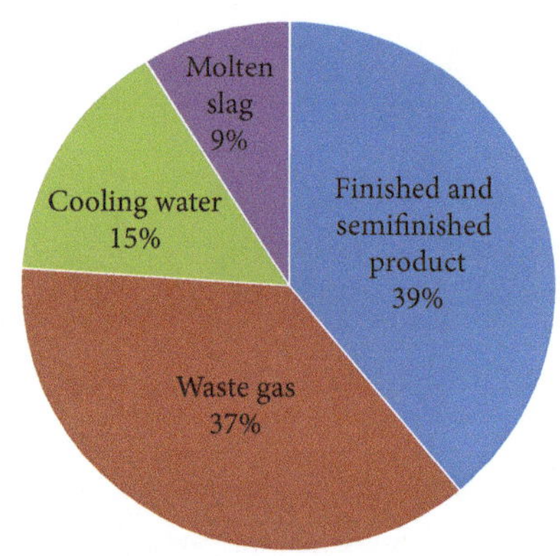

FIGURE 3: Waste heat resource structure of CSI.

2.3. Hot Connection Technology for Processes Interface. Hot connection technology is the connecting and matched physico-chemical properties control technology of material flow between adjacent processes [45, 46]. Developing hot connection technology may achieve the close connection, especially hot connection, of material and energy flow in flow rate, temperature, composition, space, time, and so forth. It also promotes the stability, continuity and high efficiency of the overall manufacturing process. CSI attaches great importance to hot connection technologies for process interfaces like BF-BOF section, BOF-continuous casting section, and continuous casting to reheating furnace section [47]. And interface technology of "route with one open ladle from BF to BOF" has been put into use in Shougang Jingtang United Iron & Steel Co., Ltd., Shagang Group, and other new established steel plants, which receives a significant energy-saving effect. The model abolishes the traditional hot metal mixer, torpedo car, and pouring in steel-making workshop. It also shortens the time of hot metal pretreatment and transfer process. And thus the general layout of BF-BOF section is compacted. Multiple superiorities appear like reducing heat dissipation, iron loss, and dust emissions. So it is the exact interface model that BF-BOF section is developed towards.

2.4. High-Temperature Air Combustion Technology. In the 1990s, CSI developed regenerative flame furnace based on HTAC technology with pure BFG as fuel [48–50]. Since 2000, many small- and medium-sized steelworks have replaced their recuperative heating furnace with regenerative furnace gradually, and then HTAC technology disseminated rapidly in China. Regenerative heating furnace takes advantage of the regenerative chamber closely connected to the furnace body or burner to recover sensible heat of the waste gas from furnace hearth [49, 51]. And that preheats BFG and combustion air to about 1000°C both. Firstly it meets the requirements of metal heating process and furnace system. Secondly it reduces the energy consumption of rolling process. Last but not the least, it eases some surplus BFG severe diffuse problems of small- and medium-sized steelworks [52]. With the expansion application scope of regenerative heating furnace and gas structure change of CSI after 2005, the number of regenerative heating furnace fueled with pure COG or mixed gas of COG and BFG is increasing year by year.

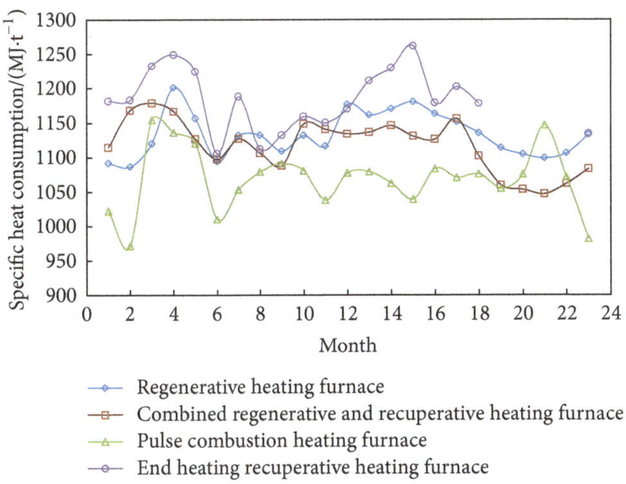

FIGURE 4: Comparison of specific heat consumption of four types of heating furnaces.

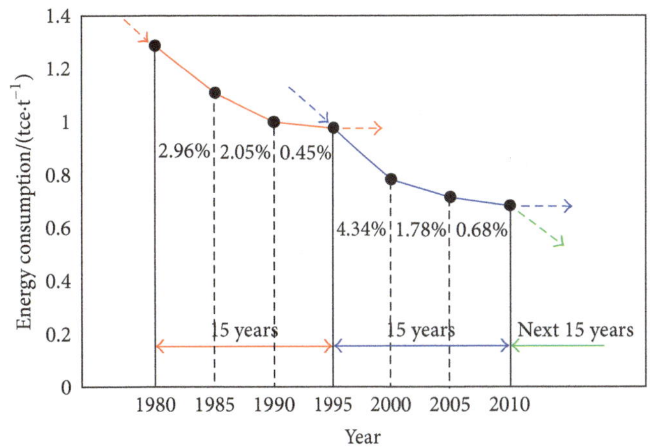

FIGURE 5: Change trend of energy consumption per ton of crude steel of CSI.

For the reason of furnace design and pyrological operating, the energy-saving effect of regenerative heating furnace is severely affected. So combined regenerative and recuperative heating furnace and intelligent furnace control technology emerged. Figure 4 gives the comparison of specific heat consumption among four types of heating furnaces. It can be seen from Figure 4 that specific heat consumption level is closely related to metal heating process and furnace working conditions. It is necessary for any type of furnace to change the operational and structural parameters when pyrological processes changes; otherwise it will not receive the expected energy-saving effect. Therefore, the type of furnaces must be diverse to fit the changes in raw material and fuel conditions and product structure.

2.5. Energy Management and Control Technology. The number of EMS has increased to 30 in CSI since Baosteel built the first one in 1985. CSI has a plan that steelworks at 2 Mt crude steel level or more than 500,000 tons of special steel will set up EMS at the end of "Twelfth Five-Year Plan" [53, 54]. The practice shows that EMS is not a simple energy management department or a single computer data acquisition system, but rather the entity for energy management and control system [55]. EMS is equipped with a variety of the digital monitoring instruments and mainframe computers, presenting coal gas, steam, electricity, technical gases (oxygen, nitrogen, argon, compressed air, etc.), water (fresh water, circulating water, soft water, and the exterior drainage), and other information on the same platform. It not only has the functions of energy flow rate monitoring, energy supply and demand display, and energy forecasting, but also can autogenerate an optimal scheduling strategy to achieve a steady supply and efficient utilization according to the change of production. Since the beginning of the new century, the majority of EMS of CSI is in the hardware construction phase of energy measurement network and energy flow network, and only a small number enters into the development period of offline decision-making and operational optimization. However, the function of software is not fully presented. Work is left to be done to realize the dynamic energy forecasting and real-time scheduling [56, 57].

3. Conclusions and Future Prospects

A dissection of the changing process of energy consumption per ton of crude steel of CSI given in Figure 2 shows not only differences but also similarities among different periods. Figure 5 shows the change trend of energy consumption per ton of crude steel of CSI from 1980 to 2010. There is a cyclical change within the 30 years with 15 years as a cycle. Expect for the different average annual energy conservation rates, the shape and downtrend of the two curves are the same. The first five years of a cycle get the fastest drop rate, followed by the middle five years; and the last five years have slowed down continuously. For instance, the average annual energy conservation rate from 1995 to 2000 is 4.34% by adjusting production structure and optimizing manufacturing process, then the energy conservation rate dropped to 1.78% during 2000–2005 and only 0.68% in the period 2005–2010.

For next the 15 years or even a longer time, there will be smaller space and more difficulties for CSI energy conservation. A third substantial drop will appear if novel theories, technologies, and management methods come out to support. In the near future, CSI can deepen energy conservation and push it to a new stage only through relying more heavily on scientific and technological progress, giving full play to the advantages of efficient energy conversion and effective energy use, and focusing on the development of key common technologies suitable for CSI.

(1) Using the concept of system and energy carrier [58], detailed analysis and solutions should be conducted on each link in the life cycle of steel products and the relationship between two different links. It makes each line meet the requirements of scientific energy utilization and systems energy conservation as much as possible. To reach the synergistic operating between material flow and energy flow, several researches should be conducted: to study how energy

consumption per ton of crude steel is influenced by ore, scrap steel, coal, electricity, and other raw materials and fuels under market economy conditions, to research the relationships and optimization of different links in production process at temperature, composition, geometry, time, and so forth, and to study the running rules of energy flow and the construction and optimization of energy flow network in steel manufacturing process.

(2) According to the quantity, quality, and user demand, CSI should maximize the recovery of waste heat from steel production processes. It is best for waste heat to be directly used in steel production processes without conversion or changed to other kinds of energy or resources. For example, surplus gas and steam could be sent to neighboring cities or other factories as energy resources, self-used for heating and/or cooling, or used for implementation of common thermal power together with the city.

(3) It is necessary to establish a new energy use evaluation criteria, evaluation method, and energy consumption indices system by applying the first and second laws of thermodynamics. That helps to study theoretical limit energy consumption per ton of crude steel, which will be used to guide CSI energy conservation in the next 15 years (2011–2025) to reach or get close to the minimum energy consumption and become the world's most advanced steel industry.

Conflict of Interests

The authors declare they have no conflict of interests.

Acknowledgment

This work is supported by the S&T research funds of the Liaoning Provincial Education Department (L2012082), China.

References

[1] J. R. Vehec, *Technology Roadmap Research Program for the Steel Industry*, American Iron and Steel Institute, Washington, DC, USA, 2010.

[2] Z. W. Lu and A. G. Xie, "Analysis of overall energy intensity of Chinese steel industry," in *Proceedings of the International Conference on Energy and Environment*, pp. 198–202, Begell House Inc., New York, NY, USA, 1996.

[3] Q. Zhai and Z. Lu, "Prediction on the energy intensity of chinese iron and steel enterprises," in *Proceedings of the International Conference on Energy and Environment*, pp. 47–51, China Machine Press, Shanghai, China, May 1998.

[4] Z. W. Lu, *Energy Utilization in Metallurgical Industry*, Metallurgical Industry Press, Beijing, China, 1986.

[5] Z. W. Lu and J. J. Cai, *Foundations of Systems Energy Conservation*, Northeastern University Press, Shenyang, China, 2010.

[6] X. P. Zhang and Q. Y. Zhang, "Development trend of world steel industry and the influence on China's steel industry," *World Regional Studies*, vol. 14, no. 2, pp. 80–86, 2005.

[7] Q. Zhang, J. J. Cai, G. J. Duan, J. J. Wang, and H. Q. Liu, "System energy conservation method and technical analyses of iron and steel industry," *Energy Conservation*, no. 1, pp. 35–37, 2006 (Chinese).

[8] J. Zhang and G. Wang, "Energy saving technologies and productive efficiency in the Chinese iron and steel sector," *Energy*, vol. 33, no. 4, pp. 525–537, 2008.

[9] Z. Lu, J. Cai, Q. Yu, and A. Xie, "The influences of materials flows in steel manufacturing process on its energy intensity," *Acta Metallurgica Sinica*, vol. 36, no. 4, pp. 370–378, 2000 (Chinese).

[10] Q. B. Yu, Z. W. Lu, and J. J. Cai, "Calculating method for influence of material flow on energy consumption in steel manufacturing process," *Journal of Iron and Steel Research International*, vol. 14, no. 2, pp. 46–51, 2007.

[11] A. Xie and Z. Lu, "Influence of structure adjustment of Chinese iron and steel industry on energy consumption," *Iron and Steel*, vol. 31, no. 11, pp. 53–70, 1996 (Chinese).

[12] G. Y. Ma, J. J. Cai, L. H. Zhang, and W. Q. Sun, "Influence of steam recovery and consumption on energy consumption per ton of steel," *Energy Procedia*, vol. 14, pp. 566–571, 2012.

[13] P. Michaelis and T. Jackson, "Material and energy flow through the UK iron and steel sector—part 2: 1994–2019," *Resources, Conservation and Recycling*, vol. 29, no. 3, pp. 209–230, 2000.

[14] Y. A. Gudim, A. A. Golubev, S. G. Ovchinnikov, and I. Y. Zinurov, "Promising technology for making steel with the use of scrap and a metallized raw material," *Metallurgist*, vol. 53, no. 3-4, pp. 196–200, 2009.

[15] H. Liu, X. Zhang, M. Wu et al., "Computational and experimental study of cooling process in coke dry Quenching experimental shaft," *Journal of Thermal Science*, vol. 11, no. 2, pp. 121–127, 2002.

[16] C. M. Macedo, R. Schaeffer, and E. Worrell, "Exergy accounting of energy and material flows in steel production systems," *Energy*, vol. 26, no. 4, pp. 363–384, 2001.

[17] B. Lin, Y. Wu, and L. Zhang, "Estimates of the potential for energy conservation in the Chinese steel industry," *Energy Policy*, vol. 39, no. 6, pp. 3680–3689, 2011.

[18] Z. W. Lu, "Analysis of the comprehensive energy consumption per ton of steel of China's iron and steel industry," *Energy for Metallurgical Industry*, vol. 11, no. 1, pp. 14–20, 1992 (Chinese).

[19] Z. W. Lu, "A brief analysis on the steel scrap resources for steel industry," in *CSM Annual Meeting Proceedings*, pp. 70–80, Metallurgical Industry Press, Beijing, China, 2001.

[20] W. Q. Sun, J. J. Cai, T. Du, and D. W. Zhang, "Specific energy consumption analysis model and its application in typical steel manufacturing process," *Journal of Iron and Steel Research International*, vol. 17, no. 10, pp. 33–37, 2010.

[21] J. J. Cai, J. C. He, and Z. W. Lu, "Analysis of energy saving and energy consumption in Chinese steel industry for last 20 years and next 5 years," *Iron and Steel*, vol. 37, no. 11, pp. 69–73, 2002 (Chinese).

[22] H. X. Jin, F. Z. Wu, and Y. H. Li, "Analysis on energy saving for blast furnace-converter process," *Modern Machinery*, no. 6, pp. 30–35, 2008 (Chinese).

[23] J. J. Cai and W. Q. Sun, "Systems energy conservation and scientific energy utilization of iron and steel industry in China," *Iron and Steel*, vol. 47, no. 5, pp. 1–5, 2012 (Chinese).

[24] Z. W. Lu, Q. G. Zhai, A. G. Xie, J. J. Cai, and Q. S. Meng, "Prediction on the energy consumption of Chinese steel industry," *Iron and Steel*, vol. 32, no. 5, pp. 69–74, 1997 (Chinese).

[25] Q. C. Bu, Z. W. Lu, and J. J. Cai, "Analysis on energy consumption of crude steel in the ninth five-year plan in JISCO," *Energy for Metallurgical Industry*, vol. 22, no. 1, pp. 10–13, 2003 (Chinese).

[26] Z. Lu, A. Xie, and D. Zhou, "More on direction and energy conservation measures of Chinese iron and steel industry," *Iron and Steel*, vol. 31, no. 2, pp. 54–58, 1996 (Chinese).

[27] Z. Liu, J. Liu, and Y. Wang, "Energy consumption in the iron and steel industry in P.R. China," *Energy for Sustainable Development*, vol. 3, no. 3, pp. 18–24, 1996.

[28] X. L. Ma, C. Zhang, J. G. Liu, and H. J. Zhao, "Practice and discussion about the high-quality burden production in Anyang I&S Co.," *Sintering and Pelletizing*, vol. 30, no. 4, pp. 56–59, 2002 (Chinese).

[29] O. Movshuk, "Restructuring, productivity and technical efficiency in China's iron and steel industry, 1988–2000," *Journal of Asian Economics*, vol. 15, no. 1, pp. 135–151, 2004.

[30] H. T. Makkonen, J. Heino, L. Laitila, A. Hiltunen, E. Pöyliö, and J. Härkki, "Optimisation of steel plant recycling in Finland: dusts, scales and sludge," *Resources, Conservation and Recycling*, vol. 35, no. 1-2, pp. 77–84, 2002.

[31] L. Song, "Practice of increasing power generation with sintering waste heat," *Sintering and Pelletizing*, vol. 33, no. 2, pp. 55–58, 2008 (Chinese).

[32] J. J. Cai, J. J. Wang, C. X. Chen, and Z. W. Lu, "Recovery of residual-heat integrated steelworks," *Iron and Steel*, vol. 42, no. 6, pp. 1–7, 2007 (Chinese).

[33] J. Y. Gao, R. Z. Wang, and M. Zhou, "Development and application of dry coke quenching technology in China," *Gas & Heat*, vol. 31, no. 1, pp. 4–7, 2011 (Chinese).

[34] D. Wang, T. Yang, Z. Wen et al., "A mathematical model for optimized operation and control in a CDQ-Boiler system," *Journal of University of Science and Technology Beijing*, vol. 12, no. 5, pp. 390–393, 2005.

[35] G. P. Ren, C. J. Ren, and Y. J. Wei, "Waste heat utilization technologies and application," *Shanghai Energy Conservation*, no. 5, pp. 2–6, 2009 (Chinese).

[36] J. J. Cai, H. Dong, T. Du, C. B. Xu, J. W. Zhou, and K. Lin, "Study on grade recovery and cascade utilization of waste heat from sintering-cooling process," *Iron and Steel*, vol. 46, no. 4, pp. 88–92, 2011 (Chinese).

[37] J. Ma, D. G. Evans, R. J. Fuller, and D. F. Stewart, "Technical efficiency and productivity change of China's iron and steel industry," *International Journal of Production Economics*, vol. 76, no. 3, pp. 293–312, 2002.

[38] Y. Q. Weng, "Further development of coal injection and charge beneficiation for BF, promotion of energy saving, consumption reduction and structural regulation of ironmaking system," *Iron and Steel*, vol. 33, no. 7, pp. 1–4, 1998 (Chinese).

[39] G. Q. Zhang, Q. F. Zhang, and Y. G. Liu, "Application of fluidized bed type coal moisture control technology with coke oven flue waste gas," *Fuel & Chemical Processes*, vol. 41, no. 6, pp. 9–11, 2010 (Chinese).

[40] H. Deng, "Application of blast furnace dewetting blast technology and development," *Metallurgical Power*, no. 4, pp. 62–64, 2009 (Chinese).

[41] L. C. Xiao, Q. Li, Z. J. Ding, L. Xiao, and B. Zhu, "Overview of dehydration of coal gas," *Journal of Hebei University (Natural Science Edition)*, vol. 27, no. 4, pp. 444–448, 2007 (Chinese).

[42] C. L. Li, "Application of dry dusting technique in blast furnace," *Metallurgical Power*, no. 1, pp. 28–29, 2010 (Chinese).

[43] J. L. Tian, "Application status of dry dust removal technology for blast furnace gas," *Energy for Metallurgical Industry*, vol. 26, no. 5, pp. 3–4, 2007 (Chinese).

[44] D. L. Zhang, Y. L. Mao, Y. L. Qu, and Z. Wang, "Dry-type deducting technology for converter gas," *World Steel*, no. 5, pp. 51–59, 2012 (Chinese).

[45] T. S. Su, D. N. Lan, R. Z. Liu, S. T. Qiu, L. W. Liu, and X. R. Zheng, "The developing direction and forecast of China's continues casting technique in 'Tenth Five-year Plan'," *Journal of East China University of Metallurgy*, vol. 17, no. 4, pp. 271–281, 2000 (Chinese).

[46] R. Y. Yin, "Discussion on high efficient and low cost clean steel production platform: one of the key technologies for steel industry in 21st century," *China Metallurgy*, vol. 20, no. 10, pp. 1–10, 2010 (Chinese).

[47] L. Tang, P. B. Luh, J. Liu, and L. Fang, "Steel-making process scheduling using Lagrangian relaxation," *International Journal of Production Research*, vol. 40, no. 1, pp. 55–70, 2002.

[48] A. H. Wang, J. J. Cai, and G. W. Xie, "Numerical study of flame properties and nitrogen oxide formation in high temperature air combustion," in *Challenges of Power Engineering and Environment: Proceedings of the International Conference on Power Engineering 2007*, pp. 1376–1379, Zhejiang University Press, Springer, Hangzhou, China, 2007.

[49] W. Dong, *Design of Advanced Industrial Furnaces Using Numerical Modeling Method*, Royal Institute of Technology, Stockholm, Sweden, 1999.

[50] W. Dong and W. Blasiak, "Study on mathematical modelling of highly preheated air combustion," in *Proceeding of the 2nd International High Temperature Air Combustion Symposium*, pp. C521–C528, Kaohsiung, Taiwan, 1999.

[51] N. Rafidi and W. Blasiak, "Heat transfer characteristics of HiTAC heating furnace using regenerative burners," *Applied Thermal Engineering*, vol. 26, no. 16, pp. 2027–2034, 2006.

[52] I. Glassman, *Combustion*, Academic Press Inc., London, UK, 1996.

[53] K. Fan, "Structure and function realization of the energy management and control center in Jinan Iron & Steel Co., Ltd.," *Metallurgical Industry Automation*, vol. 35, no. 2, pp. 24–28, 2011 (Chinese).

[54] W. M. Feng and L. Q. Cong, "Energy management system of entire iron and steel plant," *Control Engineering of China*, vol. 28, no. 3, pp. 597–600, 2005 (Chinese).

[55] J. H. Yang, W. Q. Sun, J. J. Cai, H. J. Mao, and R. Fang, "Development of supply-demand balance and distribution software of gas system for iron and steel industry," *Procedia Engineering*, vol. 15, pp. 5143–5147, 2011.

[56] Z. C. Chen and P. Robin, "Energy management and environmental awareness in China's enterprises," *Energy Policy*, vol. 28, no. 1, pp. 49–63, 2000.

[57] H. S. Wang, "Strengthening productive forces based on information to create more benefits for enterprises," *Metallurgical Industry Automation*, vol. 30, no. 1, pp. 1–5, 2006 (Chinese).

[58] J. Cai, Z. Lu, and T. Du, "The analysis of energy saving and the energy intensity of steel industry in China for last 20 years and next 5 years," in *Proceedings of the International Conference on Energy and the Environment*, pp. 37–42, Shanghai Scientific and Technical Publishers, Shanghai, China, December 2003.

4

Ecological Network Analysis for a Low-Carbon and High-Tech Industrial Park

Yi Lu,[1] Meirong Su,[1] Gengyuan Liu,[1] Bin Chen,[1,2] Shiyi Zhou,[2] and Meiming Jiang[2]

[1] *State Key Joint Laboratory of Environmental Simulation and Pollution Control, School of Environment, Beijing Normal University, Beijing 100875, China*
[2] *Beijing Development Area Co. Ltd, Beijing 100176, China*

Correspondence should be addressed to Bin Chen, chenb@bnu.edu.cn

Academic Editors: V. Kazukauskas, R. Maceiras, S. S. Martinez, S. Medved, and J. L. Míguez

Industrial sector is one of the indispensable contributors in global warming. Even if the occurrence of ecoindustrial parks (EIPs) seems to be a good improvement in saving ecological crises, there is still a lack of definitional clarity and in-depth researches on low-carbon industrial parks. In order to reveal the processes of carbon metabolism in a low-carbon high-tech industrial park, we selected Beijing Development Area (BDA) International Business Park in Beijing, China as case study, establishing a seven-compartment- model low-carbon metabolic network based on the methodology of Ecological Network Analysis (ENA). Integrating the Network Utility Analysis (NUA), Network Control Analysis (NCA), and system-wide indicators, we compartmentalized system sectors into ecological structure and analyzed dependence and control degree based on carbon metabolism. The results suggest that indirect flows reveal more mutuality and exploitation relation between system compartments and they are prone to positive sides for the stability of the whole system. The ecological structure develops well as an approximate pyramidal structure, and the carbon metabolism of BDA proves self-mutualistic and sustainable. Construction and waste management were found to be two active sectors impacting carbon metabolism, which was mainly regulated by internal and external environment.

1. Introduction

The study of ecoindustrial parks (EIPs) has assumed great deal of importance within the past ten to fifteen years. One of the best definitions of an EIP has been provided by the UESPA, which was stated as "*a community of manufacturing and service businesses seeking enhanced environmental and economic performance by collaborating in the management of environmental and reuse issues. By working together the community of businesses seeks a collective benefit that is greater than the sum of the individual benefits each company would realize if it optimized its individual performance only*" [1]. This definition mainly aims to close material cycles in the industrial chain and considers the entire life cycle from raw material production to product consumption and waste management [2].

Despite cooperation between companies to find win-win solutions [3], from the traditional perspective of EIPs, one of their strong points is basically focused on the structural planning and functional utility of industrial recycling. It means that a better design or running of EIPs' cycle have arisen to reduce waste management and disposal costs, extract with cheaper materials and energy, and gain benefits from residues [4]. However, in the more furious conflict between development of economic society and recovery of polluted ecosystem, grimmer situation and stricter requirements will be put forward. Traditional definition of EIPs cannot adapt to the social development better, especially when the whole world is paying more attention on climate change.

Climate change announces a fast global socioeconomic transition, but nobody could predict the ultimate results of the next industrial revolution. Obviously, business and government play a key role in promoting or destroying the revolution. It is undeniable that climate change affects both natural ecosystems and human societies [5–7]. Taking climate change into consideration, EIPs can be classified into green industry parks and integrated ecoindustry parks. Therein, green industry park was defined as "*a range of enterprises that use cleaner production technologies, process much of their waste and/or reduce the emissions of greenhouse gases, in situ*" [8]. In Belgium, green industry park is embodied as the "carbon neutral industrial parks," an initiative of the Flemish region, which brings in the carbon footprint of companies into the quality requirements for industrial parks to enforce carbon emission reduction and sustainable energy policy measures [7]. And some studies have suggested that it is practical to implement climate change through industrial symbiosis [9, 10] and energy innovation in EIPs [11].

An industrial ecosystem is constructed with the flows of matter, nutrients, energy, and carbon [12]. As a necessary part of industrial flows, some scientists make great contribution to stretch out the carbon cycle in different industrial ecosystem. Korhonen et al. considered the material (including carbon flow) and energy flows of forest ecosystem in Finland [12]. Liu et al. provided an outline of energy-based greenhouse gas emissions inventory in Suzhou Industrial Park, China [13]. Norstebo et al. and Midthun et al. both analyzed the taxation of CO_2 emissions and carbon capture in a case study of an extension of an EIP in Norway [14, 15]. Munir et al. presented a newly developed carbon emission Pinch Analysis technique for achieving holistic minimum carbon targets in EIPs [16]. Besides, there is still a lack of speciality of the system-wide carbon emission in an EIP.

EIPs originate from three major concepts, namely, scilicet industrial ecology, biological ecology, and the spatial perspectives based on landscape ecology [17]. In the second field of EIPs, scientists have tried to seek the inner-mimicry from biological individuals with industrial unit to natural ecosystem with industrial system. It is a metaphor running through the biological-industrial process, structure, and function. Hence the concept of industrial metabolism is first established by Ayres, which is defined as "*the whole integrated collection of physical processes that convert raw materials and energy, plus labor, into finished products and wastes*" [18, 19]. The summary of industrial metabolism is analyzing the whole material flows, extracting all possible emission sources, and assessing the influences within these flows [20]. After that, Graedel shows a good example in using statistical indexes of food web from natural ecosystems for reference in his evaluations of EIPs [21].

Nowadays, diverse analysis tools are used in the research of EIPs, which can be classified into two main trends of methodology. One of them is generally based on the inventorying of life-cycle ecological and economic input-output flows, including material flow analysis [3, 22], input-output analysis [23], life-cycle assessment [24], and structure and network analysis based on industrial metabolism [20, 25, 26]. The other method is concerned with the idea of the available solar energy.

Ecological Network Analysis (ENA) is a general metabolism-based analytical tool for studying the system connectivity and for quantifying and qualifying direct and indirect ecological flows in the system [27]. In general, the key concepts of ENA are behavior, structure, and function of a system [27]. An EIP can be extracted as a symbiotic network, where many material and energy flows link diverse compartments in ensuring a smooth running of industrial processes (both of the components' mutual interactions in integral environment and the passing relationship between integral and external environment) and system functioning. So it is appropriate to introduce ENA into EIP study. However, present attempts in seeking carbon metabolism in EIPs based on the network view are still a blank space, and challenges include defining what low-carbon high-tech industrial park is (especially in what "low-carbon" could be defined), understanding why these parks could reduce more carbon emission, as well as seeking to use an accessible tool for metabolic structure and functioning study. The adoption of ENA may provide a feasible prospect in resolving the latter two issues by evaluating the metabolic intensities, processes, structure, and control of carbon emissions.

In China, there exists a conflict among economic development, shortage of natural resource, and serious pollution suggested by the impetus for developing EIPs [28]. Hence developing high-tech industrial parks has been a promising trend of EIPs' development. These EIPs generally assemble high-tech business located in the upstream of industrial chain, such as companies in the domains of intellective, R & D, design, or head office. BDA International Business Park (BDA) is situated in Beijing High-Tech Industrial Park, which is characterized as its graceful ecological environment, low energy consumption, low carbon emission intensity, and amassing of high-tech business. As a feasible trial in developing low-carbon industrial parks, BDA covers a landscape area of $0.1735\,km^2$ and a construction area of $0.336\,km^2$. The park consists of 34 separate office buildings where 159 high-tech companies have been stationed in total. It is worth emphasizing that BDA is the first EIP considering the low-carbonic concept in its design in Beijing. This study tries to present an ENA-based methodology for carbon metabolism in low-carbon high-tech industrial parks and selects BDA as a case study for promoting carbon reduction in EIPs.

2. Materials and Methods

2.1. Ecological Network Model for Low-Carbon High-Tech Industrial Park. The essence of ecological network model is a transmitting network for materials and energy, which includes both of the components' mutual interactions in integral environment and the passing relationship between integral and external environment. For this sake, establishing the reasonable system boundary and making sure of limiting factors should be the necessary step for the ecological network model.

Even though EIPs are mainly artificially controlled, both artificial and natural processes of parks' carbon fluxes should be taken into account in the network model. The system boundary does not just coincide with the administrative boundaries, but a virtual boundary that contains metabolic processes links both inside and outside of the park. Taking carbon metabolic processes and their relationship through different compartments within the virtual boundary, a metabolic network model for low-carbon high-tech industrial parks (we might call it as Low-Carbon Metabolic Network (LCMN) as well) is established for tracking carbon flows within an low-carbon park ecosystem (Figure 1). In the LCMN, it embraces seven individual compartments: energy supporting sector (Eng), construction sector (Con), industry, business and service sector (IBS), waste management sector (Wst), green project sector (Grn), internal environment (Int), and external environment (Ext). Diverse carbon fluxes running through these compartments are identified and characterized as the compartmental interactions within LCMN, including the flows within Eng, Con, IBS, and Wst in terms of exchanging materials (both goods and wastes) and energy, flows between the park and its external environment, and the natural carbon exchange of the park. Besides, the flow of goods transporting and transport fuel induced by BDA (within and outside the park) were also involved.

2.2. Ecological Network Analysis

2.2.1. Network Utility Analysis (NUA). "Utility" is an economic conception similar to "efficiency." Since Patten [29, 30] firstly introduced the concept of NUA, it has mainly been applied to indicate both qualitative and quantitative exchange-based relationships between different compartments of a network system [27]. The relation forms within components are in variety, where one of the simplest ideas is sorted as direct and indirect (similar as a series of consequent direct transfer). In NUA, direct and integral relationships are expressed by a direct utility matrix **D** and dimensionless integral utility intensity matrix **U**, respectively. Matrix $\mathbf{D} = [d_{ij}]$ illustrates the relative strength of direct input and output control in the network [27]. d_{ij} represents the direct interactions between compartment i and j, which can be expressed as

$$d_{ij} = \frac{\left(f_{ij} - f_{ji}\right)}{T_i}, \tag{1}$$

where f_{ij} represents the metabolic flow (e.g., carbon flow) from compartment j to i; T_i is the sum of input or output flows for the ith compartment at the steady state. Then, whole-system, integral, utility-based relations are given by considering all the indirect influences in the network carried by the higher-order interactions [31]. Distinguished with the direct utility matrix **D**, integral utility intensity matrix **U** contributes to reveal the strength of the entire network organization. Compared with the sum of elements between the direct and indirect matrix, it could often find a greater contribution from indirect processes than from the direct one [27]. For revealing the net utility of each compartment to make use of materials along different-step pathways, matrix **U** is computed as

$$\mathbf{U} = \mathbf{D}^0 + \mathbf{D}^1 + \mathbf{D}^2 + \cdots + \mathbf{D}^m = (\mathbf{I} - \mathbf{D})^{-1}, \tag{2}$$

where **U** shows the utilities conveyed by pathways in different lengths $1, 2, \ldots, m$ (m is the total number of compartments, $m \geq 2$); the identity matrix **I** shows the self-feedback of flows through each compartment; the matrix $\mathbf{D}^1$ reflects the direct interactions between components; $\mathbf{D}^m$ represents the indirect relations between components along m-length-pathway. In the view of common network analysis indirect interactions $\mathbf{D}^k$ ($2 \leq k \leq m$) means specific materials convey via relative longer pathways greater than length one, which can be verified by taking the higher order powers of **D**, for example, $\mathbf{D}^2$ gives utilities conveyed along two-step pathways, $\mathbf{D}^3$ is along three-step pathways, and so on [31].

In NUA, Sign D and Sign U are introduced as two sign matrices of **D** and **U** in order to reveal the mutualism relationship between components. Compared with Sign D, Sign U gives a deeper perspective in revealing potential connections between each component. Referring to the gain (+), loss (−), or neutrality (0), interactions can be calculated by two objects: (+, +) stands for mutualistic condition, (+, −) for exploitation condition, (−, +) for exploited condition, (−, −) for competition, and (0, 0) for neutrality [31]. In these two matrices, the sign changing would affect the interactions between components, and then affect the network structure.

Fath and Patten [27] then investigated network synergism (also known as mutualism), another NUA property, to convey that positive utility was more than negative utility in quantity. There are two ways for testing the network synergism: one of them can be quantified by total utility in the dimensional utility matrix, while another one is revealed by the ratio of positive to negative utility in the network system [27]. Hence, at the level of entire system, network mutualism index (MI) and synergism index (SI) are adapted to show the fitness of the whole-system [32–34]. MI reflects the ratio of the number of positive and negative signs in the Sign U. While SI quantifies the total magnitude of the positive and negative utilities, which assess the mutualistic condition of a system in slightly different angles [35]. If MI is greater than one, or SI is greater than zero, the system mutualism could occur [27, 29, 30]. MI and SI are computed as

$$\mathrm{MI} = \frac{\mathrm{Sign}U(+)}{\mathrm{Sign}U(-)}, \tag{3}$$

$$\mathrm{SI} = \sum_{j=1}^{n}\sum_{i=1}^{n} u_{ij}, \tag{4}$$

where

$$\begin{aligned} \mathrm{Sign}U(+) &= \sum_{ij} \max\left(\mathrm{Sign}\left(u_{ij}\right), 0\right), \\ \mathrm{Sign}U(-) &= \sum_{ij} -\min\left(\mathrm{Sign}\left(u_{ij}\right), 0\right). \end{aligned} \tag{5}$$

2.2.2. Network Control Analysis (NCA). Patten [36] introduced NEA-based measures of control or dominance by using the input and output environ concept to develop a control matrix [37, 38]. Network control is based on a pair of integral flow through network flow analysis, which indicates the control from system compartments in the configuration

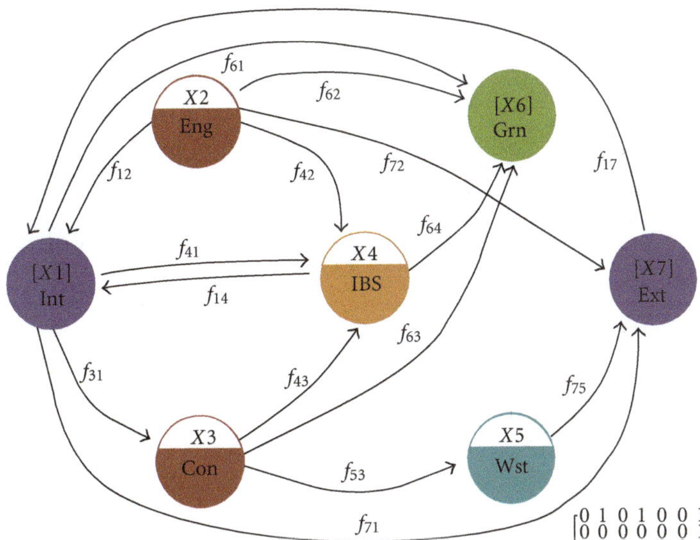

$$\begin{bmatrix} 0 & 1 & 0 & 1 & 0 & 0 & 1 \\ 0 & 0 & 0 & 0 & 0 & 0 & 1 \\ 1 & 0 & 0 & 1 & 0 & 0 & 1 \\ 1 & 1 & 0 & 0 & 0 & 0 & 1 \\ 0 & 0 & 1 & 0 & 0 & 0 & 0 \\ 1 & 1 & 1 & 1 & 0 & 0 & 1 \\ 0 & 1 & 0 & 0 & 1 & 0 & 0 \end{bmatrix}$$

Int: internal environment
Eng: energy supporting sector
Wst: waste management sector
Con: construction sector
IBS: industry, business, and service sector
Grn: green project sector
Ext: external environment
f_{ij}: carbon flow from j to i
[Xi]: stock in environment
Xi: storage of sectors

Figure 1: Metabolic network model for carbon metabolism of BDA International Business Park. The matrix at the right bottom is the adjacency matrix A of the model, where $A = [a_{ij}]$. If there exists a carbon flow from compartment j to i, $a_{ij} = 1$, else $a_{ij} = 0$.

of the whole system [35]. And network flow analysis is predicated on a conceptual flow model of a system, revealing both the structure and function of the system [39].

Be similar as the direct utility matrix D and integral utility matrix U, in control analysis (or flow analysis), flow interactions can also be divided into direct and integral (including initial input, direct, and indirect interactions) ones. Matrix $\mathbf{G}$ is the direct interaction matrix, giving the functional influences due to all paths of lengths commensurate with the power. While matrix $\mathbf{N}$ shows the indirect interaction, summing the infinite power series of the direct interaction matrix [39].

For the output environ, from the generating or flow-forward transfer efficiencies and the receiving or flow-backward transfer efficiencies $\mathbf{G} = [g_{ij}]$ and $\mathbf{G}' = [g'_{ij}]$, dimensionless integral output and input flow intensity matrices $\mathbf{N} = [n_{ij}]$ and $\mathbf{N}' = [n'_{ij}]$ can be computed as

$$\mathbf{N} = \left[n_{ij}\right] = (\mathbf{I} - \mathbf{G})^{-1}, \qquad \mathbf{N}' = \left[n'_{ij}\right] = (\mathbf{I} - \mathbf{G}')^{-1}, \tag{6}$$

where $g_{ij} = f_{ij}/T_j$, it shows the nondimensional, output-oriented, intercompartmental flows; if $g'_{ij} = f_{ij}/T_i$, it shows the input-oriented, intercompartmental flows [32, 38]. So two distributed control metrics based on these could be established to reflect the control and dependence condition, which are control allocation (CA) and dependence allocation (DA)

$$\mathrm{CA} = \left[ca_{ij}\right] \equiv \begin{cases} n_{ij} - n'_{ij} > 0, & ca_{ij} = \dfrac{n_{ij} - n'_{ij}}{\sum_{i=1}^{m} n_{ij} - n'_{ij}} \\ n_{ij} - n'_{ij} \leq 0, & ca_{ij} = 0, \end{cases} \tag{7}$$

$$\mathrm{DA} = \left[da_{ij}\right] \equiv \begin{cases} n_{ij} - n'_{ij} > 0, & da_{ij} = \dfrac{n_{ij} - n'_{ij}}{\sum_{i=1}^{m} n_{ij} - n'_{ij}} \\ n_{ij} - n'_{ij} \leq 0, & da_{ij} = 0, \end{cases} \tag{8}$$

where $0 \leq da_{ij}, ca_{ij} \leq 1$. By definition both CA and DA are calculated by the difference of two pairwise integral flows (i.e., n_{ij} and n'_{ij}). ca_{ij} reflects the degree that compartment j controls compartment i based on the controller's output environ, while da_{ij} indicates the degree that compartment j is dependent on compartment i from the observer's input environ.

Based on the network control and dependence formulation, the system-wide control condition can be revealed by the system control index (CI). CI combines control degree with dependence degree, and thus it indicates the control utility and organization capability of the whole system and can be employed to index the self-regulation of system metabolism [35] as follows:

$$\mathrm{CI} \equiv \frac{\sum_{j=1}^{m}\sum_{i=1}^{m} ca_{ij} + \sum_{j=1}^{m}\sum_{i=1}^{m} da_{ij}}{m^2}. \quad (9)$$

2.2.3. System-Wide Indicators. For the purpose of giving an overall perspective on metabolic performance of industrial park and contributing to design a both sustainable and low-carbonic park, it is necessary to define a set of indicators in addressing the system performance of the MN. Some of these indicators have already been introduced by NUA and NCA as above, while others were extracted from other researches of ENA [28]. Each indicator reflects a facet of carbon metabolism in LCMN for BDA. The formulations and short description of the whole-system indicators were illustrated (Table 1).

2.3. Data Source. The metabolism data sources were extracted from construction and operation data of BDA which were all calculated based on the IPCC recommended method and life cycle analysis. These data originated from investigations and calculation into carbon composition of artificial activities, raw materials' transportation, the relationships between these flows and stocks, and also within anthropogenic-natural processes.

3. Results

3.1. Ecological Structure of Carbon Emissions. Carbon fluxes between two compartments within LCMN of BDA are listed in Table 2. The result shows that Con (11.2%), Eng (8.2%), and Wst (8.2%) are three major carbon donors providing carbon to Con (41.4%) and Grn (24.5%), two major carbon accepters, in the form of materials, wastes, fossil fuels and machinery, and so forth. Diverse carbon exchanges between these major donors and accepters make a great contribution to support the operation of the park. Yet Eng and Wst's (both 8.2%) contribution in accepting carbon are not as great as their superior performance on the supply side. Compared with these major sectors, IBS as one of the most inactive sectors is both inferior in supplying and accepting carbon (3.9% and 6.7% resp.,). The ultimate suppliers and recipients of carbon are Int and Ext. Among which, both supplying and receipting of Int are not as much as other sectors (3.8% and 4.3% resp.,) showing a weak effect on the LCMN of BDA; Ext has a better performance in supplying (64.7%) than accepting carbon (8.9%), indicating the BDA park may be more dependent on the supply of the external environment. The biggest carbon emissions are from Eng to Int (3046 t CO_2-eq) and Wst to Ext (17547 t CO_2-eq). And IBS (6001 t CO_2-eq) and Grn (2176 t CO_2-eq) are two major sectors obtaining carbon from internal environment, which are much more than Con (13 t CO_2-eq). Similarly, Con (82509 t CO_2-eq), Grn (39620 t CO_2-eq), and Eng (12796 t CO_2-eq) are all extracting carbon greatly from the external environment, revealing that these three main sectors are more dependent on external supply again. The carbon emission (from Wst to Ext, 8.2%), extraction (from Ext to Con, 38.6%), and processing (from Con to Wst, 8.2%) are the most active and significant processes. Overall, the total carbon through flow of BDA is 213670 t CO_2-eq.

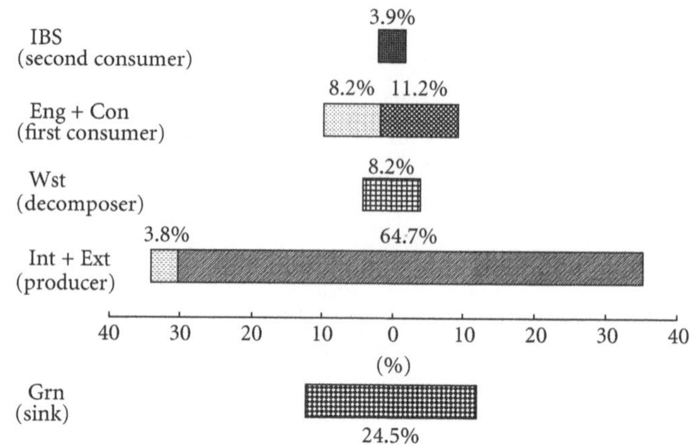

FIGURE 2: Ecological structure of carbon metabolism in BDA.

The proportion of carbon flows within each sector can reflect the ecological structure of the carbon metabolism in BDA, which forms an approximate pyramidal shape (Figure 2). In natural ecosystem, the pyramidal trophic structure based on the food web is one of the most stable structures leading an ordered and healthy ecosystem. In this sense, the carbon metabolic system in BDA is also in stable condition relatively in the role of producer (Int and Ext), decomposer (Wst), first consumer (Eng and Con), and second consumer (IBS). Besides, as a little deficiency, the decomposer cannot take full advantage thanks to the limit of waste managing technology. If we enhance the role of decomposers in the BDA's carbon metabolic network, the utilization efficiency and system stability of carbon metabolism will be improved.

Reading from left to right, the values are the total carbon inputs or outputs (Grn fits the output value) of compartments. Producer: supplying distal carbon into BDA system; decomposer: releasing carbon for producers' reuse by decomposition of the waste; first consumer: carbon agents that transfer carbon from natural environment to human society; second consumer: anthropogenic using processed carbon resources from first consumers by processes of creating products or utilizing energy; sink: eliminating carbon through photosynthesis of green trees, it does not belong to the trophic structure.

3.2. Network Mutual Relationships. Table 3, respectively, shows the direct mutual relationships (in the matrix Sign D) and integral interactions (in the matrix Sign U) between such compartments of LCMN in BDA. It is apparently that the positive/negative signs in the matrix Sign U sometimes vary from those in the matrix Sign D, which implies that both quantities and qualities of the integral relationships could alter compared with those direct ones [27]. By comparing the relation changes between direct and integral utility matrix, it is obvious that circumstances of no changes occur most

TABLE 1: System-wide indicators of metabolism model.

Indicators	Formulation	Short description
Nodes	m	Quantity of metabolic compartments, also the size of network
Links	L	Quantity of metabolic direct flows or arcs
Link density	L/m	Metabolic linking degree
Connectance	L/m^2	Metabolic connectivity, also the proportion of realized direct pathways
MI	Equation (3)	Metabolic system
SI	Equation (4)	Metabolic system synergism
CI	Equation (9)	Self-regulation of metabolism

TABLE 2: Carbon flows within the low-carbon metabolic network of BDA (unit: t CO_2-eq)[a].

Doner/accepter	Int	Eng	Con	IBS	Wst	Grn	Ext	T_j
Int	0 (0.0%)	3046 (1.4%)	0 (0.0%)	39 (0.0%)	0 (0.0%)	0 (0.0%)	6014 (2.8%)	9100 (4.3%)
Eng	0 (0.0%)	0 (0.0%)	0 (0.0%)	0 (0.0%)	0 (0.0%)	0 (0.0%)	12796 (6.0%)	12796 (6.0%)
Con	13 (0.0%)	0 (0.0%)	0 (0.0%)	6001 (2.8%)	0 (0.0%)	0 (0.0%)	82509 (38.6%)	88523 (41.4%)
IBS	6001 (2.8%)	8288 (3.9%)	0 (0.0%)	0 (0.0%)	0 (0.0%)	0 (0.0%)	23 (0.0%)	14312 (6.7%)
Wst	0 (0.0%)	0 (0.0%)	17547 (8.2%)	0 (0.0%)	0 (0.0%)	0 (0.0%)	0 (0.0%)	17547 (8.2%)
Grn	2176 (1.0%)	4644 (2.2%)	6408 (3.0%)	2234 (1.0%)	0 (0.0%)	0 (0.0%)	36920 (17.3%)	52382 (24.5%)
Ext	0 (0.0%)	1462 (0.7%)	0 (0.0%)	0 (0.0%)	17547 (8.2%)	0 (0.0%)	0 (0.0%)	19009 (8.9%)
T_i	8190 (3.8%)	17441 (8.2%)	23955 (11.2%)	8274 (3.9%)	17547 (8.2%)	0 (0.0%)	138262 (64.7%)	**213670**

[a]The numbers in parentheses mean the proportion of carbon flows, namely, the flow value divided by the total carbon flow in the whole system. T_i is the sum of flows put into the j-th compartment, and T_j is for the sum flows into the j-th compartment analogously. The T_i-T_j intersecting number in bold indicates the total carbon flow of BDA.

frequently, that is, the relationship of IBS-Int (−, +), Eng-Ext (−, +), Con-Wst (+, −), and etc. and the next is those changes to the positive side, that is, the relationship between Con and Int varies from (0, 0) in direct to (+, +) in integral, and so does between Ext-Int, Eng-Wst, IBS-Wst, and Wst-Ext. Such result indicates that relationships between compartments are more prone to system-wide stability and benefits.

Then, we tried to analyze each trophic level. There are some differences in respective mutual relations between producers (Int and Ext). Int has a variety of links and its direct and integral relationships are not unified. Both the direct and indirect linkages for Ext mainly indicate that it exploits consumers (Eng, Con, and IBS) in terms of advancing raw materials, apparatus, and energy, and then being exploited by decomposer (Wst) for carbon recycle. As the largest carbon doners, Ext dominates the contacts between producers and other compartments. The irreplaceable effect of decomposer leads to obvious direct relationships that Wst is exploited by first consumer (Con) in dealing with carbon waste and pays back to producer (Ext). Besides, there are no more direct links between other consumers and producer but diverse indirect interactions. First consumers (Eng and Con) play a significant role in transform carbon from distal environment

TABLE 3: Direct utility sign matrix (Sign D)/integral utility sign matrix (Sign$U$) of LCMN in BDA[a].

	Int	Eng	Con	IBS	Wst	Ext
Int	0/+	+/−	0/+	−/−	0/−	0/+
Eng	−/−	0/+	0/−	−/−	0/+	+/+
Con	0/+	0/−	0/+	+/+	−/−	+/+
IBS	+/+	+/+	−/−	0/+	0/+	0/−
Wst	0/+	0/+	+/+	0/+	0/+	−/+
Ext	−/+	−/−	−/−	0/−	+/+	0/+

[a]Due to the value of total supplied carbon from Grn is zero, as the carbon sink, we can consider that the mutual relationships between other compartments and Grn are exploit-exploited (including Int, Eng, Con, IBS, and Ext) or neutrality (including Wst). In this sense, we did not take into consideration Grn calculation and comparison in Table 3. In addition, the signs of direct and integral utility matrixes for each pair of compartments are separated by "/", for example, "+/−" and "−/−" between Eng and Int show that the direct mutual relationship between them is (+, −), while the integral interaction is (−, −).

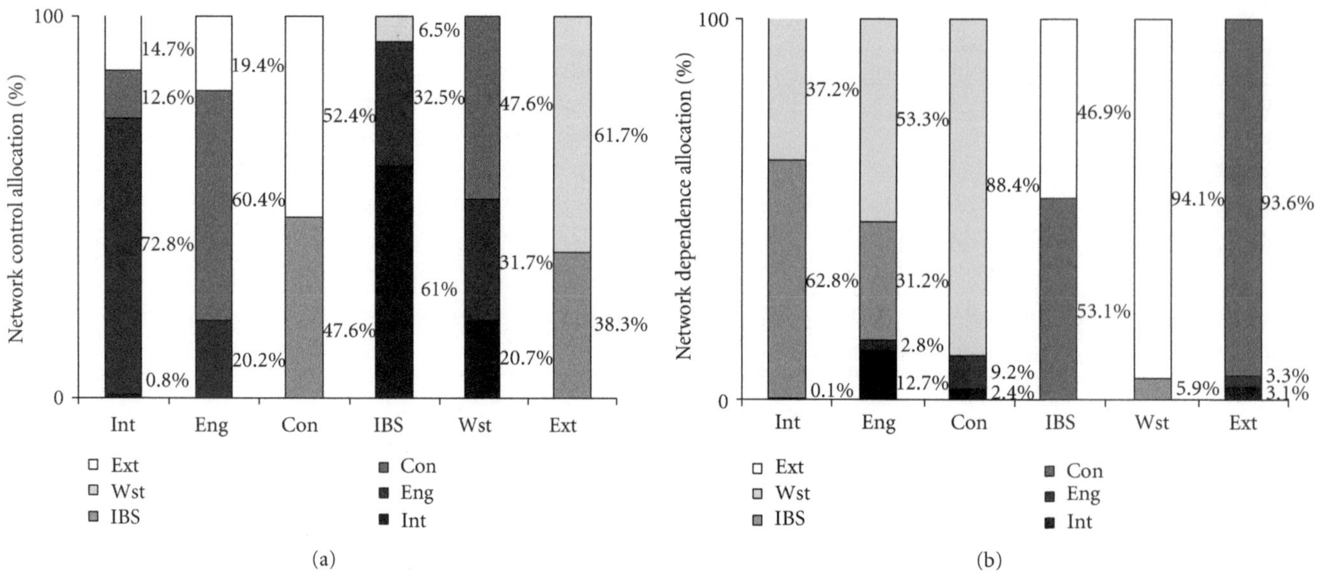

FIGURE 3: (a) Network control of carbon metabolism in BDA. (b) Network dependence of carbon metabolism in BDA. Because the value of total supplied carbon from Grn is zero, we did not consider the influence from Grn in these figures for an easy calculation.

into industrial society, so they mainly exchange carbon with producer (especially Ext) in terms of raw materials and energy. And particularly, Con is exploited by producer and second consumer (IBS) in material import and service management. Located at the highest trophic level, utility consumer (IBS) is mainly exploited by producer (Int and Ext) and first consumers (Eng and Con), for it acts better as a carbon recipient. Yet the direct and integral relationships between IBS and the decomposer (Wst) are in neutrality and mutualistic condition. That is because there is no direct carbon exchange between them, while both of them are exploited by Con as a carbon accepter in indirect ways. In a word, these compartments all play their own role in ecological structure of carbon metabolism in BDA.

3.3. Network Control Condition. Figures 3(a) and 3(b) illustrate the proportion for control and dependence condition of LCMN in BDA, respectively. These proportions are extracted from the control allocation matrix (CA) and dependence allocation matrix (DA) introduced in the preceding part, which indicates the control degree based on the controller's output environ and dependence degree from the being controlled input environ.

From the dependent perspective, Eng and Con as first consumers have some differences on dependence degree, and there exists an intrinsic linkage. Eng is more controlled by Con (60.4%), while Con is dependent on second consumer's management (IBS, 47.6%). The second consumer (IBS) generally depends on Eng's energy furnish (32.5%), whereas the influence from Wst (6.3%) is too small to mention. And at the decomposer's level (Wst), the control degree is mostly contributed by first consumers, where Con achieves 47.6% and Eng as 31.7%. The control condition shows a similarity (even more pronounced) in the systemic control for carbon metabolic in the views of controller. In addition, by observing the producers dependence degree, first consumers are more dominated by Ext (Eng is 19.4%, while Con is 52.4%), while second consumer (IBS, 61.0%) and decomposer (Wst, 20.7%) are more controlled by Int. Similar results have also been reflected in the control condition, where the control proportions from Ext to IBS and Wst reach 46.9% and 94.1%, while those from Int to Eng and Con achieve 12.7% and 2.4%. These similar results announce that the BDA International Park is further regulated by the outside world (especially by the external environ), so as to enhance the integral cooperation between those compartments, which may be a promising way to improve the utility of carbon metabolic in BDA.

TABLE 4: System-wide indicators of carbon metabolism in BDA.

Indicators	Formulation	Low-carbon metabolic network with Int and Ext	Low-carbon metabolic network without Int and Ext
Nodes	m	7	5
Links	L	15	6
Link density	L/m	2.14	1.20
Connectance	L/m^2	0.31	0.24
MI	Equation (3)	1.50	1.40
SI	Equation (4)	3.57	2.54

3.4. System Condition. The calculating values of system-wide indicators for carbon metabolism in BDA are all shown in Table 4. Apparently, compared with the carbon flows without internal and external environment, there are more connections among network compartments. Similar results are also shown in the Link Density and Connectance, where suggesting that with cooperation of Int and Ext, there are more diverse cycling ways. These results emphasize the significant affection and higher efficiency of producers (Int and Ext). Moreover, the whole system properties of LCMN in BDA have higher MI (1.50) and SI (3.57), indicating the industrial system holds more positive relationships in an integral way. Due to the fact that Mi is >1.00 both with and without Int and Ext, the carbon metabolic system in BDA can maintain self-mutualism and sustainability despite the lack of external supply. In natural ecosystem, it is quite often the case that more positive utilities than negative ones are needed to keep self-mutualism [31]. Making an analogy with natural ecosystem, these results may evince a sustainable improvement of carbon metabolism in BDA International Business Park.

4. Conclusions

Based on the methodology of ENA, we established a metabolic model for low-carbon high-tech industrial parks and analyzed the carbon metabolic system of the BDA International Business Park in this research. The results reflected the behaviors and potential linkage of system compartments and revealed the structure, function, and mutualism condition of the carbon metabolic system in the case park. In the relationship and control analyses, compartments links are quite diverse and positive, especially the Con and Wst, which play a pivotal role in exchanging carbon flows between industrial compartments and environs (both Int and Ext). Comparing the scenarios with Int and Ext or not, variations of system-wide indicators have revealed the significance of artificial control including carbon supplying and decomposing. Regarding ecological structure and function, the pyramidal ecological structure and positive indicators (MI and SI) both show a system mutualism and sustainable condition, which makes BDA Park a demonstration project in Chinese low-carbon EIPs' construction. Yet if we enhanced the effect of waste management part, different system functions would go on better.

Acknowledgments

This study was supported by Fund for Creative Research Groups of the National Natural Science Foundation of China (Grant No. 51121003), Key Program of National Natural Science Foundation (NOs. 50939001, 41271543), and Program for New Century Excellent Talents in University (NCET-09-0226). We are especially grateful for the financial support from Beijing Development Area Co. Ltd.

References

[1] B. Doyle, S. A. Martin, K. A. Weitz et al., "Eco-industrial parks: a case study and analysis of economic, environmental, technical and regulatory issues," Office of Policy, Planning and Evaluation, USEPA, Washington, DC, USA, 1996.

[2] Q. Zhu and R. P. Cote, "Integrating green supply chain management into an embryonic eco-industrial development: a case study of the Guitang Group," *Journal of Cleaner Production*, vol. 12, no. 8–10, pp. 1025–1035, 2004.

[3] C. Sendra, X. Gabarrell, and T. Vicent, "Material flow analysis adapted to an industrial area," *Journal of Cleaner Production*, vol. 15, no. 17, pp. 1706–1715, 2007.

[4] P. Desrochers, "Industrial symbiosis: the case for market coordination," *Journal of Cleaner Production*, vol. 12, no. 8–10, pp. 1099–1110, 2004.

[5] I. Allison, N. L. Bindoff, R. A. Bindschadler et al., The Copenhagen Diagnosis, The University of New South Wales Climate Change Research Centre, Sidney, Australia, 60, 2009.

[6] IPCC, "Climate Change 2007, Synthesis Report," Valencia, Spain, 2007.

[7] T. Maes, G. Van Eetvelde, E. De Ras et al., "Energy management on industrial parks in Flanders," *Renewable and Sustainable Energy Reviews*, vol. 15, no. 4, pp. 1988–2005, 2011.

[8] B. H. Roberts, "The application of industrial ecology principles and planning guidelines for the development of eco-industrial parks: an Australian case study," *Journal of Cleaner Production*, vol. 12, no. 8–10, pp. 997–1010, 2004.

[9] S. Hashimoto, T. Fujita, Y. Geng, and E. Nagasawa, "Realizing CO_2 emission reduction through industrial symbiosis: a cement production case study for Kawasaki," *Resources, Conservation and Recycling*, vol. 54, no. 10, pp. 704–710, 2010.

[10] S. Lehtoranta, A. Nissinen, T. Mattila, and M. Melanen, "Industrial symbiosis and the policy instruments of sustainable consumption and production," *Journal of Cleaner Production*, vol. 19, pp. 1865–1875, 2011.

[11] L. Sokka, S. Pakarinen, and M. Melanen, "Industrial symbiosis contributing to more sustainable energy use—an example from the forest industry in Kymenlaakso, Finland," *Journal of Cleaner Production*, vol. 19, no. 4, pp. 285–293, 2011.

[12] J. Korhonen, M. Wihersaari, and I. Savolainen, "Industrial ecosystem in the Finnish forest industry: using the material and energy flow model of a forest ecosystem in a forest industry system," *Ecological Economics*, vol. 39, no. 1, pp. 145–161, 2001.

[13] L. X. Liu, B. Zhang, J. Bi, Q. Wei, and H. Pan, "The greenhouse gas mitigation of industrial parks in China: a case strudy of Suzhou Industrial Park," *Energy Policy*, vol. 46, pp. 301–307, 2012.

[14] V. S. Norstebo, K. Midthun, and T. Bjorkvoll, "Analysis of carbon capture in an industrial park—a case study," *International Journal of Greenhouse Gas Control*, vol. 9, pp. 52–61, 2012.

[15] K. Midthun, V. S. Norstebo, G. Perez-Valdes, and T. Bjorkvoll, "Investment analysis of an integrated industrial park with carbon capture," *Journal of Natural Gas Science and Engineering*, vol. 7, pp. 44–51, 2012.

[16] S. M. Munir, Z. A. Manan, and S. R. W. Alwi, "Holistic carbon planning for industrial parks: a waste-to-resources process integration approach," *Journal of Cleaner Production*, vol. 33, pp. 74–85, 2012.

[17] T. Tudor, E. Adam, and M. Bates, "Drivers and limitations for the successful development and functioning of EIPs (eco-industrial parks): a literature review," *Ecological Economics*, vol. 61, no. 2-3, pp. 199–207, 2007.

[18] S. Anderberg, "Industrial metabolism and the linkages between economics, ethics and the environment," *Ecological Economics*, vol. 24, no. 2-3, pp. 311–320, 1998.

[19] R. U. Ayres and U. E. Simonis, *Industrial Metabolism: Restructuring for Sustainable Development*, United Nations University Press, Tokyo, Japan, 1994.

[20] L. Chen, R. S. Wang, J. X. Yang, and Y. L. Shi, "Structural complexity analysis for industrial ecosystems: a case study on

LuBei industrial ecosystem in China," *Ecological Complexity*, vol. 7, no. 2, pp. 179–187, 2010.

[21] T. Graedel, B. R. Allenby, and H. Shi, *Industrial Ecology*, Tsinghua University Press, Beijing, China, 2004.

[22] J. P. Tian, H. Shi, Y. Chen, and L. J. Chen, "Assessment of industrial metabolisms of sulfur in a Chinese fine chemical industrial park," *Journal of Cleaner Production*, vol. 32, pp. 262–272, 2012.

[23] Y. P. Fang and H. Z. Zhou, "Value flow analysis based on EAP industrial chain: case of Huaning in Xichang, Sichuan," *Journal of Cleaner Production*, vol. 17, no. 2, pp. 310–316, 2009.

[24] A. Singh, H. H. Lou, C. L. Yaws, J. R. Hopper, and R. W. Pike, "Environmental impact assessment of different design schemes of an industrial ecosystem," *Resources, Conservation and Recycling*, vol. 51, no. 2, pp. 294–313, 2007.

[25] W. Ashton, "Understanding the organization of industrial ecosystems: a social network approach," *Journal of Industrial Ecology*, vol. 12, no. 1, pp. 34–51, 2008.

[26] H. M. Zheng, Y. Zhang, and N. J. Yang, "Evaluation of an eco-industrial park based on a social network analysis," *Procedia Environment Sciences*, vol. 13, pp. 1624–1629, 2012.

[27] B. D. Fath and B. C. Patten, "Network synergism: emergence of positive relations in ecological systems," *Ecological Modelling*, vol. 107, no. 2-3, pp. 127–143, 1998.

[28] H. Shi, Y. Moriguichi, and J. Yang, "Industrial ecology in China, part I: research," *Journal of Industrial Ecology*, vol. 6, no. 3-4, pp. 7–11, 2003.

[29] B. C. Patten, "Network ecology: indirect determination of the life-environment relationship in ecosystems," in *Theoretical Studies of Ecosystems: The Network Perspective*, M. Higashi and T. P. Burns, Eds., Cambridge University Press, New York, NY, USA, 1991.

[30] B. C. Patten, "Energy, emergy and environs," *Ecological Modelling*, vol. 62, no. 1–3, pp. 29–69, 1992.

[31] B. D. Fath, "Network mutualism: positive community-level relations in ecosystems," *Ecological Modelling*, vol. 208, no. 1, pp. 56–67, 2007.

[32] B. D. Fath and S. R. Borrett, "A MATLAB function for network environ analysis," *Environmental Modelling & Software*, vol. 21, no. 3, pp. 375–405, 2006.

[33] G. Lobanova, B. D. Fath, and E. Rovenskaya, "Exploring simple structural configurations for optimal network mutualism," *Communications in Nonlinear Science and Numerical Simulation*, vol. 14, no. 4, pp. 1461–1485, 2009.

[34] Y. Zhang, Z. F. Yang, and X. Y. Yu, "Ecological network and emergy analysis of urban metabolic systems: model development, and a case study of four Chinese cities," *Ecological Modelling*, vol. 220, no. 11, pp. 1431–1442, 2009.

[35] S. Q. Chen and B. Chen, "Network environ perspective for urban metabolism and carbon emissions: a case study of Vienna, Austria," *Environmental Science and Technology*, vol. 46, no. 8, pp. 4498–4506, 2012.

[36] B. C. Patten, "Systems approach to the concept of environment," *The Ohio Journal of Science*, vol. 78, pp. 206–222, 1978.

[37] B. C. Patten and G. T. Auble, "System theory of the ecological niche," *American Naturalist*, vol. 117, pp. 893–922, 1981.

[38] B. D. Fath, "Distributed control in ecological networks," *Ecological Modelling*, vol. 179, no. 2, pp. 233–245, 2004.

[39] B. D. Fath and B. C. Patten, "Quantifying resource homogenization using network flow analysis," *Ecological Modelling*, vol. 123, no. 2-3, pp. 193–205, 1999.

5

Relationship Study on Land Use Spatial Distribution Structure and Energy-Related Carbon Emission Intensity in Different Land Use Types of Guangdong, China, 1996–2008

Yi Huang,[1,2] Bin Xia,[1,3] and Lei Yang[4]

[1] *Sustainable Development Research Center, Guangzhou Institute of Geochemistry, Chinese Academy of Sciences, Guangzhou 510640, China*
[2] *University of Chinese Academy of Sciences, Beijing 100049, China*
[3] *School of Marine Sciences, Sun Yat-Sen University, Guangzhou 510006, China*
[4] *College of Resources and Environment, Guangdong University of Business Studies, Guangzhou 510320, China*

Correspondence should be addressed to Bin Xia; adely021@163.com

Academic Editors: A. Greco, X. Qin, and F. Xiao

This study attempts to discuss the relationship between land use spatial distribution structure and energy-related carbon emission intensity in Guangdong during 1996–2008. We quantized the spatial distribution structure of five land use types including agricultural land, industrial land, residential and commercial land, traffic land, and other land through applying spatial Lorenz curve and Gini coefficient. Then the corresponding energy-related carbon emissions in each type of land were calculated in the study period. Through building the reasonable regression models, we found that the concentration degree of industrial land is negatively correlated with carbon emission intensity in the long term, whereas the concentration degree is positively correlated with carbon emission intensity in agricultural land, residential and commercial land, traffic land, and other land. The results also indicate that land use spatial distribution structure affects carbon emission intensity more intensively than energy efficiency and production efficiency do. These conclusions provide valuable reference to develop comprehensive policies for energy conservation and carbon emission reduction in a new perspective.

1. Introduction

Since energy-related carbon emission was recognized as the most important anthropogenic factor that leads to global warming [1, 2], numerous studies have focused on the mechanism of how human activities affect energy consumption. Economic development, as well as industrial structure and energy consumption structure are typically considered as primary influencing factors for energy consumption [3–7]. Actually, human activity affects the regional carbon emission through changing the land use patterns, which in turn changes the energy consumption patterns and finally influences the amount and rate of carbon emission [8–10]. However, studies on land use change-induced carbon emission are usually directed toward calculating carbon storage in soil and vegetation cover given that soil respiration also considerably accounts for carbon dioxide emission into the atmosphere [11–20]. The social attribution of land use, which affects energy consumption via the spatial distribution of land use, is always neglected.

Related research focuses on the relationship between urban form and energy consumption [21, 22]. The majority of such research suggests a strong negative correlation between urban density and energy consumption [23–28]. Concentration distribution reduces energy consumption, resulting in the emission of fewer pollutants and maintenance of sustainable urban development, particularly for traffic and residential lands [29–37]. This view is accepted by policy makers in many developed countries, and is considered as an effective way to reduce energy consumption [26, 27]. Opposing perspectives, however, also exist. For example, Mindali et al. (2004) argues that the relationship between urban

FIGURE 1: Geographic location and component cities of Guangdong Province in China.

form and energy consumption is uncertain [26]. Holden and Norland (2005) indicated that decentralized concentration may reduce energy consumption in households [38]. Ma and Jin (2011) found a weak correlation between urban density and energy consumption under high population density in China [39].

Urban form, defined on the basis of the spatial distribution of land use, encompasses urban size, population density, and other related elements [40]. Except for traffic land and residential land distribution, other types of land use are seldom regarded as influencing factors for energy consumption in existing research. Thus, the existed urban form perspective is an unsuitable approach to determining the relationship between the distribution of different land use types and energy-related carbon emission. A comprehensive relation study between the spatial distribution structure of different land use types and energy-related carbon emission should be considered at a regional scale, specific to domestic conditions in terms of the large population, defective infrastructure, and binary structures of urban and rural areas.

Some researchers analyzed the energy-related carbon emission caused by land use patterns change in China, but none of them studied on their relationships. Zhao and Huang (2010) found that the energy-related carbon emission per unit land area in Jiangsu Province increased from 8.24 t/hm^2 to 15.53 t/hm^2 from 2003 to 2007, and that the highest carbon emission intensities originated from residential and industrial land [8]. Zhao et al. (2010) studied the energy-related carbon emissions of different industrial spaces in China [41]. The results indicate that living and industrial-commercial spaces contribute the highest carbon emissions, and that the carbon emission intensities of each industrial space in Shanghai are the highest among the investigated 30 provinces. Using data envelopment analysis, You and Wu (2010) measured the provincial energy-related carbon emission efficiency of land use [42]. The results show that carbon emission per unit land area is effective in only three provinces. Zhang (2011) analyzed the land use constitution structure in Yunnan Province through information entropy method, and confirmed that the changes in land use structure cause energy-related carbon emission in the province [43].

Guangdong, which located in south China and comprises 21 cities (Figure 1), is one of the most developed provinces in the country. Its rapid development is attributed to the reform and opening policy implemented in 1978. Statistical data (taken from the Guangdong Statistical Yearbook and the Guangdong Agricultural Statistical Yearbook) show that Guangdong's GDP increased from 18.58 billion Yuan to 4601.31 billion Yuan from 1978 to 2010, and that its contribution to China's GDP increased from 5.1% to 11.47% during the same period. The proportions of output value in three major industrial sectors changed from 29.8% : 46.6% : 23.6% to 5% : 50% : 45%. The permanent population increased from 50.64 million to 104.40 million. In addition, land use spatial distribution and energy consumption significantly changed as human activities increased. Crop land decreased from 6.6416 million hectares in 1978 to 4.5245 million hectares in 2010, while terminal energy consumption increased from 18.13 million tons in 1980 to 263.4485 million in 2010. Guangdong was identified as one of the pilot provinces for national low-carbon economic development in 2010. To this end, the Guangdong government set a goal of reducing carbon dioxide emission per unit of GDP by 35% in 2015 relative to 2005 levels [44]. Land use pattern, a factor that significantly affects energy consumption, is also considered in the government's macro control policy. Therefore, ascertaining the relationship between land use spatial distribution structure and energy-related carbon emission is important for formulating land use policies, which promote efficient energy conservation and emission reduction.

This study analyzes the relationship between land use spatial distribution structure and energy-related carbon emission intensity in Guangdong from 1996 to 2008. The core components of the research are as follows: (1) The

TABLE 1: Classification of land use types and their corresponding relationship with industrial energy consumption.

Land use type	Specific types	Corresponding industrial energy consumption
Agricultural land	Cultivated land, orchard land, forest land, grassland, and other agricultural lands	Farming, forestry, animal husbandry, fishery
Industrial land	Independent industrial and mining lands	Industry
Residential and commercial land	Urban area, rural-residential areas	Construction, wholesale and retail trade, catering services, residential living
Traffic land	Traffic land	Transportation, storage, postal and telecommunications services
Other land	Unused land and other land (e.g., special designated land, lakes, and rivers)	Other industries

spatial distribution structures of different land use types in Guangdong are analyzed by applying the spatial Lorenz curve and calculating the Gini coefficient. (2) The energy-related carbon emissions in different land use types are calculated based on the IPCC accounting method. (3) The relationship between land use spatial distribution structure and energy-related carbon emission intensity is investigated by regression analysis and cointegration test.

To accurately calculate the energy-related carbon emission in each type of land use, we attempt to establish the corresponding relationships between industrial energy consumption and land use types according to the content of land use status classification in China [45]. Based on the research of Li (2009) and Zhao and Huang (2010), land uses in this study are classified into five different types: agricultural land, industrial land, residential and commercial land, traffic land, and other land (includes lakes, rivers, and special designated land; Table 1) [8, 46].

2. Methods and Data Sources

2.1. Spatial Lorenz Curve and Gini Coefficient. Numerous methods are used to study the spatial distribution structure of land use. Such approaches include information entropy theory [47–49], geographic information system method combined with remote sensing imagery [50–54], landscape ecology method [55–57], and mathematical methods, such as Markov chain modeling and linear programming [58–62]. However, most of these methods are complicated and provide abstract results [63]. Spatial Lorenz curve analysis combines location entropy with the Lorenz curve concept in economics, making it an intuitive method for studying land use distribution structures [63–65]. The spatial Lorenz curve is quantified by the Gini coefficient, whose value indicates the concentration degree of land use spatial distribution. Thus, on the basis of relevant research, we use the spatial Lorenz curve and Gini coefficient to measure the concentration degree of the spatial distribution structures of different land use types.

2.1.1. Land Use Spatial Lorenz Curve. The spatial Lorenz curve of land use can be established by citing two economic concepts, which are Lorenz curve and location quotient. The original Lorenz curve, which comprises population percentage and the corresponding income percentage, is an extensively used tool for examining national wealth distribution [66]. Lorenz curve bends indicate the concentration degree of national wealth and a straight line with no bending translates to equitably distributed wealth (Figure 2) [67]. For intuitive distribution analysis, Lorenz curve is applied to many fields by using relevant percentage variables [68–71]. To measure the spatial distribution of different land use types in this study, the spatial Lorenz curve of land use should comprise specific land percentage and regional area percentage. The specific land percentage refers to the ratio of specific land use area in each city to the whole specific land use area in the province, while the regional area percentage denotes the ratio of each city area to the province area. Actually, the relationship between specific land percentage and regional area percentage can be defined by location quotient in economics [64, 65].

location quotient is the ratio between a special factor percentage and the entire factor percentage. In this study, the special factor percentage is the specific land percentage, and the entire factor percent is the regional area percentage. These percentages are expressed as follows:

$$Q_{ni} = \frac{S_{ni}/\sum S_{ni}}{S_n/\sum S_n}, \quad (1)$$

where i and n are the indices for land use types and cities in Guangdong, respectively; Q_{ni} is the location quotient of land use type i ($i = 1, 2, 3, 4, 5$) in city n ($n = 1, 2, 3, \ldots, 21$) (Table 1 and Figure 1); S_{ni} represents the area occupied by land use type i in city n; $\sum S_{ni}$ denotes the area occupied by land use type i in Guangdong; S_n represents the area occupied by city n; and $\sum S_n$ is the area occupied by Guangdong Province. The numerator represents the area ratio of land use type i in city n, and the denominator represents the regional area percentage.

On the basis of the implications of Lorenz curve, we rank Q_{ni} from the smallest to the largest; consider an ordinate value as the accumulative area ratio of different land use types and an abscissa value as the accumulative regional area percentage; and then plot the spatial Lorenz curve (Figure 3). The curve bends indicate the concentration degree of land use spatial distribution in the studied region. A straight line with no bending corresponds to equitably distributed land use.

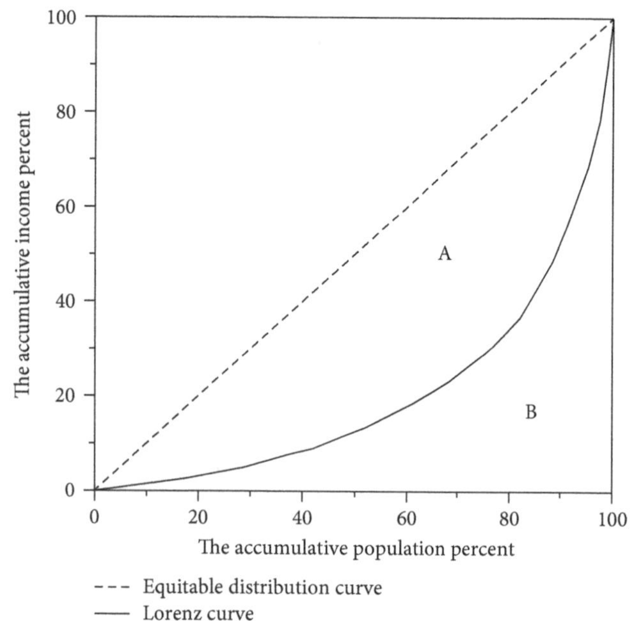

FIGURE 2: Lorenz curve.

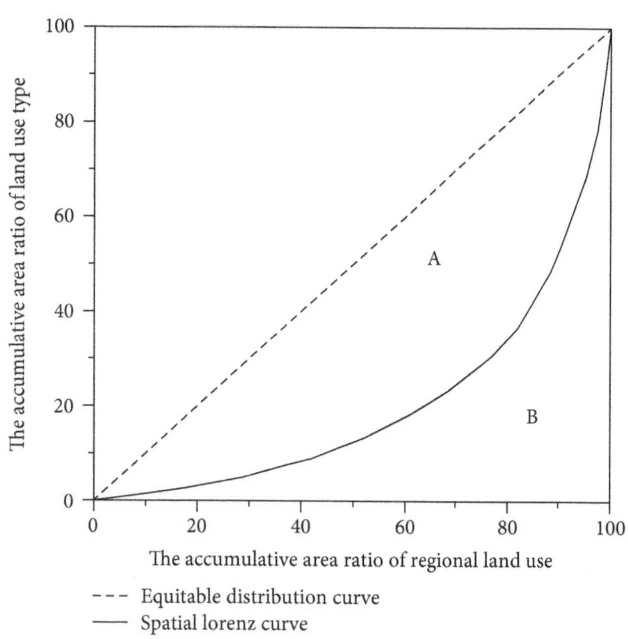

FIGURE 3: Spatial Lorenz curve.

2.1.2. Gini Coefficient. The Gini coefficient measures the land use spatial distribution based on quantifies the bends of spatial Lorenz curve [72]. It is defined as

$$G = \frac{S_A}{S_A + S_B}, \quad (2)$$

where G is the Gini coefficient, S_A represents the area between the equitable distribution curve and spatial Lorenz curve (Figure 3, area A), $S_A + S_B$ denotes the area under the equitable distribution curve (Figure 3, areas A and B). The Gini coefficient reflects deviation from the equitable distribution curve. In the spatial Lorenz curve, therefore, a high Gini coefficient represents high degree of land use concentration.

We calculate the Gini coefficient in (2) by the curve regression method [73, 74]. When the vertical and horizontal axes are both set to one unit, (2) is converted into

$$G = 1 - 2S_B = 1 - \sum (X_n - X_{n-1}) \times (Y_n + Y_{n-1}), \quad (3)$$

where X_n and Y_n are the indices for the abscissa value and ordinate value of the spatial Lorenz curve, respectively; thus, $0 < G \leq 1$. According to the commonly used division standard of the Gini coefficient, land use distribution characteristics can be estimated as follows: when $0 < G \leq 0.2$ stands for absolute decentralization, $0.2 < G \leq 0.3$ represents decentralization; $0.3 < G \leq 0.4$ indicates appropriate concentration; $0.4 < G \leq 0.5$ means concentration; and $0.5 < G \leq 1$ corresponds to absolute concentration [73].

2.2. Energy-Related Carbon Emission Intensity

2.2.1. Calculation of Energy-Related Carbon Emission. Fossil energy accounts for more than 90% of total energy consumption. It is the main source of energy-related carbon emission in China [75]. Energy-related carbon emissions, which include the carbon content of both carbon dioxide and methane, are released from the consumption of fossil and biomass energy. To calculate the fossil energy-related carbon emission both from final consumption and transformational consumption, we separately calculate the carbon emission from final fossil energy consumption and that from terminal thermal power and heating consumption because the majority of fossil energy consumption in the energy conversion process is used to produce thermal power and heating (refer to the energy balance table for Guangdong in the China Energy Statistics Yearbook). The energy-related carbon emissions are calculated as follows [8, 76]:

$$\mathrm{CT}_i = \mathrm{CR}_i + \mathrm{CE}_i + \mathrm{CH}_i + \mathrm{CB}_i, \quad (4)$$

where CT_i represents the total energy-related carbon emission of land use type i (classified in Table 1); CR_i is the carbon emission from final fossil energy consumption, excluding the thermal power and heating supply for land use type i; CE_i and CH_i are the indices for the fossil energy-related carbon emissions during the supply of thermal power and heating, respectively, in land use type i; CB_i denotes the carbon emission from noncommercial energy, which is used by the residents of residential and commercial land.

The detailed calculation methods are as follows:

$$\mathrm{CR}_i = \sum_j \mathrm{QRN}_{ij} \times \mathrm{EF}_j = \sum_j \mathrm{QR}_{ij} \times \mathrm{NCV}_j \times \mathrm{EF}_j \times 10^{-3},$$
$$\mathrm{EF}_j = A_{cj} \times B_{cj} + A_{mj} \times B_{mj}, \quad (5)$$

$$\mathrm{CE}_i = \mathrm{QE}_i \times \mathrm{EE} \times \mathrm{CF}_c, \quad (6)$$

$$\mathrm{CH}_i = \mathrm{QH}_i \times \mathrm{EH} \times \mathrm{CF}_c. \quad (7)$$

TABLE 2: Net calorific values and default carbon emission factors of the considered energy types.

Energy type	NCV_j (TJ/Gg)	EF_j (t/TJ)
Raw coal	20.91	25.64
Cleaned coal	26.34	24.89
Other washed coal	8.36	24.89
Briquettes	20.91	30.24
Coke	28.44	27.44
Coke oven gas (MJ/m^3)	16.73	12.93
Other gas types (MJ/m^3)	22.14	12.93
Crude oil	41.82	19.70
Gasoline	43.07	18.52
Kerosene	43.07	18.52
Diesel oil	42.65	19.80
Fuel oil	41.82	20.68
Liquefied petroleum gas	50.18	16.86
Refinery gas	46.06	17.84
Natural gas (MJ/m^3)	38.93	15.15
Other petroleum products	41.82	19.60
Other coking products	28.44	27.44
Other energy types	29.31	12.20
Biogas (MJ/m^3)	20.91	14.89
Crop stalk	12.55	30.57
Firewood	16.73	30.57

In (5), CR_i (10^4 t) is the carbon emission from final fossil energy consumption, excluding thermal power and heating for land use type i; QRN_{ij} (TJ) represents the combusted quantity of energy type j (j refers to the fossil energy types in Table 2) in land use type i; QR_{ij} (10^4 t (10^7 m^3)) is the consumed fossil energy type j in land use type i; NCV_j (TJ/Gg) denotes the net calorific value of fossil energy j; QRN_{ij} is the product of QR_{ij} and NCV_j; EF_j (t/TJ) is the default carbon emission factor of fossil energy j; A_{cj} and A_{mj} (t/TJ) represent the default carbon contents in carbon dioxide and methane, respectively; B_{cj} and B_{mj} denotes the indices for default oxidation carbon factors; and 10^{-3} is the coefficient of unit conversion.

The carbon emission from consumed biomass energy (CB_i), which includes biogas, crop stalk, and firewood, can also be calculated by (5). The carbon emission factors are taken from the Provincial Greenhouse Gas List Preparation Guide (Table 2) [77].

In (6) and (7), CE_i (10^4 t) and CH_i (10^4 t) are the indices for the carbon emissions from thermal power consumption and heating supply, respectively, in land use type i. The carbon emissions from thermal power and heating can be calculated by using standard coal consumption given that coal is the main source of thermal power and heating supply in China. QE_i (10^{10} kWh) and QH_i (10^{13} kJ) represent the quantities of thermal power consumption and heating supply, respectively, in land use type i; EE (g/kWh) and EH (kg/GJ) are the standard coal consumption levels per unit of thermal power and heating supply, respectively; and CF_c is the carbon emission coefficient of unit standard coal consumption. EE and EH are based on the China Power Statistical Yearbook (1997–2009) (Table 3) [78]. CF_c equals 0.67 kg C/kgce [79].

2.2.2. Energy-Related Carbon Emission Intensity in Different Land Use Types. The energy-related carbon emission intensity of land use pertains to the energy-related carbon emission in the unit area of land use [8]. It is calculated as follows:

$$CI_i = \frac{CT_i}{S_i}, \qquad (8)$$

where CI_i (t/hm^2) represents the energy-related carbon emission intensity in land use type i, CT_i (10^4 t) denotes the energy-related carbon emission in land use type i; and S_i (10^4 hm^2) is the area occupied by land use type i.

2.3. Regression Analysis and Cointegration Test. Regression analysis is commonly used in research on relationships among time series variables. Because of the relevance of time series data, the unit root always exists in time series. Time series with unit roots are called nonstationary time series, and may considerably influence the results of a regression model [80]. Although some time series are nonstationary, certain linear combinations among such series are stationary. This linear combination reflects a stable long-term relationship among variables, and suggest that cointegration among the variables exist. If the unit roots of time series variables are uniformly integrated, the cointegration relationship can be verified by the Engle-Granger testing method [81, 82]. Engle and Granger argue that the residual series of regression models should be stationary if cointegration among variables exists [83]. Thus, the cointegration relationship can be evaluated by determining the stationarity of residual series.

Eliminating inaccurate regression in relationship research entails the following steps for regression analysis and cointegration testing: first, the stationarity of variables is determined by using Augmented Dickey-Fuller (ADF) values. The variables in a uniform integration order are then chosen to build a regression model. Finally, in accordance with the Engle-Granger method, ADF values are used to test the stationarity of the residual series in the regression model. If the residual series exhibit integration of order 1, then cointegration exists in the regression variables and the regression model is reasonable [83].

Studies on the energy-related carbon emission factors in China show that energy-related carbon emission is influenced by multiple factors, including industrial structure, energy consumption structure, energy efficiency, and production efficiency [84–87]. Different land use types are classified by industries and are investigated as separate research objects. These attributes indicate that the influence of structures, including industrial structure and energy consumption structure, are insignificant. Thus, GDP per unit energy consumption (denoted as GE) and GDP per unit capital investment (denoted as GC), as well as land use distribution structure (denoted as LS), are chosen as the independent variables in this study. GE and GC represent energy efficiency and production efficiency, respectively. Considering the effect of area changes on energy-related carbon emission, we choose

TABLE 3: Standard coal consumption per unit of thermal power and heating supply.

Year	Standard coal consumption per unit of thermal power (g/kWh)	Standard coal consumption per unit of heating supply (kg/GJ)
1996	392	40.13
1997	386	40.77
1998	381	40.39
1999	372	41.22
2000	372	39.70
2001	363	40.08
2002	358	40.48
2003	358	40.64
2004	357	40.22
2005	353	40.24
2006	352	40.32
2007	343	40.50
2008	336	40.14

Data are obtained from China Power Statistical Yearbooks.

energy-related carbon emission intensity as the dependent variable, denoted as CI. To eliminate the heteroscedasticity that may exist in time series, we transform the four variables into natural logarithmic form (the logarithm, which is suitable for measuring the changes in elasticity among variables, can eliminate heteroscedasticity without affecting the relationship among variables) and denote these as lnGE, lnGC, lnLS, and lnCI. The relationships among these variables in land use types, except for other land, are determined by multiple regression analysis. The energy-related carbon emission in other land originates from other economic development, which does not relate to certain corresponding industry and value accounting. For this category, therefore, a simple linear regression model is constructed to determine the relationship between land use distribution structure and carbon emission intensity.

2.4. Data sources. Land use data are taken from the Classification Survey Statistics of Land Use for Guangdong (1997–2009), collected by the Guangdong Department of Land and Resources [88]. Energy data are obtained from China Energy Statistical Yearbook (1997–2009) [89] and China New Energy and Renewable Energy Yearbook (2009) [90]. Other data come from Guangdong Statistical Yearbook (1997–2011) [91] and Guangdong Agricultural Statistical Yearbook (1997–2011) [92].

3. Results and Discussion

3.1. Spatial Distribution Structures of Different Land Use Types. The spatial Lorenz curves of different land use types are shown in Figure 4. The spatial Lorenz curves of agricultural land are always the closest ones to the equitable distribution curve, whereas the industrial land curves are always the farthest. These characteristics indicate that agricultural land exhibits the most decentralized distribution among the five land use types, whereas industrial land shows the most concentrated distribution. Traffic land is more concentrated than residential and commercial land, as well as other land, over 1996 to 2008.

The Gini coefficients are calculated based on the spatial Lorenz curve to quantify the distribution characteristics of land use. The results are shown in Figure 5 and Table 4. The average Gini coefficient of agriculture land was 0.0491, the smallest among the Gini coefficients of the five land use types. Although the Gini coefficient of agricultural land did not visibly increase in terms of absolute value, its average annual growth rate was higher than that observed for the other land use types. These results indicate that during the study period, agricultural land was absolutely decentralized, with a trend of concentrated use. Industrial land had the highest average Gini coefficient at 0.6163, indicating absolute concentration. Although minimal fluctuation occurs, the increasing trend of industrial land concentration was clearly observable, especially from 2002 to 2004. The average Gini coefficients of residential and commercial land and other land were 0.2999 and 0.2874, respectively, indicating that these two types of land use were distributed in decentralized manners, with different change trends. Residential and commercial land tended toward appropriate concentration, whereas other land tended toward decentralization. Traffic land exhibited appropriate concentration, with an average Gini coefficient of 0.3679; however, this value increased, indicating a concentrated trend for traffic land.

3.2. Energy-Related Carbon Emission Intensity in Different Land Use Types. Using the methods discussed in Section 2, we calculate the energy-related carbon emissions in different land use types (Table 5, Figures 6 and 7). During the studied period, the total amount of energy-related carbon emission increased from 55.53 million tons to 159.33 million tons, with the annual increased rate of 9.18%. The energy-related carbon emissions in all land use types significantly increased in varying degrees. The carbon emission in industrial land increased from 32.79 million tons to 101.43 million tons. The proportion of total carbon emission in this land use type also increased from 59.06% to 63.66%, the fastest increase among all the land use types. On the basis of the increase in carbon emission, we rank the rest of the land use types as follows: residential and commercial land > traffic land > other land > agricultural land (Table 5). Although the carbon emissions in residential and commercial land and agricultural land increased 19.29 million tons and 0.49 million tons, respectively, their proportions of total carbon emission decreased by 5.86% and 2.47%, respectively. The carbon emissions in traffic land and other land increased 9.83 million tons and 5.56 million tons, respectively, and their proportions of total carbon emission increased by 2.02% and 1.71%, respectively. These results show that during the studied period, energy-related carbon emission intensively increased in all the land use types. The anthropogenic activities in the other three land use types more considerably increased than those in agricultural land and residential and commercial land.

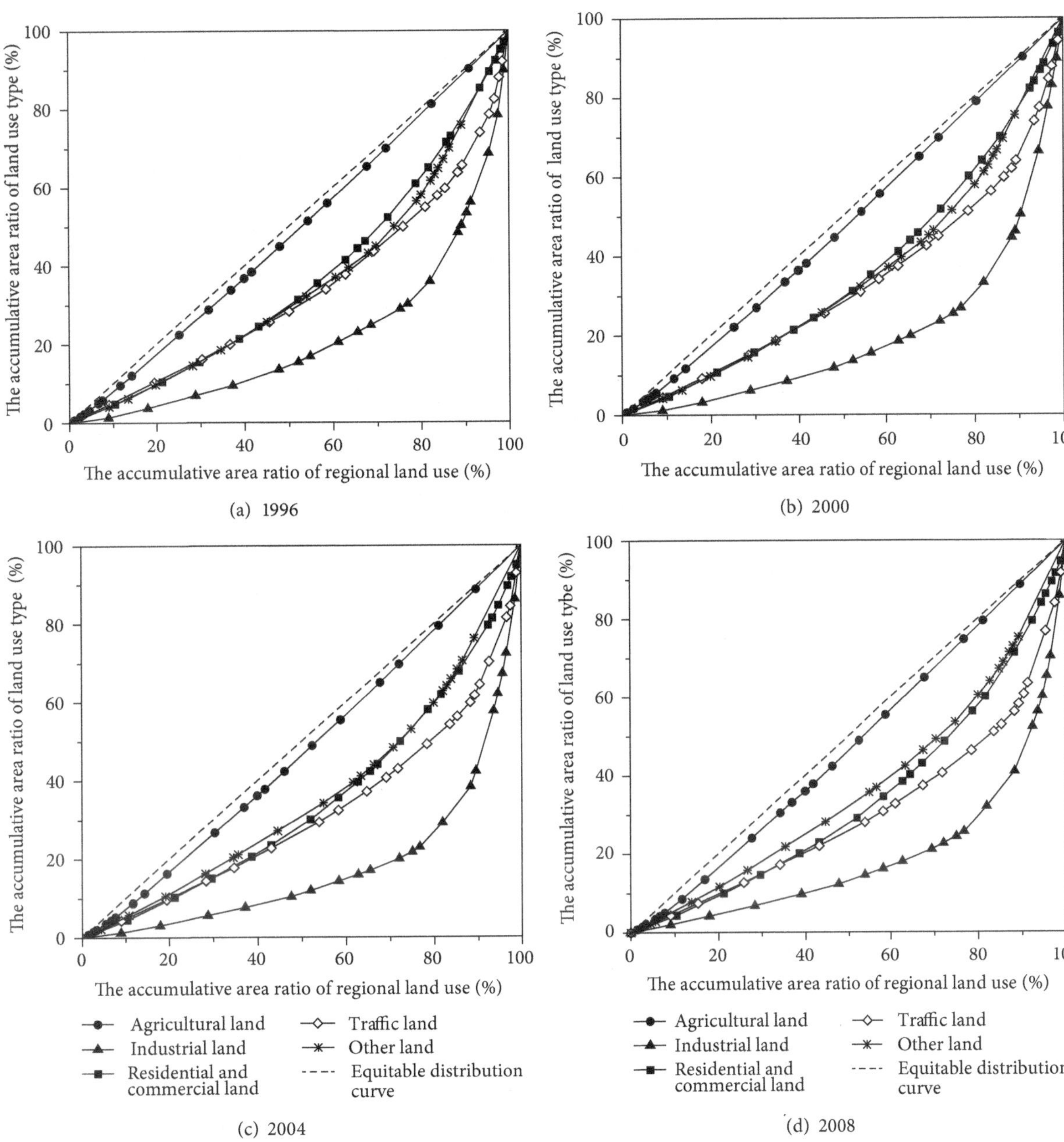

FIGURE 4: Spatial Lorenz curves of different land use types.

Energy-related carbon emission intensities are calculated according to (8). Results are shown in Table 6. For the period 1996 to 2008, the average energy-related carbon emission intensity in Guangdong increased from 3.2326 t/hm^2 to 9.2525 t/hm^2, with an annual increase rate of 9.16%. The carbon emission intensities in industrial land, traffic land, and residential and commercial land were higher than the provincial level. In addition, the carbon emission intensities in agricultural land and other land were lower than the average values, and the carbon emission intensity in agricultural land was the lowest among all the land use types. The carbon emission intensities in various land use types increased with different ratios. On the basis of annual increase rates, we rank the land use types as follows: other land > traffic land > residential and commercial land > industrial land > agricultural land. The results indicate that the anthropogenic activities per unit area in each land use type increased, and that the continued development in other land caused a high annual increase rate of energy-related carbon emission in this land use type.

3.3. Regression Analysis and Cointegration Testing of Different Land Use Types. Implementing the steps in Section 2.3, we test the stationarity of each variable using ADF values. Table 7 shows that the variables in the models of agricultural land, industrial land, residential and commercial land, and

TABLE 4: Spatial distribution structures and divisions of different land use types.

Year	Agricultural land	Industrial land	Residential and commercial land	Traffic land	Other land
1996	0.0422	0.5757	0.2799	0.3441	0.3037
1997	0.0447	0.6011	0.2817	0.3575	0.3066
1998	0.0456	0.6068	0.2819	0.3508	0.3065
1999	0.0458	0.6094	0.2802	0.3359	0.3056
2000	0.0464	0.6047	0.2867	0.3506	0.3035
2001	0.0469	0.6089	0.2891	0.3545	0.3027
2002	0.0491	0.6322	0.2995	0.3671	0.2890
2003	0.0509	0.6395	0.3047	0.3727	0.2810
2004	0.0525	0.6447	0.3087	0.3752	0.2780
2005	0.0524	0.6327	0.3201	0.3820	0.2642
2006	0.0533	0.6208	0.3213	0.3949	0.2636
2007	0.0541	0.6191	0.3222	0.3964	0.2655
2008	0.0545	0.6164	0.3223	0.4006	0.2667
Average value	0.0491	0.6163	0.2999	0.3679	0.2874
Annual change rate	2.15%	0.57%	1.18%	1.28%	−1.08%
Distribution division	Absolute decentralization	Absolute concentration	Decentralization	Appropriate concentration	Decentralization

TABLE 5: Energy-related carbon emissions in different land use types during 1996–2008 (million ton).

Year	Agricultural land	Industrial land	Residential and commercial land	Traffic land	Other land	Total
1996	2.36	32.79	15.32	3.54	1.51	55.53
1997	2.11	35.35	15.57	3.36	1.54	57.93
1998	2.25	36.30	16.54	4.08	1.71	60.88
1999	1.52	37.74	17.62	4.82	1.81	63.52
2000	2.35	40.80	18.93	5.55	2.12	69.75
2001	2.45	43.15	21.80	6.06	2.31	75.76
2002	2.29	47.67	23.27	6.58	2.54	82.36
2003	1.88	59.58	25.06	7.83	3.80	98.16
2004	2.13	64.80	26.85	9.10	3.75	106.62
2005	3.02	75.97	29.02	10.94	5.51	124.45
2006	2.91	87.17	31.57	11.27	6.12	139.04
2007	2.83	96.70	33.42	12.44	6.76	152.15
2008	2.85	101.43	34.61	13.37	7.07	159.33
Change value	0.49	68.64	19.29	9.83	5.56	103.80
Annual increased rate	1.58%	9.87%	7.03%	11.71%	13.73%	9.18%

other land are integrated in the same order. This integration indicates that stable long-term relationships exist among the variables in each of the four models. Regression models can therefore be constructed using these variables. In the traffic land model, the integration order of lnGC differs from those of the other three variables, indicating that a stable long-term relationship exists among the other three variables. Thus, the other three variables are considered in constructing the traffic land model.

Five regression models are constructed, with their coefficients shown in Table 8. All of the regression models exhibit good fit, as indicated by the large R-square values. The models also pass the F-test at highly significant levels. Table 9 shows that the residual series of the regression models pass the stationarity test. These results indicate that cointegration exists among the variables in each model, and confirm the rationality of the five regressions models.

As indicated by the five models, land use distribution structure influences energy-related carbon emission intensities at different directions and extents. Aside from industrial land, the spatial distribution Gini coefficients of the other four land use types show a significant positive correlation with

TABLE 6: Carbon intensities in different land use types from 1996 to 2008 (t/hm^2).

Year	Agricultural land	Industrial land	Residential and commercial land	Traffic land	Other land	Average value
1996	0.1604	147.8420	14.5272	45.1691	1.3999	3.2326
1997	0.1439	142.0132	14.5558	39.2389	1.4306	3.3740
1998	0.1536	138.4803	15.3286	44.6711	1.5923	3.5453
1999	0.1035	139.1317	16.1736	49.7709	1.7002	3.6992
2000	0.1608	150.8519	17.1461	54.5343	1.9975	4.0613
2001	0.1675	156.1487	19.6515	58.7774	2.1658	4.4113
2002	0.1574	156.5664	20.5831	61.7720	2.3601	4.7923
2003	0.1296	179.9739	21.9476	72.3479	3.4971	5.7068
2004	0.1467	184.5160	23.3094	82.5935	3.4396	6.1973
2005	0.2079	220.9347	24.4428	96.8572	5.1895	7.2319
2006	0.2006	237.5971	26.3965	95.8099	5.7947	8.0751
2007	0.1959	254.4186	27.7520	103.5313	6.4395	8.8357
2008	0.1969	262.7841	28.6123	110.0966	6.7675	9.2525
Annual increased rate	1.73%	4.91%	5.81%	7.71%	14.03%	9.16%

TABLE 7: ADF values and integration orders of variables in each model.

Model	Agricultural land		Industrial land		Residential and commercial land		Traffic land		Other land	
variables	ADF value	Integration order	ADF value	Integration order	ADF value	Integration order	ADF value	Integration order	ADF value	Integration order
lnGE	-7.0353^{***}	1	-3.3655^{**}	1	-4.2561^{***}	1	-3.9848^{**}	1	—	—
lnGC	-5.6262^{***}	1	-3.6838^{**}	1	-3.5406^{**}	1	-4.6038^{***}	0	—	—
lnLS	-3.3381^{**}	1	-2.7999^{*}	1	-2.7910^{**}	1	-3.6112^{**}	1	-3.8533^{*}	1
lnCI	-6.3699^{***}	1	-3.6165^{*}	1	-3.7370^{**}	1	-5.9159^{***}	1	-5.1981^{***}	1

Notes: *, **, *** denote significance at the 10%, 5%, and 1% levels, respectively.

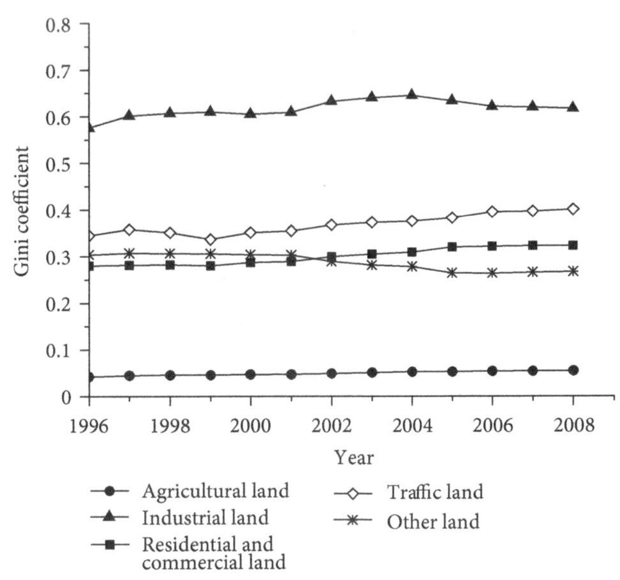

FIGURE 5: Changes in the Gini coefficients of different land use types from 1996 to 2008.

carbon emission intensities. These results indicate that concentrated land use distribution increases the carbon emission intensities in agricultural land, residential and commercial land, traffic land, and other land. Although the land structure variable is significant only at 20% level in the statistical t-test (P value, 0.1722), the model of industrial land also indicates that the Gini coefficient of spatial distribution is negatively correlated with carbon emission intensity in the long term. With respect to extent of influence, every 1% change in the spatial distribution Gini coefficients causes changes of 1.4794%, −1.4262%, 2.4352%, 3.8939%, and 9.7493% in the agricultural land, industrial land, residential and commercial land, traffic land, and other land, respectively. Among the land use types, other land is the type most strongly affected by spatial distribution structure. This result indicates that the other four types of land use, where intensive anthropogenic activities occur, are affected by other factors, such as energy efficiency and production efficiency. Meanwhile, the regression model of other land shows that the spatial distribution structure of land use strongly influences carbon emission intensity when no obvious industrial development and social activities occur in this land use type.

On the basis of the integrative effects of all the independent variables in the regression models (except for the other land model), we can analyze measures for reducing carbon emission intensity as follows. In the regression model of agricultural land, energy efficiency and product efficiency are negatively correlated with energy-related carbon emission intensity, whereas the concentration degree of spatial distribution is positively correlated with the dependent variable.

Table 8: Coefficients of the five models.

Variables	Agricultural land	Industrial land	Residential and commercial land	Traffic land	Other land
Intercept	4.3221***	5.6705***	5.4749***	7.6108***	10.8168***
lnGE	−1.0860***	1.3714***	0.7125**	−1.5946***	—
lnGC	−0.0869***	−0.3723***	0.1493	—	—
lnLS	1.4794***	−1.4262	2.4352***	3.8939***	9.7493***
R-square	0.9910	0.9354	0.9767	0.9773	0.8956
F-test value	330.3216***	43.45113***	125.6984***	215.4236***	94.4115***

Notes: *, **, *** denote significance at the 10%, 5%, and 1% levels, respectively.

Table 9: Stationarity test results for the residual series in the regression models.

Models	Agricultural land	Industrial land	Residential and commercial land	Traffic land	Other land
ADF value	−3.5780	−3.2991	−5.1956	−2.1985	−2.9509
P value	0.0019	0.0034	0.0001	0.0322	0.0069
Stationarity	Stationary	Stationary	Stationary	Stationary	Stationary

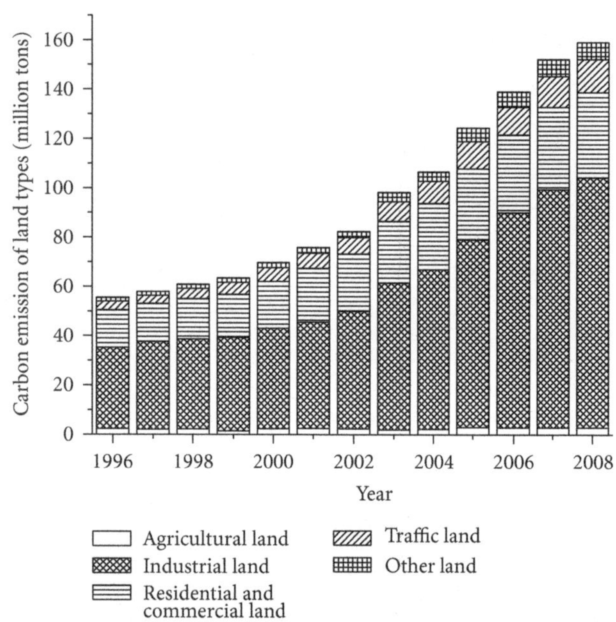

Figure 6: Energy-related carbon emissions in different land use types.

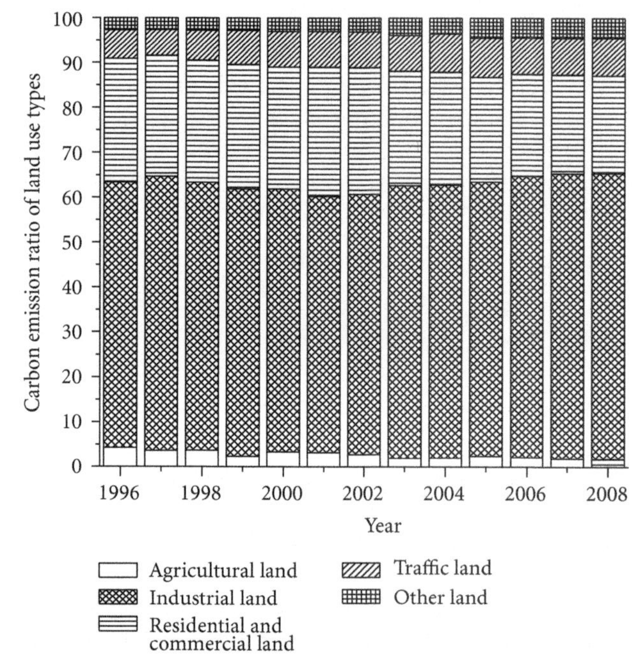

Figure 7: Composition of energy-related carbon emission.

Thus, improving the efficiency of energy use and production, as well as decentralize the agricultural land distribution can reduce the carbon emission intensity in agricultural land. The industrial land model shows that improving energy efficiency increases the carbon emission intensity in industrial land. This phenomenon also occurs in the residential and commercial land model, in which improving the efficiency of energy utilization and production increases carbon emission intensity (Table 8). Carbon emission per unit land use may increase when production per unit land use rises as a result of improved energy efficiency and production efficiency. Improving energy efficiency and production is the development trend driven by technological progress. Thus, controlling energy-related carbon emission through land use distribution structures is crucial. The coefficients of the models indicate that centralize the industrial land and decentralize the residential and commercial land can reduce the carbon emission intensity, respectively. In traffic land, energy efficiency improvement and land use decentralization both can reduce energy-related carbon emission intensity.

The analyses above indicate that adjusting the spatial distribution structure of land use is important in reducing the energy-related carbon emission intensities in all land use types. The coefficients shown in Table 8 indicate that among all the variables in each model, the spatial distribution structure of land use has the largest effect on carbon emission intensity. Thus, focusing on efficient spatial distribution structure of land use can effectively controls carbon emission.

TABLE 10: Energy consumption in different types of land use (10^4 ton standard coal equivalent).

Year	Agricultural land	Industrial land	Residential and commercial land	Traffic land	Other land
1996	368.27	4824.86	2240.35	613.05	229.93
1997	334.65	5234.19	2285.91	583.35	233.29
1998	355.70	5407.87	2414.26	699.89	258.99
1999	245.40	5709.13	2586.78	839.21	274.21
2000	370.75	6110.75	2785.05	965.75	320.64
2001	385.33	6482.84	3199.58	1054.31	349.43
2002	360.55	7177.19	3409.13	1146.83	384.88
2003	298.69	8938.86	3698.44	1351.97	573.72
2004	336.92	9938.45	3974.78	1568.56	567.12
2005	470.63	11664.64	4361.33	1922.88	836.03
2006	452.46	13445.77	4674.41	1982.17	927.34
2007	439.06	14845.45	5022.51	2186.92	1023.54
2008	441.42	15510.91	5226.68	2352.14	1069.61

4. Conclusion

The spatial distribution structure of land use influences energy consumption patterns and is one of the most important factors for regional energy-related carbon emission. However, studies on the relationship between land use spatial distribution structure and energy-related carbon emission are rare. The current work discusses this relationship for different land use types in Guangdong, an economically advanced region in China and has been chosen as a pilot province for national low-carbon development. On the basis of land use pattern and industry types, land use is classified into five types which contain agricultural land, industrial land, residential and commercial land, traffic land, and other land. The spatial distribution structures of different land use types are analyzed by using the spatial Lorenz curve and Gini coefficient. The corresponding energy-related carbon emissions are calculated with factors taken from the Provincial Greenhouse Gas List Preparation Guide [77]. In constructing regression models, the energy-related carbon emission intensities in different land use types are set as dependent variables. Energy efficiency, production efficiency, and land use spatial distribution structure are chosen as independent variables. According to the results form regression analysis and cointegration testing, we draw the following conclusions. (1) Aside from other land, the other four types of land use displayed different concentration trends during 1996 to 2008. Among the five types of land use, industrial land was the most concentrated, whereas agricultural land was the most decentralized but exhibits the fastest change rate. (2) During the study period, energy-related carbon emission in each type of land use considerably increased. The anthropogenic activities in industrial land still account for the highest energy consumption. (3) Spatial distribution concentration is negatively correlated with energy-related carbon emission intensity in industrial land in the long term; by contrast, these two variables show a positive correlation in agricultural land, residential and commercial land, traffic land, and other land. These results indicate that the centralized distribution of industrial land may reduce the corresponding carbon emission intensity, but that the centralized distribution of the other four land use types can increase carbon emission intensity. (4) Land use spatial distribution structure has a more intensive effect on carbon emission intensity than energy efficiency and production efficiency do.

Thus, adjusting the spatial distribution structure of land use is an effective way to control carbon emission. The findings provide valuable reference to develop comprehensive policies for regional energy conservation and carbon emission reduction at a new perspective. Meanwhile, some of the related suggestion can be summarized as follows. (1) Strengthening the intensive use of agricultural land by improving the input-output efficiency of energy, labor, and capital. Ensure the spatial distribution of agricultural land keep on decentralizing when the land is occupied and compensated by other land types or adjusted in internal structure. (2) Promoting industrial agglomeration and unify the layout of energy-intensive industries coordinately. Meanwhile, technological innovation application, infrastructure construction improvement, and recycling economy should be encouraged in industrial development to make sure that resource utilization and scale economy effects are sufficiently developed. (3) Guiding the residents and commercial industry expand or transfer from population intensive region to the relative sparse region, avoiding the unnecessary energy waste from intensive energy consumption and the low commute efficiency caused by excessive concentration of residential and commercial land. (4) Increasing the density of road network in underdeveloped regions of Guangdong. Adjusting the transport instrument structure and industrial management standards, improve the traffic efficiency and reduce the unnecessary repeatedly energy cost by passengers and freight transport in the region of sparse road network. (5) Improving the development efficiency in other land by formulating specification management system. Trying to develop the other land evenly in spatial distribution instead of concentrated develop.

TABLE 11: Input capital in different types of land use (100 million Yuan).

Year	Agricultural land	Industrial land	Residential and commercial land	Traffic land
1996	5.39	457.08	384.15	340.11
1997	4.71	302.15	369.33	211.97
1998	7.65	338.07	469.65	241.12
1999	7.05	393.15	558.52	270.35
2000	8.85	321.29	548.14	283.32
2001	7.47	390.22	533.44	283.36
2002	6.09	542.35	511.88	278.79
2003	9.07	685.92	767.83	351.90
2004	7.33	938.74	887.95	404.88
2005	12.29	1602.37	1132.19	498.70
2006	14.38	1853.86	1118.69	574.98
2007	21.01	1918.81	1275.85	619.81
2008	23.98	1862.23	1423.45	820.22

TABLE 12: GDP in different types of land use (100 million Yuan).

Year	Agricultural land	Industrial land	Residential and commercial land	Traffic land
1996	954.71	2788.82	2277.93	497.68
1997	999.32	3191.03	2520.99	543.15
1998	1037.09	3603.30	2782.94	579.51
1999	1077.65	3996.54	3095.60	612.74
2000	1102.55	4526.20	3504.51	718.48
2001	1126.93	5033.99	3925.78	818.73
2002	1175.49	5783.63	4416.51	870.32
2003	1201.67	7000.60	4916.92	923.46
2004	1251.06	8420.89	5507.29	1035.12
2005	1312.66	9753.01	6260.53	1210.12
2006	1367.72	11472.51	7101.13	1385.19
2007	1411.73	13508.44	8051.59	1518.51
2008	1463.96	15163.55	8782.34	1625.17

Although the results in this study are credible and representative, the study has certain limitations. Aside from the different spatial distributions of various land use types, the land use factors that affect regional energy consumption patterns also include land use composition. Thus, the result of this empirical study does not represent the common mechanism in other regions. In the future study, coordinating the internal relationship among land use factors and discussing their integrated effects on regional energy-related carbon emission are more comprehensive ways to define the mechanism of how the changes in land use influence regional energy-related carbon emission. Another limitation is the difficult access to land use classification data. Collecting land use area data on earlier years by land use classification survey is difficult, and accurately identifying different land use types through Landsat TM images with a 30 m resolution is very hard [27]. We anticipate the collection of more relevant land use data on earlier years, so that the argument in this study can be validated for longer time series.

Appendix

See Tables 10, 11, and 12.

Acknowledgments

This study has been supported by Low Carbon Development Special Fund of Guangdong Development and Reform Commission (Grant no. 2011-049). Special appreciation is also given to Guangdong Ordinary College Key Laboratory of Offshore Oil Exploration and Development for the financial support.

References

[1] C. Green, "Potential scale-related problems in estimating the costs of CO_2 mitigation policies," *Climatic Change*, vol. 44, no. 3, pp. 331–349, 2000.

[2] IPCC, *Climate Change 2007: The Physical Science Basic*, Cambridge University Press, Cambridge, UK, 2007.

[3] H. Liu, S. K. Cheng, and L. Zhang, "The international latest research of the impacts of human activities on carbon emissions," *Progress in Geography*, vol. 21, no. 5, pp. 420–429, 2002.

[4] A. Acaravci and I. Ozturk, "On the relationship between energy consumption, CO_2 emissions and economic growth in Europe," *Energy*, vol. 35, no. 12, pp. 5412–5420, 2010.

[5] B. Zhang, B. H. Peng, and M. C. Liu, "Exergetic assessment for resources input and environmental emissions by Chinese industry during 1997-2006," *The Scientific World Journal*, vol. 2012, Article ID 692746, 13 pages, 2012.

[6] X. P. Zhang and X. M. Cheng, "Energy consumption, carbon emissions, and economic growth in China," *Ecological Economics*, vol. 68, no. 10, pp. 2706–2712, 2009.

[7] L. Fei, S. C. Dong, L. Xue, Q. X. Liang, and W. Z. Yang, "Energy consumption-economic growth relationship and carbon dioxide emissions in China," *Energy Policy*, vol. 39, no. 2, pp. 568–574, 2011.

[8] R. Q. Zhao and X. J. Huang, "Carbon emission and carbon footprint of different land use types based on energy consumption of Jiangsu Province," *Geographical Research*, vol. 29, no. 9, pp. 1639–1649, 2010.

[9] J. C. Lan, W. L. Fu, B. Yuan, T. Zhang, and J. T. Peng, "Analysis of land use patterns on carbon emission and carbon footprint in Chongqing city," *Journal of Soil and Water Conservation*, vol. 26, no. 1, pp. 146–155, 2012.

[10] J. Y. Fang, A. P. Chen, C. H. Peng, S. Q. Zhao, and L. Ci, "Changes in forest biomass carbon storage in China between 1949 and 1998," *Science*, vol. 292, no. 5525, pp. 2320–2322, 2001.

[11] V. H. Dale, "The relationship between land-use change and climate change," *Ecological Applications*, vol. 7, no. 3, pp. 753–769, 1997.

[12] Y. Q. Kuang, T. P. Ouyang, Y. Zou, Y. Liu, Y. Li, and D. H. Wang, "Present situation of carbon source and sink and potential for increase of carbon sink in Guangdong Province," *China Population, Resources and Environment*, vol. 20, no. 12, pp. 56–61, 2010.

[13] R. A. Houghton and J. L. Hackler, "Emissions of carbon from forestry and land-use change in tropical Asia," *Global Change Biology*, vol. 5, no. 4, pp. 481–492, 1999.

[14] R. A. Houghton, J. E. Hobbie, J. M. Mellilo et al., "Changes in the carbon content of terrestrial biota and soils between 1860 and 1980: a net release of CO_2 to the almosphere," *Ecological Monographs*, vol. 53, no. 3, pp. 235–262, 1983.

[15] R. A. Houghton, D. S. Lefkowitz, and D. L. Skole, "Changes in the landscape of Latin America between 1850 and 1985 II. Net release of CO_2 to the atmosphere," *Forest Ecology and Management*, vol. 38, no. 3-4, pp. 173–199, 1991.

[16] C. J. E. Schulp, G. J. Nabuurs, and P. H. Verburg, "Future carbon sequestration in Europe-Effects of land use change," *Agriculture, Ecosystems and Environment*, vol. 127, no. 3-4, pp. 251–264, 2008.

[17] D. M. Howard, P. J. A. Howard, and D. C. Howard, "A Markov model projection of soil organic carbon stores following land use changes," *Journal of Environmental Management*, vol. 45, no. 3, pp. 287–302, 1995.

[18] M. Easter, K. Paustian, K. Killian et al., "The GEFSOC soil carbon modelling system: a tool for conducting regional-scale soil carbon inventories and assessing the impacts of land use change on soil carbon," *Agriculture, Ecosystems and Environment*, vol. 122, no. 1, pp. 13–25, 2007.

[19] X. Y. Zhang, X. J. Huang, X. F. Zhao, R. C. Lu, and L. Lai, "Impacts of land use change on the vegetation carbon storage in region around Taihu Lake," *Journal of NaturaL Resources*, vol. 24, no. 8, pp. 1343–1353, 2009.

[20] T. Zhou and P. J. Shi, "Indirect impacts of land use change on soil organic carbon change in China," *Advances in Earth Science*, vol. 21, no. 2, pp. 138–143, 2006.

[21] W. P. Anderson, P. S. Kanaroglou, and E. J. Miller, "Urban form, energy and the environment: a review of issues, evidence and policy," *Urban Studies*, vol. 33, no. 1, pp. 7–35, 1996.

[22] R. Camagni, M. C. Gibelli, and P. Rigamonti, "Urban mobility and urban form: the social and environmental costs of different patterns of urban expansion," *Ecological Economics*, vol. 40, no. 2, pp. 199–216, 2002.

[23] P. W. G. Newman and J. R. Kenworthy, *Cities and Automobile Dependance: A Sourcebook*, Avebury Technical Press, Avebury, UK, 1989.

[24] I. Larivière and G. Lafrance, "Modelling the electricity consumption of cities: effect of urban density," *Energy Economics*, vol. 21, no. 1, pp. 53–66, 1999.

[25] M. Jenks and R. Burgess, *Compact Cities: Sustainable Urban Forms for Developing Countries*, E&FN Spon Press, London, UK, 2000.

[26] O. Mindali, A. Raveh, and I. Salomon, "Urban density and energy consumption: a new look at old statistics," *Transportation Research A*, vol. 38, no. 2, pp. 143–162, 2004.

[27] Y. M. Chen, X. Li, Y. Zheng, Y. Y. Guan, and X. P. Liu, "Estimating the relationship between urban forms and energy consumption: a case study in the Pearl River Delta, 2005–2008," *Landscape and Urban Planning*, vol. 102, no. 1, pp. 33–42, 2011.

[28] Y. L. Yao, "Energy consumption and space density in urban area," *Energy Procedia*, vol. 5, pp. 895–899, 2011.

[29] L. Thomas and W. Cousins, "The compact city: a successful, desirable and achievable urban form," in *The Compact City: Asustainable Urban Form*, M. Jenks, E. burton, and K. Williams, Eds., pp. 53–56, E&FN Spon Press, London, UK, 1996.

[30] D. McLaren, "Compact or dispersed? Dilution is no solution," *Built Environment*, vol. 18, no. 4, pp. 268–284, 1992.

[31] M. Wachs, "Learning from Los Angeles: transport, urban form, and air quality," *Transportation*, vol. 20, no. 4, pp. 329–354, 1993.

[32] G. O. Yen and X. A. Li, "A constrained CA model for the simulation and planning of sustainable urban forms by using GIS," *Environment and Planning B*, vol. 28, no. 5, pp. 733–753, 2001.

[33] Y. O. Susilo and D. Stead, "Urban form and the trends of transportation emissions and energy consumption of commuters in the Netherlands," in *Proceedings The 87th Annual Meeting of the Transportation Research Board*, Washington, DC, USA, January 2008.

[34] D. Brownstone and T. F. Golob, "The impact of residential density on vehicle usage and energy consumption," *Journal of Urban Economics*, vol. 65, no. 1, pp. 91–98, 2009.

[35] R. Ewing and F. Rong, "The impact of urban form on U.S. residential energy use," *Housing Policy Debate*, vol. 19, no. 1, pp. 1–30, 2008.

[36] S. Ghosh, R. Vale, and B. Vale, "Domestic energy sustainability of different urban residential patterns: a New Zealand approach," *International Journal of Sustainable Development*, vol. 9, no. 1, pp. 16–37, 2006.

[37] X. C. Liu and J. Sweeney, "Modelling the impact of urban form on household energy demand and related CO_2 emissions in the Greater Dublin Region," *Energy Policy*, vol. 46, pp. 359–369, 2012.

[38] E. Holden and I. T. Norland, "Three challenges for the compact city as a sustainable urban form: household consumption of energy and transport in eight residential areas in the Greater Oslo Region," *Urban Studies*, vol. 42, no. 12, pp. 2145–2166, 2005.

[39] L. Ma and F. J. Jin, "Evaluation of Chinese urban compactness," *Progress in Geography*, vol. 30, no. 8, pp. 1014–1020, 2011.

[40] H. Y. Chen, B. S. Jia, and S. S. Y. Lau, "Sustainable urban form for Chinese compact cities: challenges of a rapid urbanized economy," *Habitat International*, vol. 32, no. 1, pp. 28–40, 2008.

[41] R. Q. Zhao, X. J. Huang, and T. Y. Zhong, "Research on carbon emission intensity and carbon footprint of different industrial spaces in China," *Acta Geographica Sinica*, vol. 65, no. 9, pp. 1048–1057, 2010.

[42] H. Y. You and C. F. Wu, "Carbon emission efficiency and low carbon optimization of land use based on the perspective of energy consumption," *Journal of Natural Resources*, vol. 25, no. 11, pp. 1875–1866, 2010.

[43] Z. Q. Zhang, *Relationship Study on the Land Use Structure and Carbon Emissions in Yunnan Province, Based on the Low Carbon Economy Development Background*, Yunnan University of Finance and Economics, Yunnan, China, 2011.

[44] GNDRC (Guangdong National Development and Reform Commission), *Guangdong Low Carbon Development Report in 2010*, GNDRC (Guangdong National Development and Reform Commission), Guangdong, China, 2011.

[45] MLR (Ministry of Land and Resources of the PRC) and MOA (Ministry of Agriculture of the PRC), *National Land Use Classification Standard (Trial Implementation)*, MLR, MOA, Beijing, China, 2002.

[46] P. Li, "The optimization research of construction land use under the condition of low carbon development-Case study in Jiangsu province," Nanjing University, Nanjing, China, 2009.

[47] Y. Z. Tan and C. F. Wu, "The laws of the information entropy values of land use composition," *Journal of Natural Resources*, vol. 18, no. 1, pp. 112–117, 2003.

[48] Y. F. Wu, P. J. Shi, and N. N. Liu, "Analysis on dynamic evolution of land use structure in mountain area of longnan," *Research of Soil and Water Conservation*, vol. 17, no. 6, pp. 134–137, 2010.

[49] M. F. Wu, M. H. Ou, and Y. J. Zhu, "Temporal-spatial changes in land use structure based on information entropy in Heshan city," *Territory & Natural Resources Study*, no. 4, pp. 25–27, 2012.

[50] E. Churieco, "Intergration of linear programming and GIS for land use Modeling," *International Journal of Geographical Information System*, vol. 7, no. 1, pp. 5–9, 1993.

[51] Q. S. Ge, M. C. Zhao, and J. Y. Zheng, "Land use change of China during the 20th century," *Acta Geographica Sinica*, vol. 55, no. 6, pp. 698–706, 2000.

[52] L. G. Jiang, X. H. Nie, and E. F. Liu, "Analysis on the spatial structure of urban land use in Jinan city," *Economic Geography*, vol. 23, no. 1, pp. 70–73, 2003.

[53] J. Chen, "Analysis of changes in land use in Putian City based on RS and GIS," *Journal of Southwest China Normal University*, vol. 30, no. 1, pp. 184–187, 2005.

[54] Z. Q. Du, J. Wang, Z. H. Chen, and Y. D. Shen, "The analysis of regional land use change and its evolving tendency based on RS and GIS," *Journal of Arid Land Resources and Environment*, vol. 21, no. 1, pp. 116–119, 2007.

[55] X. D. Li and X. L. Liu, "Landscape ecology analysis of land use structure in Gansu Province," *Journal of Desert Research*, vol. 29, no. 4, pp. 723–727, 2009.

[56] J. Zhang, L. J. Pu, and B. Z. Peng, "Variation of regional land use structure based on landscape ecology," *Resource and Environment in the Yangtze Basin*, vol. 16, no. 5, pp. 578–583, 2007.

[57] S. T. A. Pickett and M. L. Cadenasso, "Landscape ecology: spatial heterogeneity in ecological systems," *Science*, vol. 269, no. 5222, pp. 331–334, 1995.

[58] Q. Liu, M. H. Ou, and X. Y. Peng, "The forecast of land use structure based on Markov process—a case study of Kunshan city," *Journal of Nanjing Agricultural University*, vol. 28, no. 3, pp. 107–112, 2005.

[59] Y. T. Han, G. B. Wang, and S. P. Li, "Study on land utilization structure optimization of Yulin city based on the collaboration of multiple goals," *Territory & Natural Resources Study*, no. 2, pp. 24–26, 2012.

[60] A. Q. Sun and K. N. Wu, "A case study on Qinghe county: application of GLP model and AHP method in land use structural optimization," *Resources & Industries*, vol. 14, no. 2, pp. 60–65, 2012.

[61] M. J. Wang, H. H. Wang, and Q. T. Zhu, "Analysis on land use structure in Hexi Corridor based on mathematical geography methods-taking jiuquan city as an example," *Journal of Gansu Agricultural University*, vol. 47, no. 1, pp. 125–133, 2012.

[62] X. H. Cai and Y. L. Li, "Study of land use structure of Guangxi Beibu gulf economic zone," *Geomatics & Spatial Information Technology*, vol. 35, no. 4, pp. 42–45, 2012.

[63] J. W. Chen, X. B. Kong, F. R. Zhang, and G. H. Jiang, "Changes of land use structure in Beijing mountain area based on spatial Lorenz curves," *Journal of China Agricultural University*, vol. 11, no. 4, pp. 71–74, 2006.

[64] Y. F. Huang, C. M. Xu, and Y. J. Huang, "Using Lorenz curve to analyze land use in Jiangxi province," *Journal of Jiangxi Normal University*, vol. 27, no. 2, pp. 177–180, 2003.

[65] P. C. Zhang, "Spatial structure analysis of land using based on lorenz's curve-take the second research data of nanchang county as an example," *Science Mosaic*, no. 97, pp. 39–42, 2009.

[66] P. A. Samuelson and W. D. Nordhaus, *Economics*, China Machine Press, Beijing, China, 16th edition, 2005.

[67] M. O. Lorenz, "Methods of measuring the concentration of wealth," *Publications of the American Statistical Association*, vol. 9, no. 70, pp. 209–219, 1905.

[68] C. J. Groves-Kirkby, A. R. Denman, and P. S. Phillips, "Lorenz curve and gini coefficient: novel tools for analysing seasonal variation of environmental radon gas," *Journal of Environmental Management*, vol. 90, no. 8, pp. 2480–2487, 2009.

[69] L. Groot, "Carbon Lorenz curves," *Resource and Energy Economics*, vol. 32, no. 1, pp. 45–64, 2010.

[70] A. Delbosc and G. Currie, "Using Lorenz curves to assess public transport equity," *Journal of Transport Geography*, vol. 19, no. 6, pp. 1252–1259, 2011.

[71] A. Clarke-Sather, J. S. Qu, Q. Wang, J. J. Zeng, and Y. Li, "Carbon inequality at the sub-national scale: a case study of provincial-level inequality in CO_2 emissions in China 1997–2007," *Energy Policy*, vol. 39, no. 9, pp. 5420–5428, 2011.

[72] C. W. Gini, "Variability and mutability, contribution to the study of statistical distributions and relations," in *Studi Economico-Giuridici Della R. Universita De Cagliari*, 1912.

[73] J. Deng and C. T. Diao, "Changes of land use structure based on spatial Lorenz curves-take Chongqing Jiangjin for example," *Territory & Natural Resources Study*, no. 1, pp. 38–40, 2008.

[74] M. X. He, "A new method to calculate Gini coefficient," *Statistics and Decision*, vol. 27, no. 19, pp. 127–128, 2005.

[75] C. X. Li, J. Wang, X. M. Ye, and Q. Yu, "Development and prospects of new energy in China," *Electric Power Science and Engineering*, vol. 28, no. 4, pp. 1–8, 2012.

[76] IPCC, "2006 IPCC guidelines for National Greenhouse Gas Inventories," http://www.ipcc-nggip.iges.or.jp/public/2006gl/vol2.html.

[77] NDRC (National Development and Reform Commission), *Provincial Greenhouse Gas List Preparation Guide (2011)*, NDRC, Beijing, China, 2011.

[78] NBSC (National Bureau of Statistics of China), *China Power Statistical Yearbook (1997–2009)*, China Statistical Press, Beijing, China, 1997–2009.

[79] F. Chen and D. J. Zhu, "Research on the content, models and strategies of low carbon cities," *Urban Planning Forum*, no. 182, pp. 7–13, 2009.

[80] L. Gan, "Research on the relationship between electricity consumption and economic growth in China," *Journal of Zhongnan University of Finance and Economics*, vol. 18, no. 4, pp. 96–101, 2011.

[81] T. M. Gao, J. M. Wang, Y. F. Liang, and Y. H. Liu, *Method and Modeling of Econometric Analysis-Application and Case Study of Eviews*, Tsinghua University Press, Beijing, China, 2009.

[82] D. H. Yi, *Data Analysis and Eviews Application*, China Statistic Press, Beijing, China, 2002.

[83] R. F. Enlge and C. W. J. Granger, "Co-integration and error correction: representation, estimation and testing," *Econometrics*, vol. 55, pp. 251–276, 1987.

[84] X. Deng, *Decomposition Model and Empirical Study of Carbon Emission Base on LMDI Technique*, Huazhong University of Science and Technology, Wuhan, China, 2009.

[85] Y. M. Wei, C. L. Liu, Y. Fan, and G. Wu, *China Energy Report (2008): CO_2 Emissions Research*, Science Press, Beijing, China, 2008.

[86] W. X. Wang, Y. Q. Kuang, and N. S. Huang, "Study on the decomposition of factors affecting energy-related carbon emissions in guangdong province, China," *Energies*, vol. 4, no. 12, pp. 2249–2272, 2011.

[87] X. Guo, "Decomposition of china's carbon emissions: based on LMDI method," *China Population, Resources and Environment*, vol. 20, no. 12, pp. 4–9, 2010.

[88] GDLR (Guangdong Department of Land and Resources), *Classification Survey Statistics of Land Use For Guangdong (1997–2009)*, GDLR, Guangdong, China, 1997–2009.

[89] NBSC (National Bureau of Statistics of China), *China Energy Statistical Yearbook (1997–2009)*, China Statistical Press, Beijing, China, 1997–2009.

[90] CREI (Chinese Renewable Energy Institute), *China New Energy and Renewable Energy Yearbook (2009)*, China Statistical Press, Beijing, China, 2010.

[91] NBSC (National Bureau of Statistics of China), *Guangdong Statistical Yearbook (1997–2011)*, China Statistical Press, Beijing, China, 1997–2011.

[92] NBSC (National Bureau of Statistics of China), *Guangdong Agricultural Statistical Yearbook (1997–2011)*, China Statistical Press, Beijing, China, 1997–2011.

6

Estimating Elasticity for Residential Electricity Demand in China

G. Shi,[1] X. Zheng,[2] and F. Song[2]

[1] Department of Techno-Economic Research, Development Research Center of the State Council, Beijing 10010, China
[2] School of Economics, Renmin University of China, Beijing 100872, China

Correspondence should be addressed to F. Song, songfeng@ruc.edu.cn

Academic Editors: C. Bin, T. Han-Song, and C. Zhan-Ming

Residential demand for electricity is estimated for China using a unique household level dataset. Household electricity demand is specified as a function of local electricity price, household income, and a number of social-economic variables at household level. We find that the residential demand for electricity responds rather sensitively to its own price in China, which implies that there is significant potential to use the price instrument to conserve electricity consumption. Electricity elasticities across different heterogeneous household groups (e.g., rich versus poor and rural versus urban) are also estimated. The results show that the high income group is more price elastic than the low income group, while rural families are more price elastic than urban families. These results have important policy implications for designing an increasing block tariff.

1. Introduction

Since the open and reform policy initiated in 1978, China has managed to maintain rapid economic growth and emerged as the second largest economy in the world. In line with the rapid economic expansion of the past three decades, electricity demand has been growing substantially. Between 1980 and 2009, electricity consumption in China increased from 3,006 TWh to 37,032 TWh, at an annual rate of 8%. Over this period, residential electricity demand grew at a faster rate of 12%. The share of residential consumption over total electricity consumption went up from 3.5% in 1980 to 13.1% in 2009. However, China's per capita residential electricity consumption is still much lower than that of developed countries, accounting for about one-fifteenth of that of the US and one-seventh of that of Japan [1]. Given the continuing trend in income growth, modernization, and urbanization, growth in residential electricity demand is expected to remain high in China.

In the past, China's electricity industry has relied on making massive investments and increasing supply to meet the fast growing demand. The installed power capacity increased from 5.7 million kw in 1978 to 900 million kw in 2010. However, as the problems, such as energy security and environmental deterioration, become more and more severe in China, increasing energy supply is becoming more costly. To promote energy conservation and reduce the local pollutants and carbon emissions, the Chinese central government has set the national energy intensity reduction target for two consecutive five-year plans, which are 20% and 16% for the 11th and 12th five-year plans (FYPs), respectively. To reach this goal, some policies must be instigated.

Given the side effects of supply policies, demand policies have been taking increasingly important roles in controlling electricity demand. Price policy, among these tools, has a particular role to play in energy conservation. In a competitive market, a rational agent would optimally reduce energy consumption in response to a higher price. However, China's energy sector is still heavily regulated by the government. Electricity pricing is completely controlled by the government. As pointed out by Lin and Jiang [1], China's residential electricity consumption is subsidized by industrial consumption and, thus, the residential tariff is even lower than the production costs. The twisted price signal encourages overconsumption and leads to a deterioration in the environment since most of China's electricity is generated by coal. Raising the residential electricity tariff to better reflect the real cost (both production costs and external costs) and to reduce the cross subsidy becomes more and more urgent as the electricity market reform moves forward.

The policy effects of raising electricity tariffs are multifold. On one hand, raising electricity tariffs is believed to be an important measure to promote energy conservation and reduce emission. On the other hand, raising electricity tariffs will inevitably affect the welfare of the household, with differentiated effects on different social groups, such as rich versus poor or urban versus rural households. Quantitatively evaluating these policy effects needs good estimates of price and income elasticities of demand for electricity. Moreover, good estimates of price and income elasticities of demand for electricity are critical input parameters for many research studies. For example, the simulation results and subsequent conclusions of the computable general equilibrium (CGE) model, which is a popular analysis tool for evaluating energy and environmental policies, critically depends on the quality of the various energy demand elasticities.

Although estimating electricity elasticities has attracted much research effort since the first energy crisis of the 1970s, most of them focus on developed counties. J. Espey and M. Espey [2] reviewed 36 studies modeling residential electricity demand, among which only six studies focused on developing countries. Estimation of electricity elasticities based on rigid econometric analysis in China is especially scarce. The only study we found is that of Qi et al. [3], which estimated the price and income elasticities of residential electricity demand to be −0.15 and 1.06, respectively, using provincial level data between 2005 and 2007.

In this paper, price and income elasticities of Chinese residential electricity consumption are estimated for the first time based on data of a unique household survey. We specify the household electricity demand as a function of local electricity price, household income, and a number of social-economic variables (such as household size, age, and education of the household head and dwelling size). The objective of undertaking such estimation is to understand more deeply the key factors that influence electricity demand at household level in China.

We contribute to the literature in two aspects. First, to the best of our knowledge, this is the first paper that uses microlevel data to estimate the electricity demand elasticity in China. Aggregate level electricity consumption data are often employed to estimate electricity demand (e.g., among many other studies, [4–7]). However, the aggregate level data often lose much information at the individual level. Using microlevel data can better reflect the individual or household behavior and adds more details to our understanding of the consumers' response [8]. Second, we estimate the electricity elasticities across different heterogeneous household groups (e.g., rich versus poor and rural versus urban). This information contributes to the ongoing debate on how China's electricity market reform will affect different social groups and help design the new pricing scheme.

The rest of the paper is organized as follows. Section 2 describes the empirical models we employed and is followed by the data and estimation results presented in Section 3. In Section 4, the results are discussed and conclusions are given.

2. Residential Demand for Electricity

2.1. Basic Model. Residential demand for electricity is a demand derived from the demand for a well-lit house, cooked food, hot water, and so on and can be specified using the basic framework of household production theory [5]. According to the theory, households purchase input factors to produce the final goods, which appear as arguments in the household's utility function. In our specific case, a household combines electricity and capital equipment to produce a composite energy commodity.

Following the specification of [9, 10], a linear double-logarithmic form using income, electricity price, the price of alternative fuels, and a number of socioeconomic factors as independent variables is used in the empirical analysis of household electricity demand as follows:

$$\begin{aligned}\ln E_{it} = {} & \beta_0 + \beta_1 \cdot \ln\left(\text{PE\_res}_p\right) + \beta_2 \cdot \ln(\text{Income}_{it}) \\ & + \beta_3 \cdot \ln\left(\text{PE\_ng}_p\right) + \beta_4 \cdot Z_{it} + \varphi_t + \eta_{it},\end{aligned} \quad (1)$$

where E_{it} is electricity usage for household i at year t. PE_res_p is residential electricity price in province p, where household i lives. Income_{it} is real income of household i at year t. PE_ng_p is real price of liquefied natural gas in province p, which may be a complement to or substitute for electricity for households. Z_{it} is a set of control variables, including household dwelling area (HSIZE_{it}), number of family members (FSIZE_{it}), the age (AGE_{it}), and years of education (EDU_{it}) of the household head. φ_t is the year-fixed effect.

The key explanatory variables that influence household electricity demand are thus described in the model above. Household income and electricity price are two important economic variables that are assumed to determine household electricity demand. Since we adopt the double-logarithmic specification, the parameter estimates of β_1 and β_2 can be directly interpreted as price and income elasticity.

Since electricity is not the only energy source for a household, electricity demand can also be influenced by the price of other alternative fuels. Therefore, we include the price of natural gas in the estimation of the demand functions. These are also included in the model in order to test the hypothesis of whether these fuels are in anyway complementary to or substitutes for electricity.

Several variables that represent household characteristics are included in the estimation model to account for the impact of the underlying preference of consumers of different backgrounds: (HSIZE_{it}), number of family members (FSIZE_{it}), the age (AGE_{it}), and years of education (EDU_{it}). We expect that a family which has more members or whose dwelling has a larger area may consume more electricity. How the age and education level of the household head will influence electricity consumption is an empirical question. The year-fixed effect is controlled to eliminate national trends, such as business cycles, that may affect household electricity demand.

2.2. Heterogeneous Effects. Residential electricity consumption may have different patterns among households which have different incomes or live in urban or rural areas. We examine these two heterogeneous effects by estimating (2) and (3) as follows:

$$\begin{aligned}\ln E_{it} = {} & \theta_0 + \theta_1 \cdot \ln\left(\text{PE\_res}_p\right) \times \text{Rich}_{it} \\ & + \theta_2 \cdot \ln\left(\text{PE\_res}_p\right) + \theta_3 \cdot \text{Rich}_{it} \\ & + \theta_4 \cdot \ln\left(\text{PE\_ng}_p\right) + \theta_5 \cdot Z_{it} + \varphi_t + \varepsilon_{it},\end{aligned} \tag{2}$$

where Rich_{it} is a dummy variable which equals one if the income of household i in year t is greater than the sample median. The interaction term between $\ln(\text{PE\_res}_p)$ and Richit captures the difference in price elasticity of the rich compared with the poor. Taking the poor household as the benchmark, the price elasticities of the poor household and the rich household are estimated by the coefficients θ_2 and $(\theta_1 + \theta_2)$, respectively. Consider the following:

$$\begin{aligned}\ln E_{it} = {} & \gamma_0 + \gamma_1 \cdot \ln \text{PE\_res}_p \times \text{Urban}_{it} \\ & + \gamma_2 \cdot \ln \text{Income}_{it} \times \text{Urban}_{it} + \gamma_3 \cdot \ln \text{PE\_res}_p \\ & + \gamma_4 \cdot \ln \text{Income}_{it} + \gamma_5 \cdot \text{Urban}_{it} + \gamma_6 \cdot \ln \text{PE\_ng}_p \\ & + \gamma_7 \cdot Z_{it} + \varphi_t + \varepsilon_{it},\end{aligned} \tag{3}$$

where Urban_{it} is the dummy variable for urban households. Similarly, the interaction terms between the price/income variables and the urban dummy variable capture the differences in price and income elasticities of urban households compared with rural households.

2.3. Estimation Issues. Econometrically estimating the electricity demand function presents several challenges. First, simultaneity problems exist between marginal price and consumption if aggregated data are used or consumers face a nonlinear price scheme because there is reverse causality between demand and price. Fortunately, the simultaneity problem is avoided in our study since we use the household level data in which the household is clearly the price taker. In addition, electricity prices have been highly regulated in China and local sales prices were set by provincial governments based on the costs of power generation, transmission, and distribution. The increasing block tariff did not start nationwide until July, 2012. During our study period (2008-2009), consumers faced the single price scheme, which enabled us to avoid the endogeneity problem caused by the nonlinear pricing scheme. Second, energy demand is influenced by long-term household decisions over appliance purchases and dwelling characteristics. Some studies have used a system of equations where the household makes two-stage optimization decisions. In the first stage, the short-run consumption of electricity depends, among other variables, on electricity price and income. In the second stage, the purchasing decision for durable goods, such as electronic appliances, is modeled [11, 12]. Since the system of equations approach has a very high data requirement (information on holdings of household-specific appliances and residence features), single equation specifications for household energy demand are most often used in linear or logarithmic form [13–17].

There are several options for estimating (1)–(3): pooled OLS, which assumes away significant individual or temporal effects among the panel; the fixed effects or random effects models, which assume there are unobserved specific individual and temporal effects. As introduced above, electricity price has been highly regulated in China. During our study period, the prices were quite uniform within a province and remained stable over years. A lack of variation in the price variable forced us to adopt the pooled OLS technique.

3. Data and Estimation Results

3.1. Data. The household level data used in this study is provided by the China Family Panel Studies (CFPS) project, conducted annually by Peking University of China since 2008. In 2008 and 2009, the survey was carried out in Beijing, Shanghai, and Guangdong, which are located in eastern China. Stratified random sampling is applied, which ensures representativeness and randomness. A more detailed description of the CFPS project can be found in Hvistendahl [18]. The CFPS survey data contain information on various aspects of the household, such as socioeconomic information and demographic status, education information, and health information of the households. The CFPS survey data have been employed in a number of studies, such as Luo and Zhang [19] on health and labor market outcomes and Su and Heshmati [20] on gender wage difference.

A panel dataset is constructed using CFPS 2008 and 2009. In 2008, the survey covered 2,375 households, among which 1,940 households were followed up in 2009. Keeping the families that are observed in both years and with nonmissing electricity usage, the balanced panel covers 1,649 households.

Descriptive statistics are reported in Table 1. In our dataset, the average electricity consumption per capita is 37.78 kWh per month, which is 37% higher than the national average of 27.66 kWh per month [21]. (The average electricity usage for each household is 112.97 kWh per month in our dataset. On average there are 2.99 household members. So each person consumes 37.78 kWh per month.) This is because Beijing, Shanghai, and Guangdong, where the data were collected, are among the most developed provinces of China. As introduced above, a single price scheme for residential electricity was implemented in China during our investigating period. The nominal electricity prices in 2008 and 2009 for Beijing, Shanghai, and Guangdong remained unchanged at 0.47, 0.54, and 0.61 Yuan/KWH, respectively. The prices of electricity and natural gas are deflated using the provincial consumer price index (CPI).

3.2. Estimation Results. Columns (2)–(4) of Table 2 report the estimation coefficients and associated t-values for the three electricity demand models ((1)–(3)) based on the household data using the OLS technique. In model 1, the own price elasticity of electricity demand is estimated to

TABLE 1: Descriptive statistics of the household survey data.

Variable	Description	Obs.	Mean	Std. dev.	Min.	Max.
E	Electricity usages (KWh/family/month)	3,298	112.97	88.65	0	550
Income[a]	Household income (Yuan)	3,298	37,239.13	43,856.17	0	33,8000
HSIZE[a]	Dwelling area (m^2/family)	3,298	92.20	70.45	0	800
FSIZE	Family size (person/family)	3,298	2.99	1.33	1	12
PE_res	Residential electricity price (Yuan/KWh)	3,298	0.56	0.06	0.49	0.62
PE_ind	Industrial electricity price (Yuan/KWh)	3,298	0.82	0.04	0.76	0.89
PE_ng	Natural gas price (Yuan/m^3)	3,298	6.12	0.62	5.50	7.29
Age	Age of household head	2,072	59.04	13.19	17	95
Edu	Years of education of household head	2,704	8.21	4.36	1	19

Note: [a]For income, HSIZE, age, and edu, we replace missing values with 0 and create missing indicators for missing observations. Missing indicators are controlled for in the following regressions.

TABLE 2: Estimation results using OLS technique.

	Model 1: basic	Model 2: income effect	Model 3: urban versus rural
	Dependent variable: ln(residential electricity consumption)		
ln(PE_res)	−2.477***	−2.638***	−3.735***
	(0.374)	(0.493)	(0.492)
ln(Income)	0.058***	0.016**	0.063***
	(0.005)	(0.007)	(0.009)
ln(PE_ng)	−0.486**	(0.252)	0.304
	(0.247)	(0.258)	(0.246)
ln(HSIZE)	0.162***	0.103***	0.159***
	(0.021)	(0.030)	(0.029)
FSIZE	0.068***	0.069***	0.099***
	(0.013)	(0.013)	(0.012)
Age	0.002	0.000	−0.005***
	(0.001)	(0.001)	(0.001)
Edu	0.064***	0.052***	0.030***
	(0.004)	(0.005)	(0.003)
ln(PE_res)*Rich		0.482	
		(0.661)	
Rich		0.201	
		(0.291)	
ln(PE_res)*Urban			2.644***
			(0.652)
ln(Income)*Urban			−0.040***
			(0.011)
Urban			(0.240)
			(0.307)
Year FE	Y	Y	Y
	OLS	OLS	OLS
Observations	3298.000	3298.000	3298.000
R-squared	0.160	0.186	0.239

Note: Robust standard errors in parentheses; $^{***}P < 0.01$, $^{**}P < 0.05$, $^{*}P < 0.1$.

be −2.477, which carries the expected sign and is statistically significant at the 1% level. This suggests that a 1% increase in residential electricity price would result in a more than 2% decline in household electricity demand (ceteris paribus). The result implies that households are very responsive to electricity price changes. Electricity consumption is responsive to the level of income with an income elasticity of 0.058. However, since elasticity is well below unity, income growth results in a much less than proportional increase in electricity consumption.

An examination of the coefficient on the price of natural gas that is included in the model provides the cross elasticity between electricity and natural gas. The estimated coefficient is −0.486 and marginally significant. The result implies that there is a weak complementary relationship between electricity and natural gas, which reflects that electricity and natural gas have largely independent uses and limited switching. A similar complementary relationship between electricity and natural gas has been found for California residents [22] and Indian residents [23].

As we expected, larger dwelling size and more family members increase the household consumption of electricity. A 1% increase in the number of square meters results in about a 0.162% increase in the household's demand for electricity, while increasing the family number by one results in a 6.8% increase in the household demand for electricity (ceteris paribus).

Finally, the demographic characteristics may also influence the electricity consumption of the household. A one-year increase in the period of education results in a 6.4% increase in electricity consumption (ceteris paribus).

In model 2, we add a dummy variable that represents relatively rich families and associated interaction terms to control for the effect of income difference on electricity demand and the estimation results are reported in column 3 of Table 2. As indicated above, the estimated coefficient of the interaction term ln(PE_res)*Rich can be interpreted as the difference in price responsiveness of electricity demand between the rich household and poor household. The estimated coefficient is small and statistically insignificant, which means that the rich family and the poor family respond similarly to electricity price changes. The estimated coefficients of other variables in model 2 remain stable compared with model 1.

In model 3, we add a dummy variable that represents whether the family lives in an urban area or a rural area and the associated interaction terms controlling for the effect of living location on electricity demand. The estimation results are reported in Column 4 of Table 2. The price elasticity for rural and urban households is −3.735 and −1.091, respectively, while the income elasticity is 0.063 and 0.023, respectively. These results indicate that urban residential electricity consumption is less elastic to both price and income. The estimates provide supportive evidence that urban household electricity consumption is less sensitive to price and income changes. Raising electricity price will be more detrimental to rural families. The income increase of the rural households will lead to more residential electricity demand. However, the overall effect is mild.

4. Conclusions

The paper provides the results of the econometric estimation of the electricity demand model using a dataset consisting of information at the individual household level in China. The basic model is used to determine the responsiveness of electricity consumption to its own price, income, price of alternative fuel, and variables relating to the social-economic characteristics of the household. The estimated models demonstrate the importance of including the household level information, which is impossible using aggregate level data. Demand elasticities for the heterogeneous household level are also examined.

We found that residential demand for electricity responds rather sensitively to its own price in China. There might be two reasons to explain the seemingly high estimate of our study. First, the estimates of electricity price elasticity depend on the model specification and data. Taylor [24] reviewed many studies and found electricity to be much more price elastic in the long run. Our estimated results, using a pooled two-year dataset and OLS technique, are consistent with his conclusion. Second, China's economy has undergone rapid development and modernization during the past 30 years. Most Chinese residents have experienced the dramatic change from lacking the basic living necessities before the 1980s to, today, being able to afford the costly electrical appliances (such as air conditioners, shower heaters, and washing machines) which represent the modern life style. However, for the Chinese, the virtue of thrift and the preference to save have hardly changed. It could be expected that an increase in the price of electricity might result in many households cutting back the use of electric appliances, for instance, they may choose to wash their clothes by hand, instead of using a washing machine, or to turn on the air-conditioner for only a few hours on a hot day instead of running it all day. Supportive evidence is that high estimates of the price elasticity of residential electricity demand were found in studies in the US before the 1970s [25, 26] and in some developing countries or regions, such as Taiwan [27] and Honduras [28].

From the policy point of view, the high price elasticity of residential electricity demand implies that there is the potential to use the price instrument to conserve electricity consumption. Electricity pricing schemes, such as the increasing block tariff, seasonal pricing, and/or residential time-of-use pricing, will increase the overall electricity price level and thus can effectively curb the demand of the electricity.

Since July 1st, 2012, China has begun to implement the increasing block tariff nationwide. How to design an effective tariff structure to ensure that the new pricing scheme can improve efficiency and equity is an essential question. The keys to the increasing block tariff design include the number of blocks and the volume and price in different blocks. The examination of the electricity demand elasticity of different groups shows that the high income group is more price elastic than the low income group, while rural families demonstrate more price elasticity than urban families. These results have three implications for designing the increasing block tariff. First, to ensure equity and satisfy the basic needs of consumers, the electricity volume of the first block should be set large enough to cover the usage of most residents. Second, the price of the last block, which targets the high income households, should be set high to ensure an effective cut in consumption. Third, since our results also show that raising electricity prices will be more detrimental to rural families, it is necessary to separate the rural and urban households and design different volumes and prices in each block for each group.

Finally, as would be expected, the estimates for income elasticities show that electricity is a necessity. The relatively low value of this elasticity means that residential electricity consumption would grow moderately with further growth in the Chinese residents' incomes.

Acknowledgment

The research support from the funds of Renmin University of China (11XNL009) is gratefully acknowledged.

References

[1] B. Lin and J. Jiang, "Designation and influence of household increasing block electricity tariffs in China," *Energy Policy*, vol. 42, pp. 164–173, 2012.

[2] J. Espey and M. Espey, "Turning on the light: a meta-analysis of residential electricity demand elasticities," *Journal of Agricultural and Applied Economics*, vol. 36, pp. 65–81, 2004.

[3] F. Qi, L. Zhang, B. Wei, and G. Que, "An application of Ramsey pricing in solving the cross-subsidies in Chinese electricity tariffs," in *Proceedings of the 3rd International Conference on Deregulation and Restructuring and Power Technologies (DRPT '08)*, pp. 442–447, April 2008.

[4] S. Balabanoff, "The dynamics of energy demand in Latin America," *OPEC Review*, vol. 18, no. 4, pp. 467–488, 1994.

[5] M. Filippini, "Swiss residential demand for electricity," *Applied Economics Letters*, vol. 6, no. 8, pp. 533–538, 1999.

[6] M. Bernstein and J. Griffin, "Regional differences in price-elasticity of demand for energy," Tech. Rep., National Renewable Energy Laboratory, Santa Monica, Calif, USA, 2005, Rand Infrastructure, Safety, and Environment.

[7] A. Alberini and M. Filippini, "Response of residential electricity demand to price: the effect of measurement error," CEPE Working Paper 75, Centre for Energy Policy and Economics, Zurich, Switzerland, 2010.

[8] P. Holtedahl and F. L. Joutz, "Residential electricity demand in Taiwan," *Energy Economics*, vol. 26, no. 2, pp. 201–224, 2004.

[9] M. Beenstock, E. Goldin, and D. Nabot, "The demand for electricity in Israel," *Energy Economics*, vol. 21, no. 2, pp. 168–183, 1999.

[10] M. P. Clements and R. Madlener, "Seasonality, cointegration, and forecasting UK residential energy demand," *Scottish Journal of Political Economy*, vol. 46, no. 2, pp. 185–206, 1999.

[11] J. T. Bernard, D. Bolduc, and D. Bélanger, "Quebec residential electricity demand: a microeconometric approach," *Canadian Journal of Economics*, vol. 29, no. 1, pp. 92–113, 1996.

[12] R. Nesbakken, "Price sensitivity of residential energy consumption in Norway," *Energy Economics*, vol. 21, no. 6, pp. 493–515, 1999.

[13] G. S. Donatos and G. J. Mergos, "Residential demand for electricity: the case of Greece," *Energy Economics*, vol. 13, no. 1, pp. 41–47, 1991.

[14] E. D. Tserkezos, "Forecasting residential electricity consumption in Greece using monthly and quarterly data," *Energy Economics*, vol. 14, no. 3, pp. 226–232, 1992.

[15] J. I. Silk and F. L. Joutz, "Short and long-run elasticities in US residential electricity demand: a co-integration approach," *Energy Economics*, vol. 19, no. 4, pp. 493–513, 1997.

[16] M. P. Clements and R. Madlener, "Seasonality, cointegration, and forecasting UK residential energy demand," *Scottish Journal of Political Economy*, vol. 46, no. 2, pp. 185–206, 1999.

[17] P. Tiwari, "Architectural, Demographic, and Economic Causes of Electricity Consumption in Bombay," *Journal of Policy Modeling*, vol. 22, no. 1, pp. 81–98, 2000.

[18] M. Hvistendahl, "Survey to reveal true face of chinese society," *Science*, vol. 328, no. 5978, pp. 554–555, 2010.

[19] M. Luo and C. Zhang, "Non-linear relationship between body massindex and labor market outcomes: new evidence from China," MPRA Paper 38936, 2011, http://mpra.ub.uni-muenchen.de/38936/.

[20] B. Su and A. Heshmati, "Analysis of gender wage differential in China's urban labor market," IZA Discussion Paper 6252, 2012, http://ftp.iza.org/dp6252.pdf.

[21] National Bureau of Statistics (NBS), *China Statistical Yearbook 2010*, NBS, Beijing, China, 2011.

[22] L. M. Garcia-Cerrutti, "Estimating elasticities of residential energy demand from panel county data using dynamic random variables models with heteroskedastic and correlated error terms," *Resource and Energy Economics*, vol. 22, no. 4, pp. 355–366, 2000.

[23] M. Filippini and S. Pachauri, "Elasticities of electricity demand in urban Indian households," *Energy Policy*, vol. 32, no. 3, pp. 429–436, 2004.

[24] L. D. Taylor, "The demand for electricity: a survey," *Bell Journal of Economics*, vol. 6, no. 1, pp. 74–110, 1975.

[25] H. Houthakker and L. Taylor, *Consumer Demand in the United States*, Harvard University Press, Cambridge, Mass, USA, 1970.

[26] R. Halvorsen, "Residential demand for electric energy," *Review of Economics and Statistics*, vol. 57, pp. 12–18, 1975.

[27] G. Hondroyiannis, "Estimating residential demand for electricity in Greece," *Energy Economics*, vol. 26, no. 3, pp. 319–334, 2004.

[28] L. Hunt, C. Salagado, and A. Thrope, "The policy of power and power of policy in Honduras," *Journal of Energy and Development*, vol. 25, pp. 1–36, 1999.

7

Emergy Analysis of Biogas Systems Based on Different Raw Materials

Yang Wang,[1] Cong Lin,[1] Jing Li,[2] Na Duan,[1] Xue Li,[1] and Yanyan Fu[1]

[1] *College of Water Resources and Civil Engineering, China Agricultural University, Beijing 100083, China*
[2] *China Zhongyuan Engineering Corporation, Beijing 100191, China*

Correspondence should be addressed to Cong Lin; lincong@cau.edu.cn

Academic Editors: B. Chen and Z.-M. Chen

Environmental pollution and energy crisis restrict the development of China, and the utilization of renewable technology is an effective strategy to alleviate the damage. Biogas engineering has rapidly developed attributes to solve environmental problems and create a renewable energy product biogas. In this paper, two different biogas plants' materials were analyzed by emergy method. One of them is a biogas project whose degraded material is feces (BPF system), and the other is the one whose degraded material is corn straw (BPC system). As a result, the ecological-economic values of BPF and BPC are $28,300/yr and $8,100/yr, respectively. Considering currency, environment, and human inputs, both of the biogas projects have the ability of disposing waste and potential for development. The proportion of biogas output is much more than fertilizer output; so, fertilizer utilization should be emphasized in the future. In comparison, BPF is better than BPC in the aspects of ecological-economic benefits, environmental benefits, and sustainability. The reason is the difficulty of corn straw seasonal collection and degradation. Thus it is proposed that BPC should be combined with the other raw materials.

1. Introduction

China is the world's largest developing country with a population over than 1.3 billion. With the development of economy and industrialization, the problems of excessive energy consumption significantly affect the future of China. Finding new energy sources to replace fossil fuels is an urgent task for China [1–5]. Anaerobic digestion is one of the most appropriate technologies to solve these problems. There are large biomass resources in China, especially crop straw, forest residue, livestock and poultry manure, and various kinds of municipal and industrial organic wastes and waste water [2]. Such sufficient materials promote the development of biogas, especially the large-and medium-sized biogas projects.

In 2003, the "agricultural ecological" program was carried out, which created a favorable condition for the development of biogas projects. According to statistics (shown in Figure 1), biogas yield in 2009 is 1.75 times more than 2003. In addition, the development of biogas projects also positively influences related industries, such as processing manufacturing, construction materials, and construction engineering. All of them have achieved good economic, social, and other comprehensive benefits [6].

With the number of biogas projects increasing, biogas ecological systems have attracted more and more attention. Biogas ecological system is a complexity composed of agricultural, environment, energy, society, and other relevant sectors. The aims of a biogas ecological system are producing an energy carrier from renewable resources and achieving multiple environmental benefits. Some research has been taken on biogas ecological systems, but mostly, the focus have been on the technical evaluation, economic benefit, and energy flows [7–13]. The research methods of assessing biogas ecological system include exergy accounting, energy analysis, life cycle assessment, and emergy analysis which have been developed in the last 30 years. Emergy is chosen in this paper because it is a particularly appropriate tool to evaluate the agricultural production system. Emergy focuses on the compound ecosystem at the interface between natural and human systems, which involves many subject areas, not only the systems of ecology, ecosystem ecology, energetic, resource science, environmental science, systematic, earth

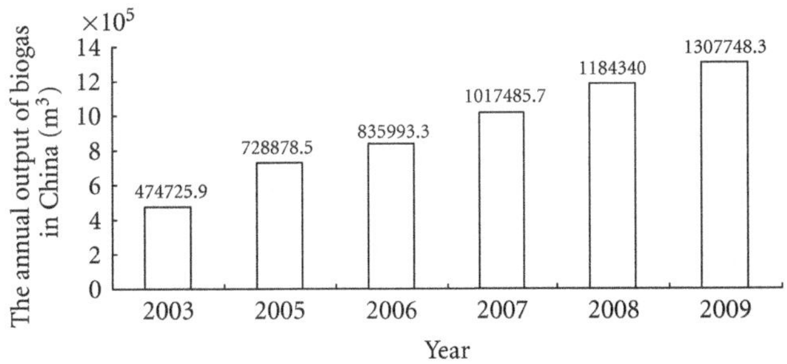

FIGURE 1: Bar chart of the number of biogas projects in China.

science, and other natural sciences, but also those relating to economics, sociology, forecasting and the other humanities [14, 15]. Emergy method unifies all the measures, and the whole inputs indices along the forming process are converted into solar emergy (sej) and solar transformation (sej/J). So, both the environmental values and the economic values can be calculated, which simplifies the assessment process.

All the research mentioned earlier is focused on one kind of biogas plant, and the material used as input is not differentiated. However, with the development of biogas technology, there are several biogas materials species. At present, the main anaerobic fermentation materials are straw and feces, and their fermentation process and parameters (temperature, concentration, refluxing ratio, stir mode, etc.) are very different. Therefore, only by considering the different characters of materials, biogas project system analysis could be comprehensive and correct. In this paper, two representative biogas projects are chosen to be compared, and emergy is used to analyze their characteristics. The aim of this study is to assess the ecological-economic benefit, environmental benefit, and sustainable development capability of two different raw materials (corn straw and feces) in the biogas generation system and lay a theoretical foundation for the generalization of the ecological biogas production system. All the calculations aim to be as transparent as possible in order to make the results useful for future analyses.

2. Materials and Methods

2.1. Study Site. The two biogas projects studied in this paper are different in degraded materials; one is corn straw, and the other is feces. All the data come from the results of site investigation among the whole year.

2.1.1. BPF System. The biogas project, where degraded material is feces (BPF), is located in the west of Hegezhuang village of Chaoyang district in Beijing. The total investment of biogas plant is RMB 2,600,000 Yuan, occupying 2930 m^2. BPF began operating in October of 2010, processing feces from three villages, Naidong, Hegezhuang, and Naixi. This biogas project solved the problem of feces pollution and improved the quality of village life.

The technology of BPF is continuous stirred tank reactor (CSTR), and the material of the tank is enameled pressed steel. The capacity of the BPF is 400 m^3, which has a liquid-gas storage integrated tank, including 224 m^3 liquid storage volume and 200 m^3 gas storage volumes. The biogas project can digest 10~20 t solid manure and 25 t domestic sewage per day, and its biogas yield is 650 m^3/d. In routine management, the fermented concentration is maintained to 6%~10% by the way of biogas slurry refluxing. A solar system is one of the most important parts of the whole biogas plant, because it can increase feeding materials temperature to 5°C in winter. Moreover, some of the output biogas is heated to maintain fermented temperature. In spring and autumn, 250 m^3 biogas will be heated, and the output is 400 m^3. In summer, because of high temperature, it is not necessary to heat the biogas tank; so, the entire biogas of 650 m^3 will be exported. In winter, there is 350 m^3 biogas heated and 300 m^3 biogas exported. In addition, 113 kg coal will be supplied per day in winter which will last for 4 months; so, the quantity of the coal is 13.6 t/yr.

2.1.2. BPC System. The other biogas project, where degraded material is BPC, is located in the Dongyaozhuang village of Qing country in Hebei province, where crop planting industry is well developed. Because of this, there are a number of types of waste crops straw stalks, and the environment has been polluted by stacking and willfully burning straws. In 1999, the first 1000 m^3 CSTR biogas underground tank was built. However, during that period, the technology was immature; so, the biogas tank could not be put into operation. After that, a brick-concrete structure biogas tank of 400 m^3 was constructed successfully in 2005, which was the first in China. However, due to limitations of building materials and technology, that biogas project is out of service now. Through summarizing construction operation experiences and improving technology, the new biogas tank of 400 m^3 was constructed in 2006. By debugging and running, it was in formal operation in 2007 and has become the star exemplary biogas project in China.

The technology of BPC is up-flow solids reactor (USR), and the material of the tank is welded steel plate. The whole volume of BPC is 400 m^3, with treatment capacity of 900 kg corn straw every day. To maintain the feeding concentration, 2~3 t water should be added per day. The fermentation temperature is 44~55°C, and the biogas production capacity is 480 m^3/d. The whole biogas system needs 200 m^3 of biogas heated per day in winter and 120 m^3/d in spring and autumn. The biogas net output amount is 280 m^3/d in winter and 360 m^3/d in spring and autumn. Because it is not necessary to heat it, the biogas net output amount is 470 m^3/d in summer. The same as BPF, 150 kg/d coal is used to heat the biogas tank which lasts about 4 months, and the amount of coal is 18 t/yr.

2.2. Emergy Accounting. The input and output raw data of the two biogas systems are recorded for the whole year of 2011. The details of emergy accounting can be referred to through the works of Odum [16]. The process of emergy accounting consists of two steps. The first is to define the boundary of the system and then make emergy evaluation tabulation. The next step is to sum all the emergy contributions from the

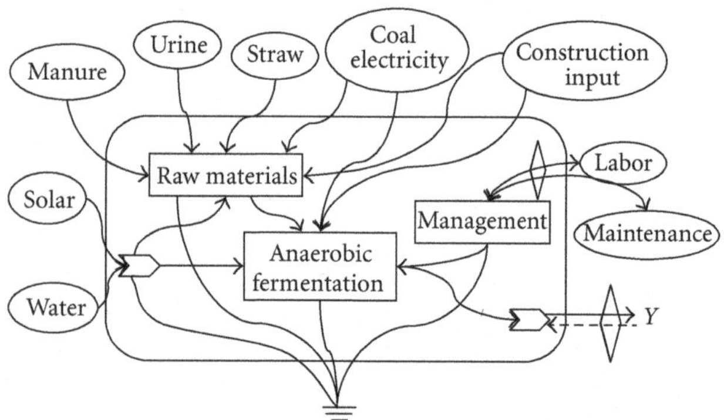

FIGURE 2: Emergy system diagram of the biogas project.

independent inputs and biogas systems evaluation as shown in the established emergy indicator framework.

The emergy system is shown in Figure 2. Input resource is divided into three parts: free local renewable resources (RRs) embracing raw materials (feces/corn straw), solar energy, and underground watering, renewable purchased (RP) inputs includes human labor, and nonrenewable purchased (NP) inputs covering construction inputs, coal and electricity. All the products including biogas and biogas residues are system yield (Y). In the counting process, the whole economic values are converted into solar emergy (sej). And then solar emergy turns into emergy-monetary value (Em$) through an emergy-monetary ratio. In this paper, emergy-monetary value (Em$) represents not only macroeconomy value of emergy flow, but also ecological-economic value, which assesses the eco-efficiency, environmental impact, and the sustainable capacity of the system.

2.3. Emergy Evaluation Index. To analyze these two types of biogas systems for the aspects of ecological-economic benefit, environmental benefit, and sustainable development ability, ten evaluation indices are applied in this paper.

2.3.1. Purchase Emergy Ratio (PER). It is a ratio of input emergy social economy feedback and total input emergy. It depends on system's dependent degree to external resources.

2.3.2. Natural Emergy/Purchase Emergy. This ratio is the emergy from natural resources divided by input emergy social economy feedback, which shows the condition of industrial competitiveness.

2.3.3. Emergy Investment Ratio (EIR). It's a ratio of input emergy social economy feedback and input emergy of natural resources, which shows the cost of the system.

2.3.4. Emergy Yield Ratio (EYR). It is a ratio of the total output emergy and input emergy social economy feedback, which represents the situation of the system production.

2.3.5. Emergy Self-Sufficiency Ratio (ESR). ESR shows the self-maintenance ability of the system. It is the emergy from nature divided by the total output emergy of the system.

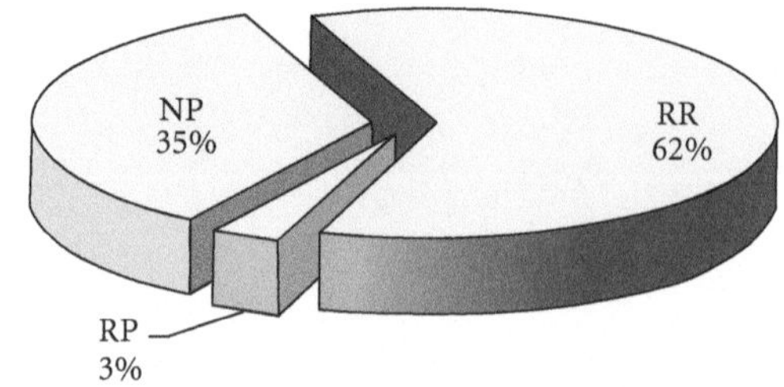

FIGURE 3: Emergy input ratio of BPC.

2.3.6. Ratio of Waste Treatment (%W). It is a ratio of the emergy of waste (for a biogas project, it means the raw materials, such as corn straw and feces) treatment and the total output emergy of the system. This value shows the ability of the treatment of waste of the whole system.

2.3.7. Environmental Loading Ratio (ELR). This is an important index of environment, [which means the pressure to environment from the system.] It is equal to the ratio of total nonrenewable emergy divided by renewable emergy.

2.3.8. Feedback Yield Ratio (FYR). This ratio means the emergy of system self-feedback (for biogas system, it is the part of biogas that is heated) divided by economy feedback emergy, which shows the self-organizing ability of the system.

2.3.9. Renewable Ratio (%R). It is a ratio of renewable emergy and input emergy of the system, which means the system's renewable property.

2.3.10. Emergy Sustainability Ratio (ESR). This evaluation index shows the situation of the system sustainability, which is equal to EYR divided by ELR.

3. Results and Discussion

3.1. Emergy Accounting. As shown in Table 1, the total emergy input of the BPC system amounts to 2.07×10^{17} sej/yr, of which the RR, RP and NP contribute 62%, 3%, and 35%, respectively (shown in Figure 3). The emergy yield of the biogas system consists of biogas and biogas residues, and the biogas residues consist of three parts: nitrogen fertilizer, phosphate fertilizer, and potash fertilizer. As shown in Figure 4, biogas is the most important production, the yield of which amounts to 6.92×10^{17} sej/yr and makes up 99% of the total emergy yield. The part of the biogas residues is only 1%.

Based on the data in Table 2, the total emergy input of BPF system amounts to 6.42×10^{17} sej/yr, of which RR, RP, and NP contribute 88%, 1%, and 11%, respectively (Figure 5). As to RR, solid manure and urine (including flushing sewage) account for 46.50% and 53.50%, respectively. The same as BPC system, the emergy yield of the biogas system also consists of biogas and biogas residues. As shown in Figure 6, biogas is the most important production, the yield of which

TABLE 1: Energy accounting for BPC system.

No.	Item[a]	Units	Raw date	Transformity (sej/unit)	References	Solar emergy (sej/yr)	Em ($)
				Local renewable resources			
1	Underground water	J	3.65*E* + 09	4.10*E* + 04	Odum, 1996 [16]	1.50*E* + 14	0
2	Corn straw	J	4.72*E* + 12	2.70*E* + 04	Odum, 1996 [16]	1.27*E* + 17	3180
	Total (RR)					1.28*E* + 17	3180
				Renewable purchased inputs			
3	Human labor	J	1.84*E* + 10	3.80*E* + 05	Xi and Qin, 2006 [17]	6.99*E* + 15	5804
	Total (RP)					6.99*E* + 15	5804
				Nonrenewable purchased inputs			
4	Steel plate[b]	US $	1.75*E* + 06	1.40*E* + 09	Brown and Ulgiati, 2004 [18]	2.45*E* + 15	846
5	Equipments, PE pipes[b]	US $	5.44*E* + 03	4.94*E* + 12	Brown and Ulgiati, 2004 [18]	2.69*E* + 16	5441
6	Civil work cost[b]	US $	6.05*E* + 02	4.94*E* + 12	Xi and Qin, 2006 [17]	2.99*E* + 15	605
7	Maintenance cost	US $	6.05*E* + 02	4.94*E* + 12	Xi and Qin, 2006 [17]	2.99*E* + 15	605
8	Coal	J	4.82*E* + 11	4.00*E* + 04	Odum, 1996 [16]	1.93*E* + 16	2177
9	Electricity	J	1.08*E* + 11	1.59*E* + 05	Xi and Qin, 2006 [17]	1.72*E* + 16	1886
10	Protein powder	US $	9.27*E* + 01	4.94*E* + 12	Xi and Qin, 2006 [17]	4.58*E* + 14	931
	Total (NP)					7.22*E* + 16	12491
	Total input					2.07*E* + 17	21475
				System feedback			
11	Biogas (heated)	J	8.28*E* + 11	2.48*E* + 05	Bastianoni and Marchettini, 2000 [15]	2.05*E* + 17	7666
	Total (F)					2.05*E* + 17	7666
				Yield			
12	Biogas (output)	J	2.79*E* + 12	2.48*E* + 05	Bastianoni and Marchettini, 2000 [15]	6.92*E* + 17	28996
13	Nitrogen fertilizer	g	6.58*E* + 05	4.62*E* + 09	Zhang et al., 2005 [19]	3.04*E* + 15	145
14	Phosphate fertilizer	g	3.29*E* + 05	1.78*E* + 10	Zhang et al., 2005 [19]	5.86*E* + 15	218
15	Potash fertilizer	g	4.94*E* + 05	2.96*E* + 09	Zhang et al., 2005 [19]	1.46*E* + 15	193
	Total yield (Y)					7.02*E* + 17	29553

[a] Local nonrenewable resources can be neglected.
[b] Items have been divided by a lifetime of 20 years.

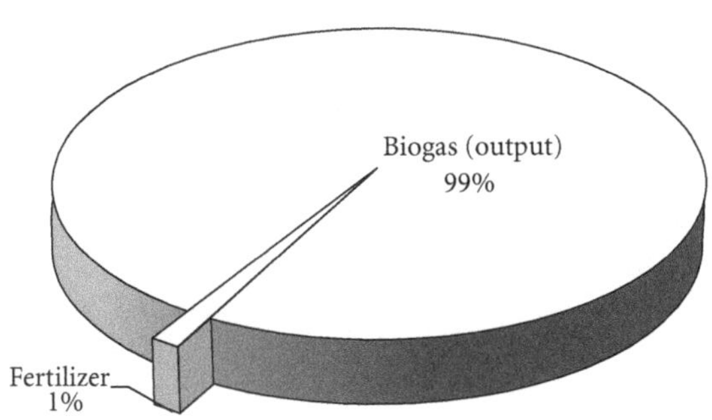

FIGURE 4: Emergy output ratio of BPC.

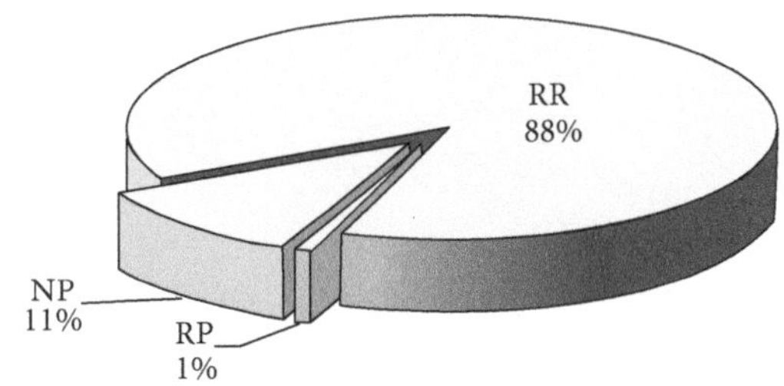

FIGURE 5: Emergy input ratio of BPF.

amounts to 8.16×10^{17} sej/yr and makes up 75% of the total emergy yield. The remaining part of the biogas residues is divided into three parts: nitrogen fertilizer (3%), phosphate fertilizer (17%), and potash fertilizer (5%).

In comparison to BPC system, the input emergy of BPF system is 4.38×10^{17} sej/yr more than BPC, because RR is a large proportion of BPF system. Otherwise, as to output emergy of BPC system, biogas output is 99%, and biogas manure is 1%; concerning output emergy of BPF system, biogas output is 75%, and biogas manure is 25%. The reason is the difference of the nature of digestion materials, construction scale, and digestion technology. For the BPC system, the greatest goal is biogas production; so, we call this kind of biogas project energy-ecology type. On the other hand, BPF system realizes the target of dealing with manure

TABLE 2: Energy accounting for BPF system.

No.	Item[a]	Units	Raw date	Transformity (sej/unit)	References	Solar emergy (sej/yr)	Em ($)
				Local renewable resources			
1	Sunlight	J	1.26*E* + 11	1.00*E* + 00	Odum, 1996 [16]	1.26*E* + 11	0
2	Solid manure	J	9.76*E* + 12	2.70*E* + 04	Odum, 1996 [16]	2.64*E* + 17	0
3	Feces	J	7.98*E* + 10	3.80*E* + 06	Odum, 1996 [16]	3.03*E* + 17	0
	Total (RR)					5.67*E* + 17	0
				Renewable purchased inputs			
4	Human labor	J	1.38*E* + 10	3.80*E* + 05	Xi and Qin, 2006 [17]	5.24*E* + 15	4353
	Total (RP)					5.24*E* + 15	4353
				Nonrenewable purchased inputs			
5	Civil work cost[b]	US $	2.96*E* + 03	4.94*E* + 12	Xi and Qin, 2006 [17]	1.46*E* + 16	2963
6	Equipments, PE pipes[b]	US $	5.20*E* + 03	4.94*E* + 12	Xi and Qin, 2006 [17]	2.57*E* + 16	5200
7	Appurtenant work[b]	US $	7.10*E* + 02	4.94*E* + 12	Xi and Qin, 2006 [17]	3.51*E* + 15	713
8	Maintenance cost	US $	6.05*E* + 02	4.94*E* + 12	Xi and Qin, 2006 [17]	2.99*E* + 15	605
9	Coal	J	3.64*E* + 11	4.00*E* + 04	Odum, 1996 [16]	1.46*E* + 16	738
10	Electricity	J	7.28*E* + 10	1.59*E* + 05	Xi and Qin, 2006 [17]	1.16*E* + 16	1173
	Total (NP)					7.29*E* + 16	11391
	Total input					6.45*E* + 17	15744
				System feedback			
11	Biogas (heated)	J	1.60*E* + 12	2.48*E* + 05	Bastianoni and Marchettini, 2000 [15]	3.97*E* + 17	13881
	Total (F)					3.97*E* + 17	13881
				Yield			
12	Biogas (output)	J	3.29*E* + 12	2.48*E* + 05	Bastianoni and Marchettini, 2000 [15]	8.16*E* + 17	28561
13	Nitrogen fertilizer	g	8.63*E* + 06	4.62*E* + 09	Zhang et al., 2005 [19]	3.99*E* + 16	1838
14	Phosphate fertilizer	g	1.02*E* + 07	1.78*E* + 10	Zhang et al., 2005 [19]	1.82*E* + 17	6844
15	Potash fertilizer	g	1.73*E* + 07	2.96*E* + 09	Zhang et al., 2005 [19]	5.12*E* + 16	6796
	Total yield (Y)					1.09*E* + 18	44039

[a]Local nonrenewable resources can be neglected.
[b]Items have been divided by a lifetime of 20 years.

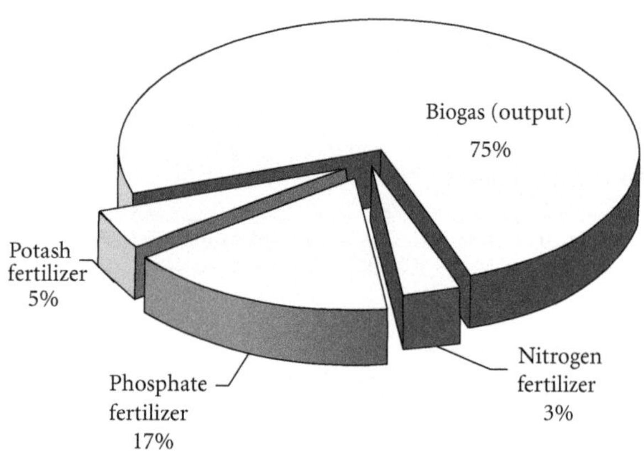

FIGURE 6: Emergy output ratio of BPF.

and wastewater efficiently, which is inclined to the energy-environmental type.

3.2. Ecological-Economic Analysis. A biogas project is one of the circular agriculture ecology patterns dealing with agricultural wastes, and the ecological-economic benefits could be measured by emergy-monetary value, which is a specific form of emergy-monetary value reflected in the economic market.

In Table 3, the total investment funds of BPF and BPC are $15,700/yr and $21,500/yr, respectively. The BPF resources (feces) do not need to be bought; so, investment funds of the BPF system include RP and NP, but those of BPC system include RR, RP, and NP. Thus, the BPF system can alleviate the government pressure and be worthy of widely promotion. In contrast, the BPC resources (corn straw) need to be bought, because of its extensive application, such as fuel to heat water and briquetting forming fuel. In addition, straw must be crushed before anaerobic digestion. And supplementary additives, such as protein powder, are required in the anaerobic digestion process. So, the RP and NP of the BPC system are both higher than those of the BPF. Moreover, the BPF system has solar heating system which decreases the quantity of coal and electricity; so, the whole input energy is less than BPC.

The outputs of BPF and BPC system are $44,000/yr and $29,600/yr, respectively. The main reason is the large gap of output economic value of biogas residues, which causes

TABLE 3: Table of emergy monetary value \$/yr.

Item	BPC	BPF
RR	3.18×10^3	/
RP	5.80×10^3	4.35×10^3
NP	1.25×10^4	1.14×10^4
Total input	2.15×10^4	1.57×10^4
Biogas (output)	2.90×10^4	2.86×10^4
Nitrogen fertilizer	1.45×10^2	1.84×10^3
Phosphate fertilizer	2.18×10^2	6.84×10^3
Potash fertilizer	1.93×10^2	6.80×10^3
Total yield	2.96×10^4	4.40×10^4
Ecological-economic value	0.81×10^4	2.83×10^4

differences between the two biogas projects technologies. The ecological-economic values of BPF and BPC system are \$28,300/yr and \$8,100/yr, respectively. Because both of the materials are renewable natural resources, the two biogas projects have high ecological-economic values and significant ecological environmental benefits. Most biogas from the project can meet the energy needs; biogas manure can be used as organic manure for crops, vegetables, and fruits. So they all improve the ecological-economy value.

3.3. Emergy Evaluation Index Analysis. All the emergy evaluation indices are classified by ecological-economic benefit, environmental benefit, and sustainable development ability (shown in Table 4).

3.3.1. Ecological-Economic Benefit. PER, natural emergy/purchase emergy, EIR, and ESR of BPF system are 0.32, 4.49, 0.22, and 1.42 times more than BPC system, respectively. This indicates that the BPF system depends less on renewable resources and has greater competitiveness and higher natural resources utilization efficiency and self-sufficiency ability than BPC systems. This is because BPC's material, corn straw, needs to be purchased, but BPF's material, feces, does not need fund support.

As to the net contribution of economic, although the feedback emergy of the two systems appears to be the same, the EYR of BPF system is about 1.57 times more than that of BPC systems, because the BPF system has much more yield emergy, which means that the BPF system has low cost, good production efficiency, high emergy utilization efficiency, and competitiveness.

Considering all the previous indices, BPF system has higher ecological-economic benefit than BPC system.

3.3.2. Environmental Benefit. The %W of BPF system is about 1.43 times more than that of BPC systems. Raw materials of BPF system are livestock manure, poultry and urine. And the total fermentation volume is 624 m^3, which consists of a primary fermentation reactor (400 m^3) and a secondary fermentation reactor (224 m^3). On the other hand, the raw material of BPC system is only straw, with fermentation volume of 480 m^3. The wastes treatment quantities of the BPF system and BPC system converted into solar energy value are 5.67×10^{17} sej/yr and 1.27×10^{17} sej/yr, which means that BPF system, has higher degradation efficiency.

ELR is an important index which reflects the degree of influence of the system on the environment. The ELRs of BPF and BPC system are 0.13 and 0.53, which are both less than 1. The results show that the two systems have a small impact on environment. However, owing to the higher renewable emergy ratio and more wastes treatment quantity, BPF system has much higher environmental benefit.

3.3.3. The Analysis of Sustainable Development Ability. To analyze sustainable development ability, there are three parts to be considered: FYR, %R, and ESI.

The first index is FYR. FYR is the ratio of the emergi of system self-feedback (for a biogas system, it is the part of biogas heated) divided by economy feedback emergy. The economic feedback emergies of BPF and BPC are 7.81×10^{16} sej/yr and 7.92×10^{16} sej/yr, while the self-feedback emergies are 3.97×10^{17} sej/yr and 2.05×10^{17} sej/yr. The previous results are the result of the quantity of energy used to heat 400 m^3 primary fermentation reactor and 224 m^3 secondary fermentation reactors. The FYR of BPF is 1.96 times more than BPC, which means that the ability of BPF's self-organizing is much better than BPC's.

Another index is %R, which indicates the renewable character of the biogas system. The renewable energy and the emergy devoted to BPF are both more than the BPC system. At the same time, %R of BPF is 1.36 times more than the BPC system. The results show that the ability of the system renewable character of BPF is better.

The last index, ESI, states the relation between the environment and the emergy produced. A perfected biogas system has not only high output, but also a far-reaching influence. The ESI of BPF is 6.61 times more than the BPC system, which shows that the BPF system has bright sustainable development possibilities.

4. Conclusion

The large- and medium-sized biogas projects where materials are corn straw and feces both have high ecological-economic benefit. In this research, the ecological-economic values of BPF system and BPC system are \$28,300/yr and \$8,100/yr, respectively. According to the monetary, environment, and human production factors, both of the two types of biogas projects own great development potential.

As to the whole input emergy, renewable natural resources take the largest percentage, which means that both of the two systems have good treatment ability of agricultural waste (feces and crop straw). In the BPC system, the amount of biogas output is much more than biogas residues. With the rapid development of biogas projects, recycling treatment of waste and sewage has become a hot topic, and the comprehensive utilization of biogas residues has become a primary goal. Therefore, technology needs to be improved in order to utilize biogas residues efficiently.

TABLE 4: The comparative table of energy index.

Item	Evaluation index	BPC	BPF
Ecological-economy benefit	Purchased emergy ratio (PER)	0.38	0.12
	Natural emergy/purchased emergy	1.62	7.26
	Emergy investment rate (EIR)	0.62	0.14
	Emergy yield ratio (EYR)	8.86	13.95
	Emergy self-sufficiency ratio (ESR)	0.62	0.88
Environmental benefit	Waste processing ratio (%W)	61.30	87.89
	Environmental load ratio (ELR)	0.53	0.13
Sustainable development	Feedback yield ratio (FYR)	2.59	5.08
	Renewable ratio (%R)	65.15	88.70
	Sustainability index (ESI)	16.57	109.50

From the analysis of the point of emergy, BPF system is superior to BPC system in the aspects of ecological economic benefit, environmental benefit and the sustainable development ability. This result is due to problems of the difficulty of corn straw seasonal collection and degradation. As to this situation, BPC system can be considered if anaerobic digestion is combined with other raw materials, such as poultry and livestock manure, urine and household garbage.

Appendices

A.

Notes to Table 1:

(1) Underground water

Number: 2500 kg/d
Standard energy value: 4 kJ/ kg
Energy: 2500 kg/d $\times$ 4 $\times$ 10^3 J/kg $\times$ 365 d/yr = 3.65 $\times$ 10^9 J/yr

(2) Corn straw

Number: 900 kg/d
Standard energy value: 14355.72 kJ/kg
Energy: 900 kg/d $\times$ 14355.72 $\times$ 10^3 J/kg $\times$ 365 d/yr = 4.72 $\times$ 10^{12} J/yr

(3) Human labor

Days: 4 $\times$ 365 d/yr = 1460 d/yr
Standard energy value: 12600 kJ/d
Energy: 1460 d/yr $\times$ 12600 kJ/d = 1.84 $\times$ 10^{10} J/yr

(4) Steel plate

Number: 3.5 $\times$ 10^7 g/20 yr = 1.75 $\times$ 10^6 g/yr

(5) Equipments, PE pipes

Number: (900000/20 yr) RMB/8.27 = 5.44 $\times$ 10^3 (2000 US$)

(6) Civil work cost

Number: (100000/20 yr) RMB/8.27 = 6.05 $\times$ 10^2 (2000 US$)

(7) Maintenance cost

Number: 5000 RMB/8.27 = 6.05 $\times$ 10^2 (2000 US$)

(8) Coal

Number: 18000 kg/yr
Standard energy value: 26777.6 kJ/kg
Energy: 18000 kg/yr $\times$ 26777.6 kJ/kg = 4.82 $\times$ 10^{11} J/yr

(9) Electricity

Number: 30000 kw·h/yr
Standard energy value: 3598.24 kJ/kw·h
Energy: 30000 kw·h/yr $\times$ 3598.24 kJ/kw·h = 1.08 $\times$ 10^{11} J/yr

(10) Protein powder

Number: (182.5 kg/yr $\times$ 4.2 RMB/kg)/8.27= 9.27 $\times$ 10^1 (2000 US$)

(11) Biogas (heated)

Number: 39600 m^3/yr
Standard energy value: 20920 KJ/m^3
Energy: 39600 m^3/yr $\times$ 20920 KJ/m^3 = 8.28 $\times$ 10^{11} J/yr
Number: 76500 m^3/yr
Standard energy value: 20920 kJ/m^3
Energy: 76500 m^3/yr $\times$ 20920 kJ/m^3 = 1.60 $\times$ 10^{12} J/yr

(12) Biogas (output)

Number: 133200 m^3/yr
Standard energy value: 20920 kJ/m^3

Energy: $133200\,m^3/yr \times 20920\,kJ/m^3 = 2.79 \times 10^{12}$ J/yr
Number: $157500\,m^3/yr$
Standard energy value: $20920\,kJ/m^3$
Energy: $157500\,m^3/yr \times 20920\,kJ/m^3 = 3.29 \times 10^{12}$ J/yr

(13) Nitrogen fertilizer

Number: 6.58×10^5 g/yr

(14) Phosphate fertilizer

Number: 3.29×10^5 g/yr

(15) Potash fertilizer

Number: 4.94×10^5 g/yr

B.

Notes to Table 2:

(1) Solar energy

Standard energy value: 4.1868×10^3 J/kg·k
Energy: $50\,m^3/d \times 10^3\,Kg/m^3 \times 120\,d/yr \times 4.1868 \times 10^3\,J/(Kg{\cdot}k) \times 5\,k = 1.26 \times 10^{11}$ J/yr

(2) Solid manure

Number: 4.015×10^6 kg/yr
Total solid: 18%
Standard energy value: 13500 kJ/kg
Energy: $4.015 \times 10^6\,kg/yr \times 18\% \times 13500\,kJ/kg = 9.76 \times 10^{12}$ J/yr

(3) Urine (including flushing sewage)

Number: $12410\,m^3/yr$
Standard energy value: $6.43 \times 10^6\,J/m^3$
Energy: $12410\,m^3/yr \times 6.43 \times 10^6\,J/m^3 = 7.98 \times 10^{10}$ J/yr

(4) Human labor

Days: 3×365 d/yr = 1095 d/yr
Standard energy value: 12600 kJ/d
Energy: $1095\,d/yr \times 12600\,kJ/kg = 1.38 \times 10^{10}$ J/yr

(5) Civil work cost

Number: (734000/30 yr)RMB/8.27= 2.96×10^3 (2000 US$)

(6) Equipments, PE pipes

Number: (1289000/30 yr)RMB/8.27 = 5.20×10^3 (2000 US$)

(7) Appurtenant work

Number: (176000/30 yr)RMB/8.27 = 7.10×10^2 (2000 US$)

(8) Maintenance cost

Number: 5000 RMB/8.27 = 6.05×10^2 (2000 US$)

(9) Coal

Number: 13600 kg/yr
Standard energy value: 26777.6 k J/kg
Energy: $13600\,kg/yr \times 26777.6\,kJ/Kg = 3.64 \times 10^{11}$ J/yr

(10) Electricity

Number: 20221 kw·h
Standard energy value: 3598.24 kJ/kw·h/yr
Energy: $20221\,KW{\cdot}h/yr \times 3598.24\,kJ/kw{\cdot}h = 7.28 \times 10^{10}$ J/yr

(11) Biogas (heated)

Number: $76500\,m^3/yr$
Standard energy value: $20920\,kJ/m^3$
Energy: $76500\,m^3/yr \times 20920\,kJ/m^3 = 1.60 \times 10^{12}$ J/yr

(12) Biogas (output)

Number: $157500\,m^3/yr$
Standard energy value: $20920\,kJ/m^3$
Energy: $157500\,m^3/yr \times 20920\,kJ/m^3 = 3.29 \times 10^{12}$ J/yr

(13) Nitrogen fertilizer

Number: 8.63×10^3 kg/yr

(14) Phosphate fertilizer

Number: 1.02×10^4 kg/yr

(15) Potash fertilizer

Number: 1.73×10^4 kg/yr.

Acknowledgment

The research work is supported by the National Key Technology R&D Program of China (nos. 2008BADC4B03 and 2009BAC64B08).

References

[1] H. Ma, L. Oxley, and J. Gibson, "China's energy situation in the new millennium," *Renewable and Sustainable Energy Reviews*, vol. 13, no. 8, pp. 1781–1799, 2009.

[2] H. Ma, L. Oxley, J. Gibson, and W. Li, "A survey of China's renewable energy economy," *Renewable and Sustainable Energy Reviews*, vol. 14, no. 1, pp. 438–445, 2010.

[3] N. Zhang, N. Lior, and H. Jin, "The energy situation and its sustainable development strategy in China," *Energy*, vol. 36, no. 6, pp. 3639–3649, 2011.

[4] P. Crompton and Y. Wu, "Renewable energy policy and electricity market reforms in China," *Energy Economics*, vol. 27, no. 1, pp. 195–208, 2005.

[5] N. Zhou, M. D. Levine, and L. Price, "Overview of current energy-efficiency policies in China," *Energy Policy*, vol. 38, no. 11, pp. 6439–6452, 2010.

[6] P. Vassallo, S. Bastianoni, I. Beiso, R. Ridolfi, and M. Fabiano, "Emergy analysis for the environmental sustainability of an inshore fish farming system," *Ecological Indicators*, vol. 7, no. 2, pp. 290–298, 2007.

[7] S. Q. Chen, B. Chen, and D. Song, "Life-cycle energy production and emissions mitigation by comprehensive biogas-digestate utilization," *Bioresource Technology*, vol. 14, pp. 357–364, 2012.

[8] F. Wang and G. Wang, "Econometric analysis of the impacts of "four-in-one" household biogas construction on farmers' planting behaviors," *Transactions of the Chinese Society of Agricultural Engineering*, vol. 22, no. 3, pp. 116–120, 2006.

[9] B. L. Duan, F. Yang, Z. L. Li et al., "The application and benefit of rural energy four in one model," *Renewable Energy*, vol. 6, pp. 43–44, 2002.

[10] M. Pöschl, S. Ward, and P. Owende, "Evaluation of energy efficiency of various biogas production and utilization pathways," *Applied Energy*, vol. 87, no. 11, pp. 3305–3321, 2010.

[11] C. Coskun, E. Akyuz, Z. Oktay et al., "Energy analysis of hydrogen production using biogas-based electricity," *International Journal of Hydrogen Energy*, vol. 17, no. 36, pp. 11418–11424, 2011.

[12] J. Richard Ciotola a, L. Stephanie, and J. F. Martina, "Emergy analysis of biogas production and electricity generation from small-scale agricultural digesters," *Ecological Engineering*, vol. 37, pp. 1681–1691, 2011.

[13] D. F. Shen, P. D. Zhang, M. Zhong et al., "Study on the energy flowing of the four in one eco-agriculture model of rural courtyardin the middle region of Gansu," *China Biogas*, vol. 23, no. 1, pp. 43–46, 2005.

[14] S. Bastianoni, F. M. Pulselli, C. Castellini, C. Granai, A. Dal Bosco, and M. Brunetti, "Emergy evaluation and the management of systems towards sustainability: a response to Sholto Maud," *Agriculture, Ecosystems and Environment*, vol. 120, no. 2–4, pp. 472–474, 2007.

[15] S. Bastianoni and N. Marchettini, "The problem of co-production in environmental accounting by emergy analysis," *Ecological Modelling*, vol. 129, no. 2-3, pp. 187–193, 2000.

[16] H. T. Odum, *Environmental Accounting: Emergy and Environmental Decision Making*, John Wiley & Sons, New York, NY, USA, 1996.

[17] Y. Xi and P. Qin, "Emergy value evaluation on rice-duck organic farming mode," *Chinese Journal of Applied Ecology*, vol. 17, no. 2, pp. 237–242, 2006.

[18] M. T. Brown and S. Ulgiati, "Energy quality, emergy, and transformity: H.T. Odum's contributions to quantifying and understanding systems," *Ecological Modelling*, vol. 178, no. 1-2, pp. 201–213, 2004.

[19] D. Zhang, F. Ling, L. Zhang, S. Yang, X. Liu, and W. Gao, "Emergy analysis of planting system at Gongzhuling county in the main grain production region in Northeast China Plain," *Transactions of the Chinese Society of Agricultural Engineering*, vol. 21, no. 6, pp. 12–17, 2005.

8

An Adaptive Framework for Selecting Environmental Monitoring Protocols to Support Ocean Renewable Energy Development

Emily J. Shumchenia,[1] Sarah L. Smith,[2] Jennifer McCann,[2] Michelle Carnevale,[2] Grover Fugate,[3] Robert D. Kenney,[1] John W. King,[1] Peter Paton,[4] Malia Schwartz,[1] Malcolm Spaulding,[5] and Kristopher J. Winiarski[4]

[1] *Graduate School of Oceanography, University of Rhode Island, South Ferry Road, Narragansett, RI 02882, USA*
[2] *Coastal Resources Center and Rhode Island Sea Grant, University of Rhode Island, Narragansett, RI 02882, USA*
[3] *Rhode Island Coastal Resources Management Council, South Kingstown, RI 02879, USA*
[4] *College of the Environmental and Life Sciences, University of Rhode Island, Kingston, RI 02881, USA*
[5] *College of Engineering, University of Rhode Island, Narragansett, RI 02882, USA*

Correspondence should be addressed to Emily J. Shumchenia, emily@gso.uri.edu

Academic Editors: A. Azzellino and J. P. Kofoed

Offshore renewable energy developments (OREDs) are projected to become common in the United States over the next two decades. There are both a need and an opportunity to guide efforts to identify and track impacts to the marine ecosystem resulting from these installations. A monitoring framework and standardized protocols that can be applied to multiple types of ORED would streamline scientific study, management, and permitting at these sites. We propose an adaptive and reactive framework based on indicators of the likely changes to the marine ecosystem due to ORED. We developed decision trees to identify suites of impacts at two scales (demonstration and commercial) depending on energy (wind, tidal, and wave), structure (e.g., turbine), and foundation type (e.g., monopile). Impacts were categorized by ecosystem component (benthic habitat and resources, fish and fisheries, avian species, marine mammals, and sea turtles) and monitoring objectives were developed for each. We present a case study at a commercial-scale wind farm and develop a monitoring plan for this development that addresses both local and national environmental concerns. In addition, framework has provided a starting point for identifying global research needs and objectives for understanding of the potential effects of ORED on the marine environment.

1. Introduction

At present, there is a great need to better understand the potential effects of offshore renewable energy developments (ORED) on the marine environment [1–3]. While the development of commercial-scale ORED in the United States has lagged well behind development in Europe [4], construction on multiple projects is likely to begin within the next few years in U.S. waters [5]. Because of differences between Europe and the U.S. in terms of regulatory requirements, environmental settings, and species present at development sites [6], there is a considerable need for U.S.-specific guidance to ensure thorough data collection as ORED projects develop to evaluate effects and assess potential impacts. In this paper, we use the word "effect" to refer to a change in the environment without respect to magnitude or direction (i.e., moderate or severe, positive or negative [1]). When the effect is better characterized and a magnitude and direction can be assigned, we refer to it as an "impact." A scientific framework exists for detecting and characterizing effects, but more work is needed in order to describe and assess impacts [1].

Data collection at ORED sites should be guided by a framework that identifies the unique impacts associated with ORED [7], then selects and helps implement a suite of standardized environmental monitoring protocols relevant to each development type. Standardized protocols improve impact assessment by following a single methodology at multiple sites, permitting comparison and aggregation of data [8]. Building a uniform database of environmental impacts will allow us to better refine our understanding of drivers and stressors acting at ORED sites, improve the Environmental Impact Statement (EIS) process, and refine monitoring requirements in the future as certain impacts are suggested to be either negligible or worthy of concern. This knowledge can be used to encourage development in areas where known impacts are expected to be minimal. In addition, reducing the existing uncertainty about environmental impacts related to ORED will likely ease public concern about development and therefore improve the siting process. Such a framework should also assist in answering regulatory questions about siting and scale by ensuring that relevant data are collected, therefore reducing uncertainty in decision making.

Monitoring an effect or an impact means that the monitoring protocol must be designed to measure change against some baseline condition or management objective [9, 10]. This change may be measured temporally, as in between seasons or years, before and after construction, or spatially, including measuring differences between an affected and a control area. Designing a study that can both provide conclusive evidence of an impact (or lack thereof) and separate this impact from the noise of seasonal or interannual environmental variability can be problematic [11]. Furthermore, not all of the potential impacts will be directly observable. For example, observations of underwater ORED structures will not likely be continuous. Therefore, in order to determine if a foundation has an impact on seabed structure, for example, discrete measurements of seabed volume will be made and compared through time. In this example, measured changes in seabed volume serve as a representative of the impact of the foundation on the seabed and in this way are considered to be an indicator of that impact. In general, indicators can be used as a means to quantitatively track change in the context of ecosystem-based management goals [12–14]. Developing indicators of change at ORED sites would be immensely helpful to impact assessments and, if developed across disciplines (e.g., biology, geology, physical oceanography), enable an overall assessment of the condition of the ecosystem. Building support for an indicator as a representative of an ecosystem attribute or function is an iterative process that can be conducted as monitoring data are collected at ORED sites. At first, where few data exist, qualitative "reference directions" can be used to track change [10]. As a database is built, changes can be quantified and thresholds can be identified relative to impacts at particular developments. In rare cases where the natural variability of a parameter has already been characterized, statistical tools may be used to determine appropriate thresholds or even the sampling protocols themselves (e.g., power analyses [15–17]). In most cases, however, very little is known about natural variability and environmental monitoring efforts will be measuring natural change commingled with the effects of ORED. A monitoring framework that considers all of these concerns is essential.

To address these concerns, an adaptive, rather than a static, monitoring framework for ORED is most appropriate. Firstly, there are many points of weakness in the general understanding of the impacts of ORED on marine resources that could greatly change monitoring needs and/or requirements. For example, the impacts of indirect effects (e.g., alteration to food webs) and wholly unanticipated effects are unknown [1]. Data regarding these points may only become available at a later stage of ORED maturity, but current monitoring protocols and regulations should be prepared in anticipation of these types of effects. Next is the current understanding of linkages between effects and indicators. We can agree conceptually that certain environmental/biological parameters are indicative of an ecosystem change, but in many cases we have no estimate of thresholds of concern for these parameters (e.g., how much of a reduction can occur in a bird population before mitigation needs to take place?). Just as experience in ecosystem-based fisheries management has helped propose appropriate thresholds for indicators of fisheries status [9], experience in managing OREDs will help clarify the assumptions made between effects and indicators. An adaptive framework is also essential in a field where new technologies are developing and emerging at a rapid pace.

In this paper we propose an adaptive monitoring framework based on indicators of the likely changes to the ecosystem due to ORED. We developed the framework to be used by offshore renewable energy developers and U.S. management and regulatory agencies in order to standardize the design and methodologies used to collect data at ORED sites. The framework and protocols were developed to be scientifically valid and easy to understand and follow by nonscientists. In order to do so, we reduced the complexity of impact assessment by developing decision-support tools that guide users towards monitoring objectives. As scientists, we were challenged with the task of maintaining scientific rigor in the framework while simultaneously acknowledging the pressures on management and regulatory agencies to encourage developments and keep costs low. The resulting framework is adaptive, flexible and can be implemented at any ORED site in U.S. waters.

2. Methods

Offshore renewable energy development is here defined as the construction and operation of one or more devices designed to harness power from the marine environment (wind, tidal, and wave power considered here) and includes any necessary infrastructure, including subsea cables, the vessels necessary to construct or install an ORED, and the footprint of a project. This paper considers the effects of ORED on the benthic habitat and resources, marine mammals, sea turtles, fish, and avian species. We also considered the effect of ORED on one human use—fishing activity—because of the inextricability of the effects on fishing activity from effects on fish themselves, and the resulting concerns of fishermen about potential effects on their livelihood. We examined renewable energy developments at two scales, 1 = "demonstration" and 2 = "commercial/multiple commercial." At Scale 1, three or fewer devices are part of a "farm"; Scale 2 constitutes a farm or farms of around 100 devices and greater.

A literature review was conducted of potential positive and negative environmental effects (see Appendix A, supplementary material available online at doi:10.1100/2012/450685 for works cited) using the Programmatic Environmental Impact Statement (PEIS) for Alternative Energy Development developed by the Minerals Management Service in 2007 as a reference point [18]. Potential impacts were categorized by the five affected ecosystem components, the anticipated level of effect (minor, moderate, major), and the level of certainty (high, medium, low) at each scale of development and for each technology type within an ORED "impact matrix" (Supplementary Appendix B). The descriptions and thresholds for impact levels were derived from the definitions used in the PEIS [18]: minor—should not influence or have only small impacts on the affected resource, activity, or community; moderate—impacts could moderately influence the resource, activity, or community, generally or for particular species; major—impacts could significantly influence the resource, activity, or community, generally or for particular species. Here, we used the word "certainty" to refer to the amount of evidence available from studies conducted on a particular effect. High certainty indicates that there was a large body of literature documenting or studying an impact. It is important to note that "certainty" does not refer to the chance that an impact will occur. The chance of an impact occurring is more appropriately described as likelihood, a concept that was not addressed in this study. Therefore, where we describe an effect with a high certainty of major impact, this can be interpreted as "*if* the named effect occurs, *then* the magnitude of the impact on environment will be major."

Suites of similar potential impacts were aggregated by energy resource, foundation type, and scale of development in order to define "Impact scenarios." An Impact scenario is applicable to multiple development situations in order to distill and focus the monitoring and management actions required for ORED. Impact scenarios describe the major and moderate negative impacts of ORED on five ecosystem components—benthic habitat and resources, marine mammals, sea turtles, fish, and avian species.

We developed two types of decision trees to serve as decision-support tools, to help users determine which impacts are relevant when their development criteria are implemented, and to evaluate alternatives. The first decision tree, the "Impact decision tree," determines the approximate magnitude of impacts from ORED on each ecosystem component considering three factors—energy type, foundation type, and development scale. The second type, "Component decision trees," is a suite of finer-scale decision trees for each of the ecosystem components that determine which monitoring protocols are recommended given a more specific suite of characteristics related to the development type (e.g., stage of development). We took this approach because each ecosystem component experiences different levels of impact due to different drivers. For example, different foundation types would differentiate several types of effect for benthic habitat and resources but are not likely to do the same for avian species.

A chart-based/graphic format was rejected in favor of a text-based/key format. In the adopted format, the user answers a series of questions about the development project and is guided through the decision tree and toward an eventual "answer" based on the responses to the questions. For the Impact decision tree, the "answer" is an Impact scenario, an associated list of the ecosystem components that may experience major and moderate negative (i.e., adverse) impacts from ORED, a short description of the type of impacts, and an estimate of the certainty regarding these impacts. For each scenario, the lists of major negative potential impacts were ranked by proportion and magnitude of total impacts so that #1 reflects the component with the most negative impacts. To provide more detail on potential adverse impacts, all moderate impacts and levels of certainty are also provided for each scenario. The lists of moderate impacts are not prioritized or ranked and are listed as they appeared in the impact matrix (Supplementary Appendix B).

For the Component decision trees, the "answer" is a list of monitoring objectives. At the University of Rhode Island we have developed a series of monitoring protocols to address each of these objectives. We also developed a case study in order to demonstrate how a manager or regulator may use these tools to develop an ORED monitoring plan. The case study consists of a Scale 2 wind-turbine farm composed of around 200 jacketed structures. We describe the resulting Impact scenario and list the major and moderate impacts that monitoring should address. To demonstrate the next step, we present an example monitoring protocol that addresses monitoring of benthic habitat and resources.

3. Results

From the renewable energy impact matrix, literature review, and expert judgment, we assembled a short list of the potential effects for each ecosystem component considered to be of greatest importance. For each effect, we propose an indicator that is recommended for use as a monitoring target (Table 1).

3.1. The Monitoring Framework. We developed an adaptive and reactive monitoring framework that incorporates the use of environmental indicators to track change (Figure 1). Currently, we have the ability to characterize a baseline condition and assign reference directions to indicators; for example, increases in sediment grain size at every turbine should accelerate monitoring for scour. Reference directions are useful when data are insufficient to establish more quantitative reference levels, but they only provide an indication of a trend and do not specify when a threshold of irreversible harm has been reached [10]. In an adaptive monitoring framework, data are synthesized to produce more quantitative metrics and thresholds for environmental indicators of ORED effects. In a reactive monitoring framework, evidence of an effect should be used to accelerate study of that effect, perhaps by multiple methodologies (refer to Figure 1). Suites of ORED effect indicators would not only provide a clearer

Table 1: Potential effects of offshore renewable energy developments in the United States considered to be of greatest importance based on results of a literature review and expert judgment (see Supplementary Appendix A). Effects are organized by ecosystem component and are paired with a proposed indicator of that effect.

	Impact/monitoring objective	Indicator
Benthic habitat and resources	Changes to seafloor morphology and structure (compared to preconstruction)	Increase or decrease in seabed volume
	Changes in median grain size, or organic content	(i) Deposition: decrease in median grain size, increase in organic content, increase in seabed volume (ii) Scour: increase in median grain size, decrease in organic content, decrease in seabed volume
	Turbidity during construction/decommissioning	Change in water column turbidity
	Change in target species abundance and distribution (e.g., species of importance)	Change in abundance, diversity, % cover, multivariate community composition
	Current speed/direction inside and outside farm	Change in residual flow rates
	Reef effects, colonization on foundations	Increase in % cover, biomass of epifaunal organisms; increase in presence of nonnative species
	Change in density, diversity, dominance structure of infauna	Change in abundance, diversity, % cover, multivariate community composition
Fish	Reef or aggregation effects	Increase in fish abundance around devices, shift in species composition, increase in presence of nonnative species
	Changes to abundance/distribution caused by disturbance or habitat alteration	Increase or decrease in fish abundance; increase or decrease in target species; shift in species composition; change in density, diversity, and dominance structure of fish species; increase in presence of nonnative species
	Blade strikes/pressure gradients (tidal power)	Observation of blade strike incidents
	EMF effects	Not feasible to monitor directly—changes in fish abundance, behavior, or species composition are indicators
	Installation or operational noise effects	Not feasible to monitor directly—changes in fish abundance, behavior, or species composition are indicators
Fisheries	Catchability (catch per unit effort) during construction	Catch per unit effort increases or decreases for target species
	Catchability (catch per unit effort) during operation	Catch per unit effort increases or decreases for target species
	Loss of access to grounds	Changes in numbers of vessels fishing near or inside of the renewable energy area, change in the presence of fixed fishing gear inside of or around a renewable energy installation
	Changes in species distribution	Shift in species composition, increase in presence of nonnative species
	Reef effects (aggregation)	Increase in fish abundance around devices; shift in species composition; increase in presence of nonnative species
Avian	Displacement/attraction	Increase or decrease in avian species-specific densities postconstruction in development area
	Barrier effects—effects on foraging, roosting, migratory movements	Migrating or commuting birds avoiding developed areas
	Collision mortality	Birds found dead or injured due to direct collision with infrastructure above the water
Marine mammals and sea turtles	Vessel strikes	Detection of dead or injured animals
	Noise generated during construction	Detection of dead or injured animals; changes in distribution, abundance, or behavior of populations
	Disturbance or injury during all stages of development, including from vessels	Detection of dead or injured animals; changes in distribution, abundance, or behavior of populations
	Noise generated during operation	Changes in distribution, abundance, or behavior of populations

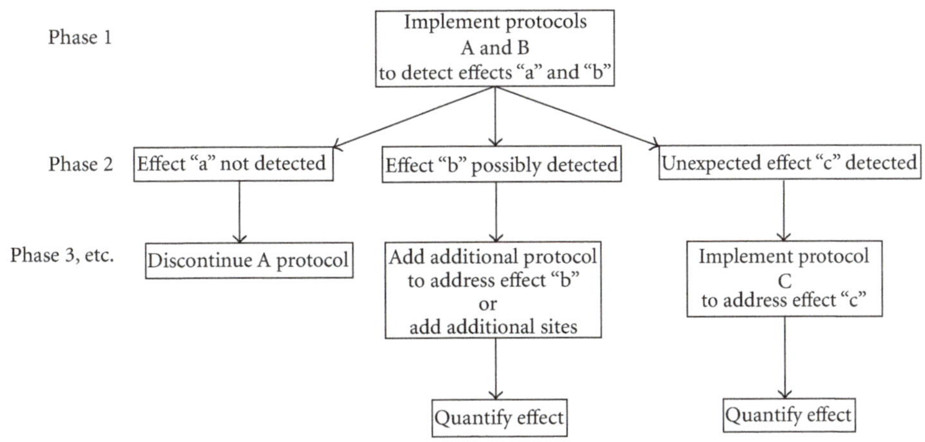

FIGURE 1: Proposed adaptive and reactive monitoring framework.

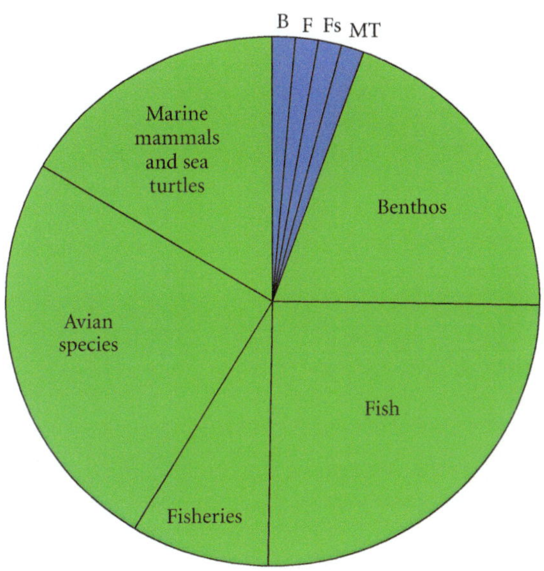

FIGURE 2: Pie chart representing the proportion of impacts for each ecosystem component for Impact scenario i1, demonstration-scale projects, categorized by whether the effect is positive (blue), minor negative (green), moderate negative (yellow), or major negative (red). B = Benthic habitat and resources; F = Fish; Fs = Fisheries; A = Avian species; MT = Marine mammals and sea turtles.

path for goal setting for developers but would encourage regulatory monitoring protocols to contribute to our general understanding of the natural variation of marine ecosystems and how human activities can be integrated and harmonized.

3.2. Impact Decision Tree. The Impact decision tree determines the approximate magnitude of impacts from ORED on each ecosystem component for a broad range of development types and scenarios. For each combination, a bar graph shows the relative number of potential impacts and their magnitudes (Table 2). Even though there are 39 possible scenarios that result when combining the three factors, our decision tree reduces these to six main Impact scenarios.

3.3. Impact Scenarios. Impact scenarios are very brief descriptions of the major environmental concerns regarding categories of ORED with similar environmental impacts. Below, the Impact scenarios are accompanied by pie charts representing the total number of impacts for each scenario, categorized by whether the impact is positive (blue), minor negative (green), moderate negative (yellow), or major negative (red). Each of these sections of the pie is further broken down by ecosystem component.

(i1) All Demonstration Scale Projects (Figure 2, Table 3). These projects are described as "Scale 1". The current literature suggests that any renewable energy development, if completed at the demonstration scale, will not have moderate or major impacts on the ecosystem components examined here. Therefore, we list the potential minor impacts and their certainty in the Impact decision tree. Of the suite of minor impacts, benthic habitat and resources, avian species, and fish species share an equally high proportion. Across ecosystem components, impacts with the highest certainty tend to be physical and chemical disturbances, such as disturbance from device installation, attraction to devices, or chemical spills. Impacts with low certainty include noise (except for marine mammals and sea turtles where the certainty for this impact is high), changes to energy regimes, and changes in organism energetic expense. Electromagnetic field (EMF) impact is the only impact that has low certainty consistently across all ecosystem components. Only those potential impacts with high certainty are listed in the decision tree; where certainty is low, it may be impossible to detect any impact at this magnitude.

(i2) Wind Turbine Developments Involving Pile Driving (Figure 3, Table 4). This scenario includes monopile wind turbine developments and jacketed- or tripod-mounted turbines at development Scale 2. If the proposed development will not utilize pile driving to install the jacketed or tripod structures, then Impact scenario i3 is more appropriate. The impacts that make this scenario unique are the presence of turbines above the water surface, the piles drilled into the seabed, and the noise associated with this activity. Therefore,

Table 2: The impact decision tree. Each step of the key is followed until an Impact scenario is reached, denoted by "i#". A bar graph is shown with each Impact scenario displaying the proportion of impacts, color-coded by direction (positive or negative) and magnitude (none, minor, moderate major). Data is derived from the literature review (Supplementary Appendix A) and renewable energy impact matrix (Supplementary Appendix B).

Determine which ecosystem components may experience impacts from renewable energy development:		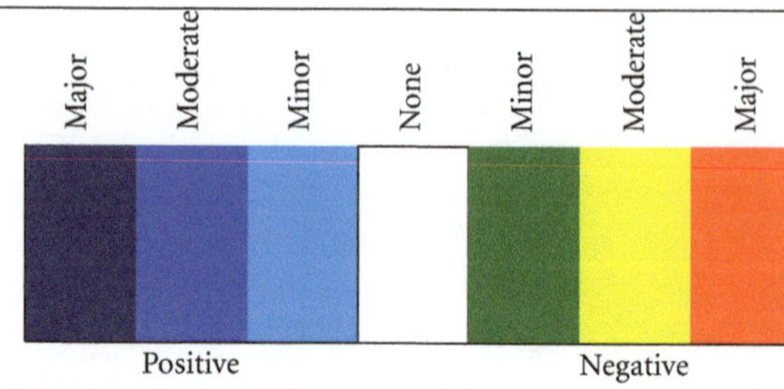
The energy resource is		
Wind		Go to A
Tidal		Go to B
Waves		Go to C
(A) The wind turbine foundation is		
Monopile		Go to A1
Gravity		Go to A2
Tripod/lattice		Go to A3
Floating mooring		Go to A4
(A1) The monopile wind turbine project scale is		
Demonstration	Go to i1	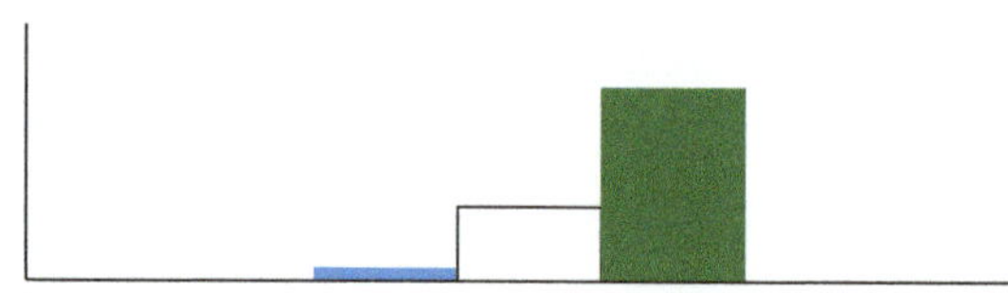
Commercial/multiple commercial	Go to i2	
(A2) The gravity wind turbine project scale is		
Demonstration	Go to i1	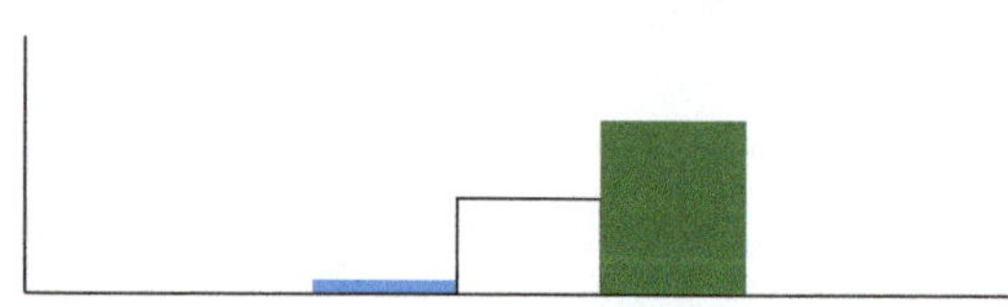
Commercial/Multiple commercial	Go to i3	
(A3) The tripod/lattice wind turbine project scale is		
Demonstration	Go to i1	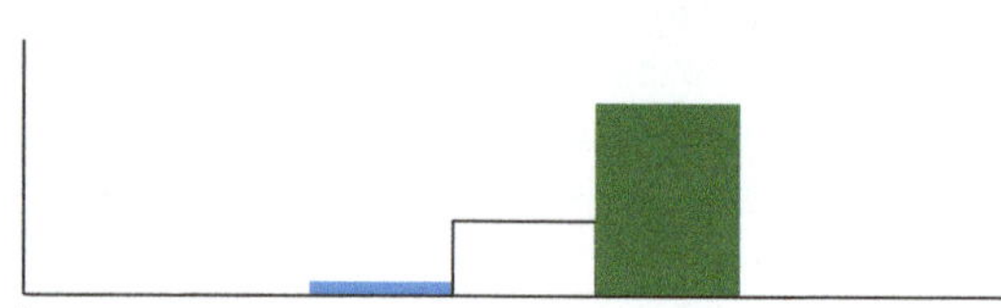
Commercial/Multiple commercial	Go to i2	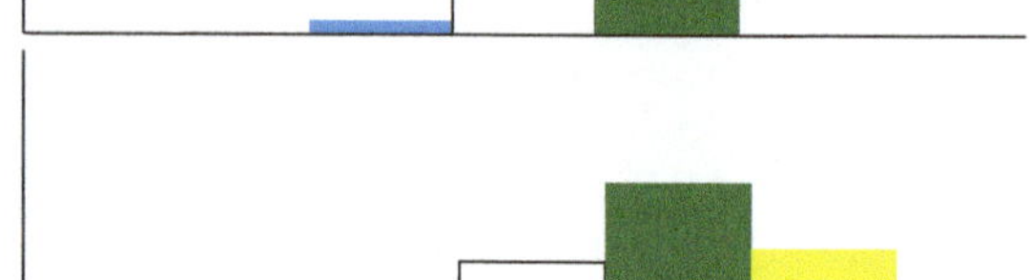
(A4) The floating mooring wind project scale is		

Table 2: Continued.

Determine which ecosystem components may experience impacts from renewable energy development:		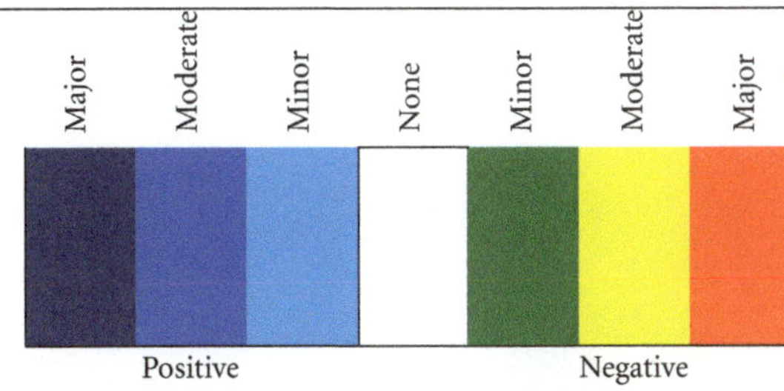
Demonstration	Go to i1	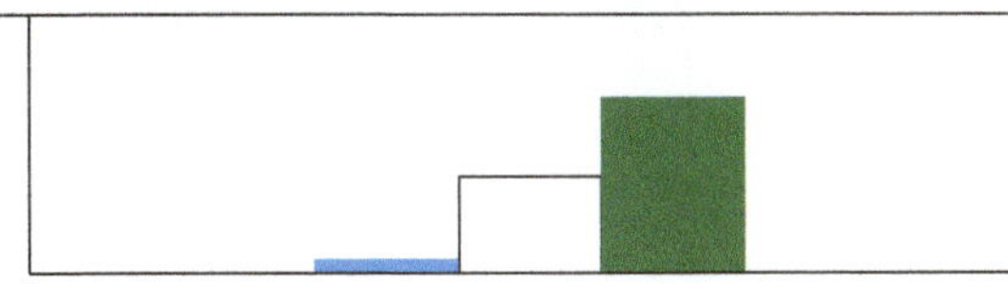
Commercial/Multiple commercial	Go to i3	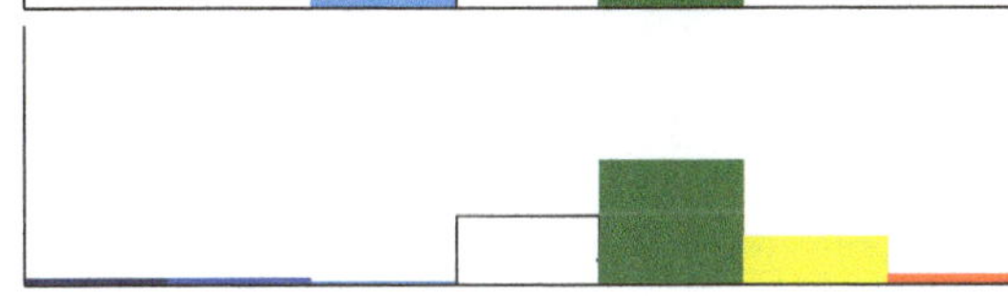
(B) The tidal turbine type is		
Open, bottom mounted		Go to B1
Open, floating mooring		Go to B2
Shrouded, bottom mounted		Go to B3
Shrouded, floating mooring		Go to B4
(B1) The open rotor, bottom-mounted tidal turbine project scale is		
Demonstration	Go to i1	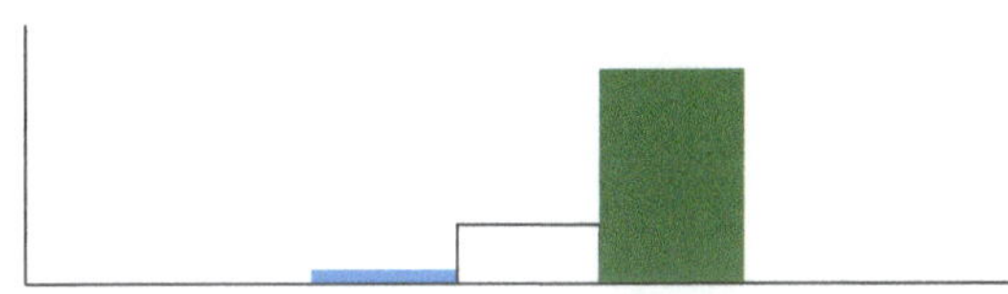
Commercial/Multiple commercial	Go to i4	
(B2) The open rotor, floating mooring tidal turbine project scale is		
Demonstration	Go to i1	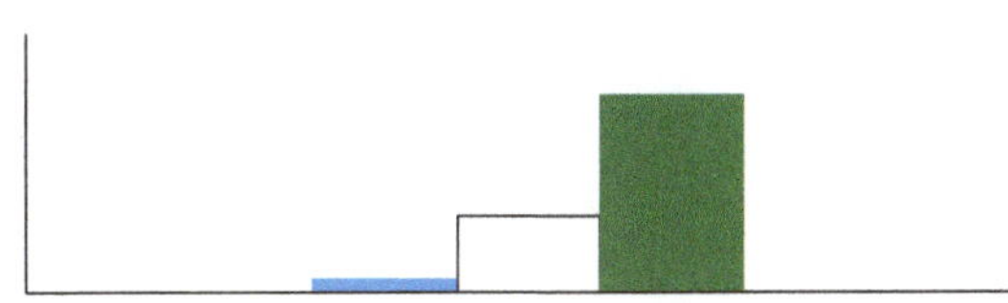
Commercial/Multiple commercial	Go to i5	
(B3) The shrouded rotor, bottom-mounted tidal turbine project scale is		
Demonstration	Go to i1	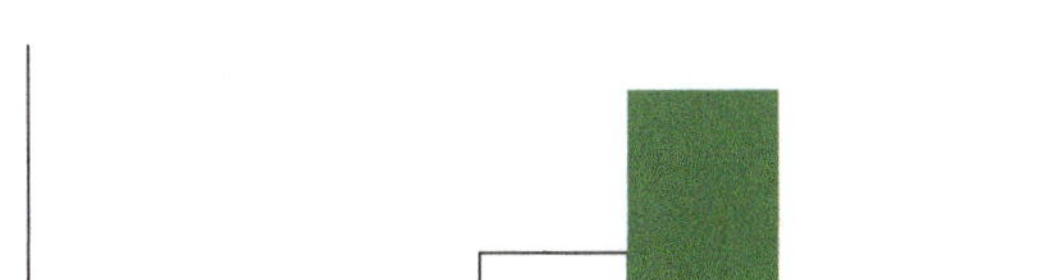
Commercial/Multiple commercial	Go to i4	

TABLE 2: Continued.

Determine which ecosystem components may experience impacts from renewable energy development:		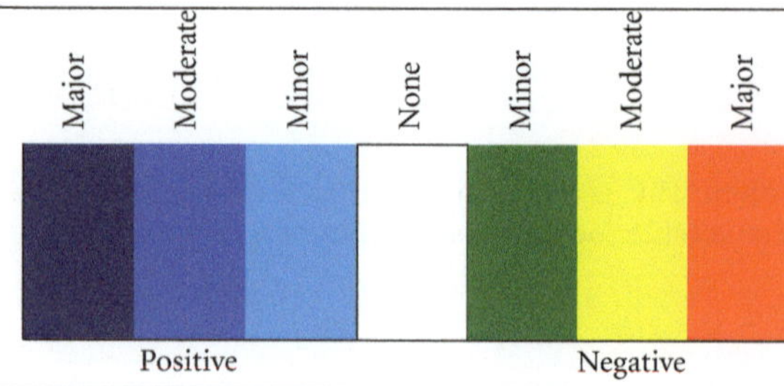
(B4) The shrouded rotor, floating mooring tidal turbine project scale is		
Demonstration	Go to i1	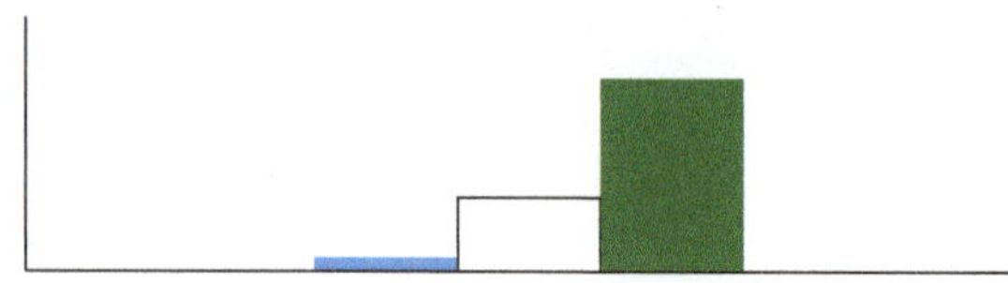
Commercial/Multiple commercial	Go to i5	
(C) The wave device type is		
Point absorber		Go to C1
Wave attenuator		Go to C2
Oscillating water column		Go to C3
Oscillating wave surge converter		Go to C4
Overtopping		Go to C5
(C1) The point absorber project scale is		
Demonstration	Go to i1	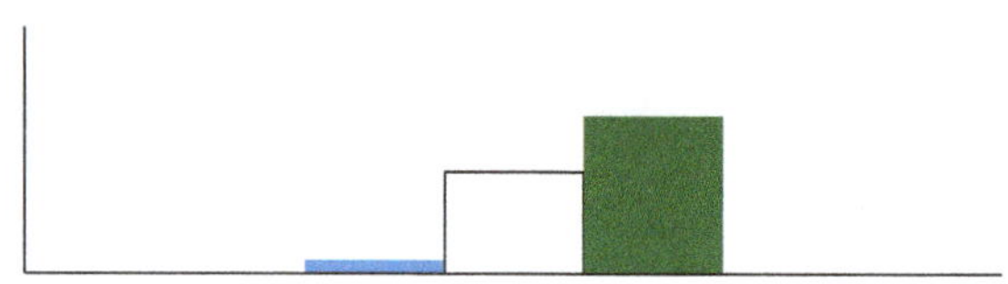
Commercial/Multiple commercial	Go to i6	
(C2) The wave attenuator project scale is		
Demonstration	Go to i1	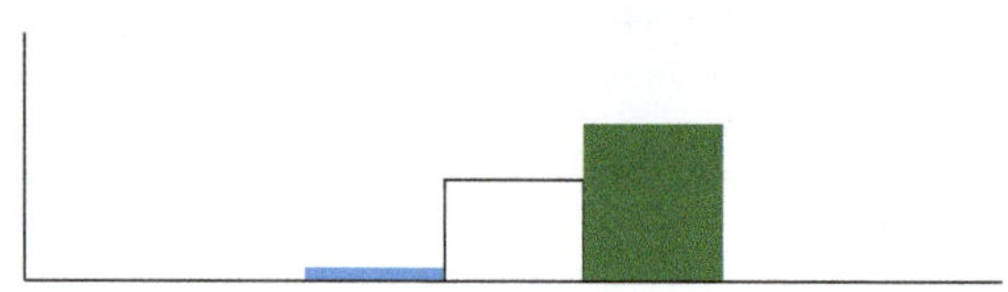
Commercial/Multiple commercial	Go to i6	
(C3) The oscillating water column device project scale is		
Demonstration	Go to i1	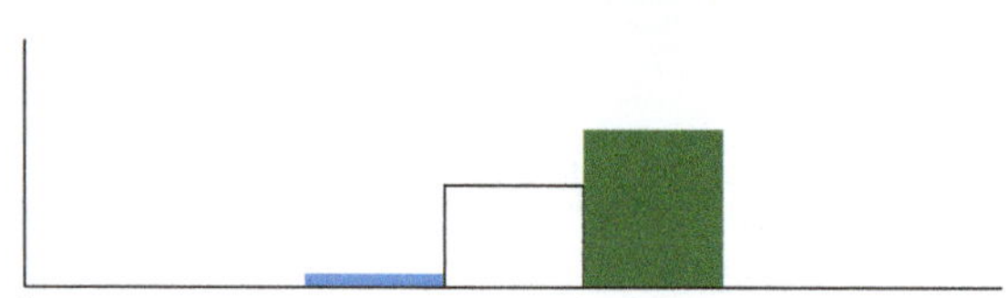
Commercial/Multiple commercial	Go to i6	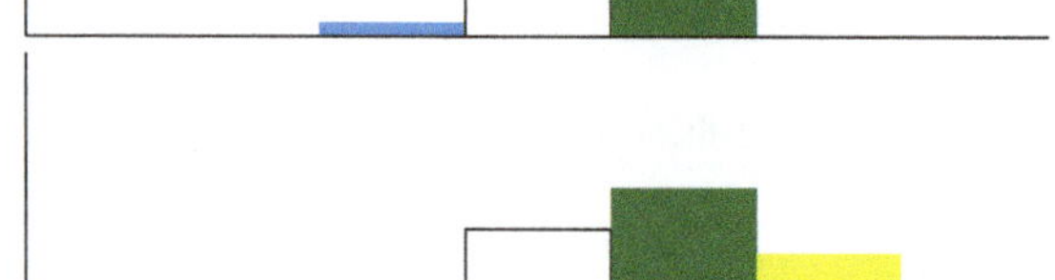

Table 2: Continued.

Determine which ecosystem components may experience impacts from renewable energy development:		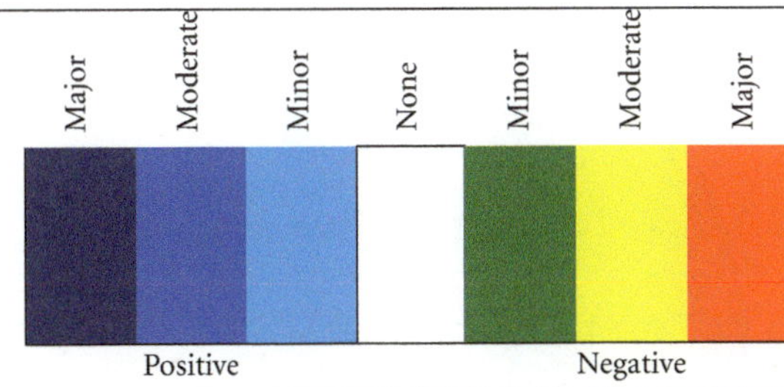
(C4) The oscillating wave surge converter project scale is		
Demonstration	Go to i1	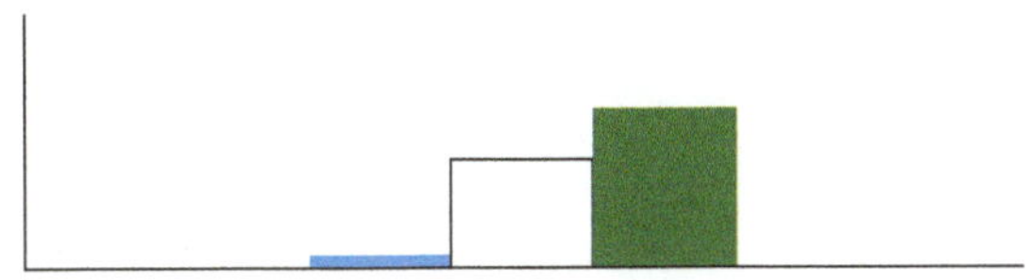
Commercial/Multiple commercial	Go to i6	
(C5) The overtopping device project scale is		
Demonstration	Go to i1	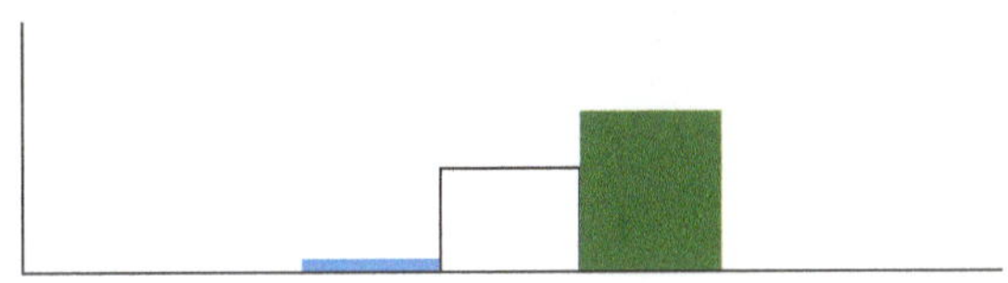
Commercial/Multiple commercial	Go to i6	

Table 3: A description of the potential impacts resulting from demonstration scale projects. See text, Section 3.3, for a narrative summary of this development scenario. Minor impacts are listed in the order they appear in the renewable energy impact matrix.

(i1) Demonstration scale projects		
Component (not ranked)	Minor impact	Certainty
Benthic habitat and resources	(i) Disturbance from installation/removal of device (including turbidity) (ii) Disturbance from installation or removal of power cable (including trenching) (iii) Scour around structures (iv) Smothering by excavated sediments (v) Reef effects (vi) Diffusion/flaking of marine coating (vii) Chemicals discharged during installation or removal (viii) Resuspension of pollutants in sediments	(i) High (ii) High (iii) High (iv) High (v) High (vi) High (vii) High (viii) High
Fish species and fishing activity	(i) Disturbance from installation or removal of device (ii) Disturbance from installation or removal or power cable (iii) Reef effects (iv) Loss of access to grounds during construction (v) Loss of access to grounds during operation	(i) High (ii) High (iii) High (iv) High (v) High
Avian species	(i) Displacement or attraction to structure above surface of the water (wind turbines) (ii) Displacement or attraction to structure below the surface of the water (iii) Disturbance from installation of device or transmission cable (iv) Collision with rotating turbine blades	(i) High (ii) High (iii) High (iv) High

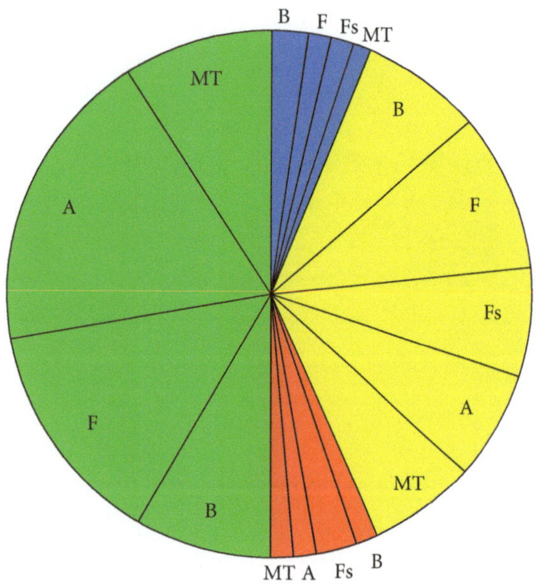

FIGURE 3: Pie chart representing the proportion of impacts for each ecosystem component for Impact scenario i2, wind turbine projects involving pile driving, categorized by whether the effect is positive (blue), minor negative (green), moderate negative (yellow), or major negative (red). B = Benthic habitat and resources; F = Fish; Fs = Fisheries; A = Avian species; MT = Marine mammals and sea turtles.

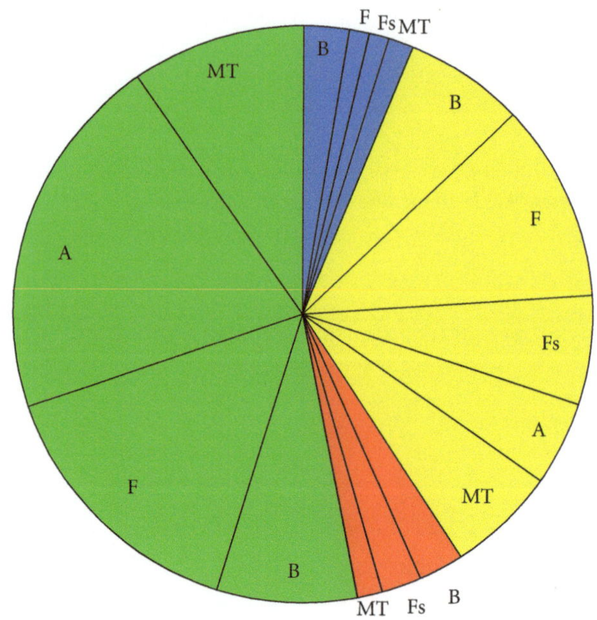

FIGURE 5: Pie chart representing the proportion of impacts for each ecosystem component for Impact scenario i4, bottom-mounted tidal turbine projects, categorized by whether the effect is positive (blue), minor negative (green), moderate negative (yellow), or major negative (red). B = Benthic habitat and resources; F = Fish; Fs = Fisheries; A = Avian species; MT = Marine mammals and sea turtles.

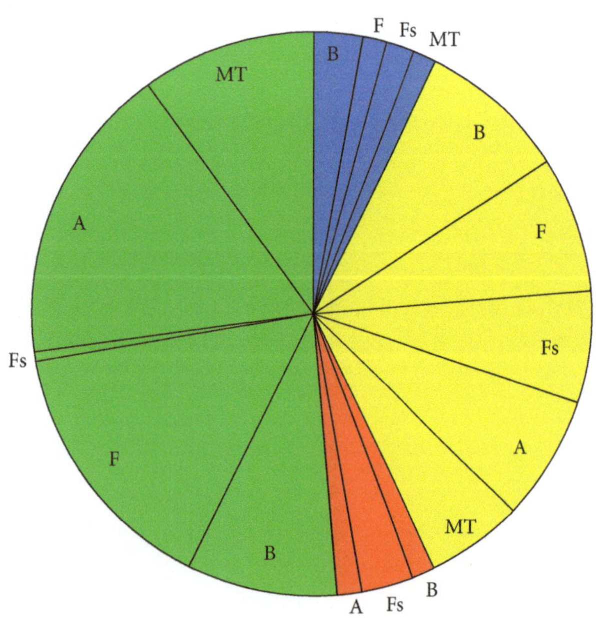

FIGURE 4: Pie chart representing the proportion of impacts for each ecosystem component for Impact scenario i3, wind turbine project involving no pile driving, categorized by whether the effect is positive (blue), minor negative (green), moderate negative (yellow), or major negative (red). B = Benthic habitat and resources; F = Fish; Fs = Fisheries; A = Avian species; MT = Marine mammals and sea turtles.

the expected major impacts include noise, scour and/or deposition around the structures, displacement or attraction to structures, and loss of access to mobile-gear fishing grounds. Notable moderate impacts include resuspension of pollutants, loss of access to recreational and fixed gear fishing grounds, decreased catchability (fisheries), damaged/lost fishing gear, and collisions and strikes for avian species, marine mammals, and sea turtles. Reef effects are likely for benthic habitat and resources and fish species at these developments.

(i3) Wind Turbine Developments Involving No Pile Driving (Figure 4, Table 5). Floating mooring or gravity-base foundations present a different suite of impacts for wind turbine developments at Scale 2. A major impact in scenario i2—noise during construction—is now absent. The suite of remaining negative impacts for each ecosystem component is very similar to i2, with the exception of benthic habitat and resources. Gravity-base foundations incur a moderate negative impact through physical disturbance to the sediment, where in i2 this impact is classified as minor. Reef effects are likely for benthic habitat and resources and fish species at these developments.

(i4) Bottom-Mounted Tidal Turbine Projects (Figure 5, Table 6). For tidal turbine developments, the profile of impacts tended to differ more based on the foundation type than on whether the rotor is shrouded or open.

Table 4: A description of the potential impacts resulting from wind turbine projects involving pile driving. See text, Section 3.3, for a narrative summary of this development scenario. Ecosystem components are ranked by their proportion of the major impacts. Moderate impacts are listed in the order they appear in the renewable energy impact matrix.

(i2) Wind turbines involving pile driving		
Priority	Major impacts	Certainty
(1) Fish species and fishing activity	Loss of access to grounds during construction and operation (mobile gear)	High
(2) Avian species*	Displacement or attraction to structure above water surface	High
(3) Benthic habitat and resources	Scour and/or deposition	High
(4) Marine mammals and sea turtles*	Noise from pile driving	Medium
Component	Moderate impacts	Certainty
Benthic habitat and resources	(i) Resuspension of pollutants in sediments	(i) High
	(ii) Chemical spills, discharge	(ii) Medium
	(iii) Disturbance from installation of cable	(iii) Medium
	(iv) Changes to current/wave regime	(iv) Medium
Fish species and fishing activity	(i) Chemical spills	(i) Medium
	(ii) Operational noise	(ii) Medium
	(iii) Noise from preconstruction seismic surveys	(iii) Medium
	(iv) Noise from pile driving	(iv) Medium
	(v) Noise from pile cutting during device removal	(v) Medium
	(vi) EMF	(vi) Low
	(vii) Habitat/community composition alteration	(vii) Medium
	(viii) Decreased catchability during construction and operation	(viii) Medium
	(ix) Loss of access to grounds during construction and operation (fixed gear and recreational)	(ix) High
	(x) Changes in species distribution	(x) Low
	(xi) Damaged/lost gear	(xi) High
Avian species	(i) Displacement or attraction to structure below water surface	(i) Medium
	(ii) Collision with rotating turbine blades	(ii) High
	(iii) Pressure gradients around rotor	(iii) Medium
	(iv) Leakage of lubricants/fluids, release of maintenance chemicals	(iv) Medium
	(v) Large chemical spills	(v) High
Marine mammals and sea turtles	(i) Entanglement with mooring lines or cables	(i) Medium
	(ii) Strikes with installation or support vessels	(ii) High
	(iii) Operational noise	(iii) Medium
	(iv) Noise from pile cutting during device removal	(iv) High

*Denotes that higher priority may be given to this component due to national/regional/local regulatory objectives and obligation.

Potential major impacts at these developments include changes to hydrodynamics, scour and/or deposition around devices/moorings, loss of access to mobile-gear fishing grounds, and noise from pile driving. If the proposed development will not utilize pile driving to install the tidal turbines, then Impact scenario i5 is more appropriate. Notable moderate impacts include physical disturbance to the sediment; collisions/strikes to rotor blades for fish species, avian species, marine mammals, and sea turtles; the effects of rotor wake/pressure gradients to fish and avian species; collisions/strikes with construction or support vehicles for marine mammals and sea turtles; decreased catchability and damaged/lost gear for fisheries.

(i5) Floating Mooring Tidal Turbine Projects (Figure 6, Table 7). Floating mooring foundations present a different suite of impacts for tidal turbine developments at Scale 2. A major impact in scenario i4—noise during construction—is now absent. The suite of remaining negative impacts for each ecosystem component is very similar to i4 with exceptions for benthic habitat and resources and fisheries. The impacts from sediment disturbance in this scenario are downgraded to minor, as are the impacts surrounding decreased catchability.

(i6) Wave Energy Projects (Figure 7, Table 8). In general, wave energy developments are not as well studied as tidal

TABLE 5: A description of the potential impacts resulting from wind turbine projects involving no pile driving. See text, Section 3.3, for a narrative summary of this development scenario. Ecosystem components are ranked by their proportion of the major impacts. Moderate impacts are listed in the order they appear in the renewable energy impact matrix.

(i3) Wind turbines involving no pile driving		
Priority	Major impacts	Certainty
(1) Fish species and fishing activity	Loss of access to grounds during construction and operation (mobile gear)	High
(2) Avian species*	Displacement or attraction to structure above water surface	High
(3) Benthic habitat and resources	Scour and/or deposition	High
Component	Moderate impacts	Certainty
Benthic habitat and resources	(i) Resuspension of pollutants in sediments	(i) High
	(ii) Disturbance from installation/removal of device (turbidity)	(ii) Medium
	(iii) Chemical spills, discharge	(iii) Medium
	(iv) Disturbance from installation of cable	(iv) Medium
	(v) Changes to current/wave regime	(v) Medium
Fish species and fishing activity	(i) Chemical spills	(i) Medium
	(ii) Operational noise	(ii) Medium
	(iii) Noise from pre-construction seismic surveys	(iii) Medium
	(iv) Noise from pile cutting during device removal	(iv) Medium
	(v) EMF	(v) Low
	(vi) Habitat/community composition alteration	(vi) Medium
	(vii) Decreased catchability during construction and operation	(vii) Medium
	(viii) Loss of access to grounds during construction and operation (fixed gear and recreational)	(viii) High
	(ix) Changes in species distribution	(ix) Low
	(x) Damaged/lost gear	(x) High
Avian species	(i) Displacement or attraction to structure below water surface	(i) Medium
	(ii) Collision with rotating turbine blades	(ii) High
	(iii) Pressure gradients around rotor	(iii) Medium
	(iv) Leakage of lubricants/fluids, release of maintenance chemicals	(iv) Medium
	(v) Large chemical spills	(v) High
Marine mammals and sea turtles	(i) Entanglement with mooring lines or cables	(i) Medium
	(ii) Strikes with installation or support vessels	(ii) High
	(iii) Operational noise	(iii) Medium

*Denotes that higher priority may be given to this component due to national/regional/local regulatory objectives and obligation.

or wind developments. Therefore, we caution against the interpretation that the pie chart suggests that wave energy developments have a lower proportion of potential major and moderate impacts than any other development type. Major impacts at Scale 2 wave energy projects are changes in hydrodynamics, scour, and/or deposition around devices and loss of access to mobile-gear fishing grounds. Notable moderate impacts include loss of access to fixed-gear and recreational fishing grounds, damaged/lost fishing gear, chemical spills, and collisions/strikes with construction or support vehicles for marine mammals and sea turtles. Specifically, oscillating wave surge converters have higher potential impacts over the other types (moderate versus minor) for operational noise on fish species, marine mammals, and sea turtles, and for sediment disturbance on benthic habitat and resources. Overtopping devices pose increased potential impacts over other types (moderate versus minor) on avian species for displacement or attraction to the device because of the above-water structure.

3.4. Component Decision Trees. The Component decision trees take component-specific concerns into consideration and terminate with a manageable number of recommended monitoring protocols and templates (Tables 9, 10, 11, 12, and 13). For example, the benthic environment decision tree (Table 9) describes 24 total monitoring scenarios but condenses them into a maximum of four monitoring templates. We have two separate decision trees for marine mammals and sea turtles, but one set of monitoring protocols, as in many cases a single protocol can be used to monitor both components. After working through the Impact decision

Table 6: A description of the potential impacts resulting from bottom-mounted tidal turbine projects. See text, Section 3.3, for a narrative summary of this development scenario. Ecosystem components are ranked by their proportion of the major impacts. Moderate impacts are listed in the order they appear in the renewable energy impact matrix.

(i4) Bottom-mounted tidal turbine projects		
Priority	Major impacts	Certainty
(1) Benthic habitat and resources	(i) Changes in hydrodynamics (ii) Scour and/or deposition	(i) Medium (ii) High
(2) Fish species and fishing activity*	Loss of access to grounds during construction and operation (mobile gear)	High
(3) Marine mammals and sea turtles*	Noise from pile driving	Medium
Component	Moderate impacts	Certainty
Benthic habitat and resources	(i) Resuspension of pollutants in sediments (ii) Disturbance from installation/removal of device (turbidity) (iii) Chemical spills, discharge (iv) Disturbance from installation of cable (v) Changes to current/wave regime	(i) Low (ii) High (iii) Medium (iv) Medium (v) Medium
Fish species and fishing activity	(i) Collision/blade strike (ii) Pressure gradients around rotor (iii) Chemical spills (iv) Operational noise (v) Noise from pre-construction seismic surveys (vi) Noise from pile driving (vii) Noise from pile cutting during device removal (viii) EMF (ix) Habitat/community composition alteration (x) Decreased catchability during construction and operation (xi) Loss of access to grounds during construction and operation (fixed gear and recreational) (xii) Changes in species distribution (xiii) Damaged/lost gear	(i) Medium (ii) Medium (iii) Medium (iv) Medium (v) Medium (vi) Medium (vii) Medium (viii) Low (ix) Medium (x) Medium (xi) High (xii) Low (xiii) High
Avian species	(i) Collision with rotating turbine blades (ii) Pressure gradients around rotor (iii) Leakage of lubricants/fluids, release of maintenance chemicals (iv) Large chemical spills	(i) Medium (ii) Medium (iii) Medium (iv) High
Marine mammals and sea turtles	(i) Entanglement with mooring lines or cables (ii) Strikes with installation or support vessels (iii) Operational noise (iv) Noise from pile cutting during device removal	(i) Medium (ii) High (iii) Medium (iv) High

*Denotes that higher priority may be given to this component due to national/regional/local regulatory objectives and obligation.

tree, the user should then select the recommended Component decision trees in order to determine specifically which monitoring objectives will apply to that technology type.

3.5. Monitoring Protocols. The Component decision trees terminate with a total of 30 monitoring objectives across all ecosystem components. Monitoring protocols have been developed to address each of these objectives at the University of Rhode Island, but for brevity, the benthic habitat and resources monitoring protocol for seabed scour and/or deposition is presented.

4. Case Study: Commercial Wind Farm

This test case was conducted using a hypothetical wind farm in the Wind Energy Area defined by BOEM in federal waters off the Massachusetts and Rhode Island coasts. A wind farm being planned for this area may include around 200 turbines with jacketed structures. Ecological concerns for this development include the sensitivities of local bird populations (scoters, red-throated loons), important commercial fisheries (demersal fish, lobsters), and the occasional presence of endangered marine mammals (particularly North Atlantic right whales).

TABLE 7: A description of the potential impacts resulting from floating mooring tidal turbine projects. See text, Section 3.3, for a narrative summary of this development scenario. Ecosystem components are ranked by their proportion of the major impacts. Moderate impacts are listed in the order they appear in the renewable energy impact matrix.

(i5) Floating mooring tidal turbine projects		
Priority	Major impacts	Certainty
(1) Benthic habitat and resources	(i) Changes in hydrodynamics (ii) Scour and/or deposition	(i) Medium (ii) High
(2) Fish species and fishing activity*	Loss of access to grounds during construction and operation (mobile gear)	High
Component	Moderate impacts	Certainty
Benthic habitat and resources	(i) Resuspension of pollutants in sediments (ii) Chemical spills, discharge (iii) Disturbance from installation of cable (iv) Changes to current/wave regime	(i) Low (ii) Medium (iii) Medium (iv) Medium
Fish species and fishing activity	(i) Collision/blade strike (ii) Pressure gradients around rotor (iii) Chemical spills (iv) Operational noise (v) Noise from pre-construction seismic surveys (vi) EMF (vii) Habitat/community composition alteration (viii) Loss of access to grounds during construction and operation (fixed gear and recreational) (ix) Changes in species distribution (x) Damaged/lost gear	(i) Medium (ii) Medium (iii) Medium (iv) Medium (v) Medium (vi) Low (vii) Medium (viii) High (ix) Low (x) High
Avian species	(i) Collision with rotating turbine blades (ii) Pressure gradients around rotor (iii) Leakage of lubricants/fluids, release of maintenance chemicals (iv) Large chemical spills	(i) Medium (ii) Medium (iii) Medium (iv) High
Marine mammals and sea turtles	(i) Entanglement with mooring lines or cables (ii) Strikes with installation or support vessels (iii) Operational noise	(i) Medium (ii) High (iii) Medium

*Denotes that higher priority may be given to this component due to national/regional/local regulatory objectives and obligation.

4.1. Potential Impacts and Monitoring Plan. The Impact decision tree identified this installation as an i2 Impact scenario (Table 4). We recommend that monitoring plans for the four impacts listed as potentially major and negative (Loss of access to grounds for commercial mobile-gear fishermen during construction and operation, displacement or attraction to a device for avian species, seabed scour and/or deposition, and noise from pile driving for marine mammals) be required by federal permitting organizations. An additional 24 impacts were identified as potentially moderate and negative; a subset of these should be considered as part of the permitting requirements but could also serve to inform additional monitoring that might be conducted by other federal or state agencies to address local environmental or stakeholder concerns.

By working through the Component decision trees, we identified nineteen monitoring objectives/protocols that are applicable to this development (Table 14), demonstrating that protocols can be developed to track single and multiple impacts. For example, a sampling protocol for benthic community analysis can also be used to collect information about sediment grain size. Based on the magnitudes of impacts and the local concerns for this wind farm, we identified a subset of eight monitoring objectives/protocols that should be implemented by developers, federal agencies, or both (Table 14). These include monitoring for seabed scour and/or deposition; ventless trap surveys for lobster and trawl surveys for demersal fish; an examination of the spatial use of fishing activity; aerial surveys using both high-definition video and still photography for avian monitoring; and visual surveys, passive acoustic monitoring, and marine mammal observers to track impacts to marine mammals and sea turtles. At a larger (regional) scale, or in an area where perhaps less is known about the local biological resources, a monitoring plan could begin with all of the recommended objectives/protocols and gradually decrease this effort through time as protocols fail to be relevant or detect ecosystem change.

TABLE 8: A description of the potential impacts resulting from projects harnessing wave energy. See text, Section 3.3, for a narrative summary of this development scenario. Ecosystem components are ranked by their proportion of the major impacts. Moderate impacts are listed in the order they appear in the renewable energy impact matrix.

(i6) Wave energy projects		
Priority	Major impacts	Certainty
(1) Benthic habitat and resources	(i) Changes in hydrodynamics	(i) Medium
	(ii) Scour and/or deposition	(ii) High
(2) Fish species and fishing activity*	Loss of access to grounds during construction and operation (mobile gear)	High
Component	Moderate impacts	Certainty
Benthic habitat and resources	(i) Resuspension of pollutants in sediments	(i) Low
	(ii) Chemical spills, discharge	(ii) Medium
	(iii) Disturbance from installation of cable	(iii) Medium
	(iv) Changes to current/wave regime	(iv) Medium
Fish species and fishing activity	(i) Chemical spills	(i) Medium
	(ii) Operational noise	(ii) Medium
	(iii) Noise from pre-construction seismic surveys	(iii) Medium
	(iv) EMF	(iv) Low
	(v) Habitat/community composition alteration	(v) Medium
	(vi) Decreased catchability during construction/operation	(vi) Medium
	(vii) Loss of access to grounds during construction and operation (fixed gear and recreational)	(vii) High
	(viii) Changes in species distribution	(viii) Medium
	(ix) Damaged/lost gear	(ix) High
Avian species	(i) Displacement/attraction to structure above water surface	(i) Medium
	(ii) Leakage of lubricants/fluids, release of maintenance chemicals	(ii) Medium
	(iii) Large chemical spills	(iii) High
Marine mammals and sea turtles	(i) Entanglement with mooring lines or cables	(i) Medium
	(ii) Strikes with installation or support vessels	(ii) High
	(iii) Operational noise	(iii) Medium

*Denotes that higher priority may be given to this component due to national/regional/local regulatory objectives and obligation.

4.2. Example Monitoring Protocol. Finally, we present an example protocol to address seabed scour and/or deposition (Table 15). Similar protocols for each impact should be developed to implement a consistent and robust monitoring plan. Protocols such as this are intended to provide guidance to regulators and developers on the most suitable methods for detecting the indicators of impacts but, like the overall framework, are designed to be adaptive to regulatory needs and site-specific concerns. In our example, we provided estimates of cost for two different monitoring strategies so that monitoring activities might be prioritized based on ecological and financial factors. We recommend that similar protocols be developed based on best practices identified in the literature and do not aim to be definitive or to provide comprehensive lists of all available methodologies.

5. Discussion

This paper presents a system for selecting both the priority impacts of ORED to be monitored and the appropriate monitoring protocols to address these impacts for a given project. One of the lessons learned from this effort was the importance of a monitoring program that is adaptive to both regulatory needs and local concerns (e.g., [19]). In drafting a set of monitoring objectives we attempted to account for variability in regions, target species, and so forth, but decisions about the most appropriate ways to monitor ORED will still have to be made on a case-by-case basis. Drivers, impacts, indicators, and technologies available to use in a monitoring program are all expected to be site specific. Our framework is flexible enough to address these concerns and to be useful for developing monitoring plans on scales ranging from local to national.

5.1. Development of Monitoring Protocols. It would be impractical to monitor every interaction that could potentially result in an effect on an organism or abiotic component of the ecosystem. Protocols should target priority impacts for monitoring given the ecological or societal importance of the resource, activity, or community and the certainty of their magnitude of impact. Monitoring protocols should test a particular regulatory question that is linked to one or more of these potential effects. In this study, we assumed that regulators would want and need to focus on the major

Table 9: Component decision tree for impacts on benthic habitat and resources. The key is followed to obtain a recommended suite of monitoring protocols.

Determine which impacts to the benthic environment need to be monitored:	
The energy resource is	
Wind	Go to A
Waves	Go to B
Tidal	Go to C
(A) The wind turbine foundation is	
Monopile OR tripod OR lattice	Go to A1
Gravity	Go to A2
Floating mooring	Go to A3
(A1) The stage of the monopile, tripod, or lattice wind turbine project is	
Construction	Z1, Z2
Operation	Z1, Z2, Z3
Decommissioning	Z1, Z2
(A2) The stage of the gravity wind turbine project is	
Construction	Z1, Z2, Z3
Operation	Z1, Z2, Z3
Decommissioning	Z1, Z2, Z3
(A3) The stage of the floating mooring wind project is	
Construction	Z1, Z2
Operation	Z1, Z2
Decommissioning	Z1, Z2
(B) The tidal turbine type is	
Open OR shrouded bottom mounted	Go to B1
Open OR shrouded floating mooring	Go to B2
(B1) The stage of the bottom-mounted tidal turbine project is	
Construction	Z1, Z2
Operation	Z1, Z2, Z3, Z4
Decommissioning	Z1, Z2
(B2) The stage of the floating mooring tidal turbine project is	
Construction	Z1, Z2
Operation	Z1, Z2, Z4
Decommissioning	Z1, Z2
(C) The wave device type is	
Point absorber OR wave attenuator OR oscillating water column	Go to C1
Oscillating wave surge converter	Go to C2
Overtopping	Go to C3
(C1) The stage of the point absorber OR wave attenuator OR oscillating water column project is	
Construction	Z1, Z2
Operation	Z2, Z4
Decommissioning	Z1, Z2
(C2) The oscillating wave surge converter project scale is	
Construction	Z1, Z2
Operation	Z1, Z2, Z3, Z4
Decommissioning	Z1, Z2
(C3) The overtopping device project scale is	
Construction	Z1, Z2
Operation	Z1, Z2, Z4
Decommissioning	Z1, Z2

Recommended protocols:
Z1: seabed scour and/or deposition.
Z2: changes in benthic community composition.
Z3: increase in hard bottom habitat.
Z4: changes in hydrodynamics.

Table 10: Component decision tree for impacts on fish species and fisheries. The key is followed to obtain a recommended suite of monitoring protocols.

Determine which impacts to fisheries resources and fishing activity need to be monitored:	
The energy resource is	
Wind	Go to A
Tidal	Go to B
Waves	Go to C
(A) The wind turbine foundation is	
Monopile OR tripod OR lattice OR gravity	Go to A1
Floating mooring	Go to A3
(A1) The stage of the monopile, tripod, lattice, or gravity wind turbine project is	
Construction	X1, X2, X5
Operation	X1, X2, X3, X5
Decommissioning	X1, X2
(A2) The stage of the floating mooring wind project is	
Construction	X1, X2, X5
Operation	X3, X4, X5
Decommissioning	X1, X2
(B) The tidal turbine type is	
Open OR shrouded bottom mounted	Go to B1
Open OR shrouded floating mooring	Go to B2
(B1) The stage of the bottommounted tidal turbine project is	
Construction	X1, X2, X5
Operation	X1, X2, X3, X4, X5
Decommissioning	X1, X2
(B2) The stage of the floating mooring tidal turbine project is	
Construction	X1, X2, X5
Operation	X1, X2 X3, X4, X5
Decommissioning	X1, X2
(C) The wave device type is	
Point absorber OR wave attenuator OR oscillating water column OR overtopping	Go to C1
Oscillating wave surge converter	Go to C2
(C1) The stage of the point absorber OR wave attenuator OR oscillating water column OR overtopping project is	
Construction	X1, X2, X5
Operation	X1, X2, X5
Decommissioning	X1, X2
(C2) The oscillating wave surge converter project scale is	
Construction	X1, X2, X5
Operation	X1, X2, X3, X5
Decommissioning	X1, X2

Recommended protocols:
X1: mesoscale changes to abundance and distribution (disturbance).
X1a: the species of concern are finfish.
X1b: the species of concern are crustaceans or rock fish.
X2: habitat alteration/community composition: microscale changes to abundance and distribution—finfish.
X3: reef effects.
X4: blade strikes.
X5: spatial use of fishing activity.

negative potential impacts of ORED. However, we identified a number of potential positive impacts that are also worthy of research and monitoring. The positive impacts, such as reef effects, should be considered in any impact assessment for the potential value they may provide to the wider marine environment [2].

Protocols may need to be developed with the assumption that there are no or insufficient existing data on the relevant species to establish reference levels prior to monitoring. In some cases, the data may exist but not at a scale appropriate for integration into monitoring efforts. In other circumstances, baseline data will exist that can and should be

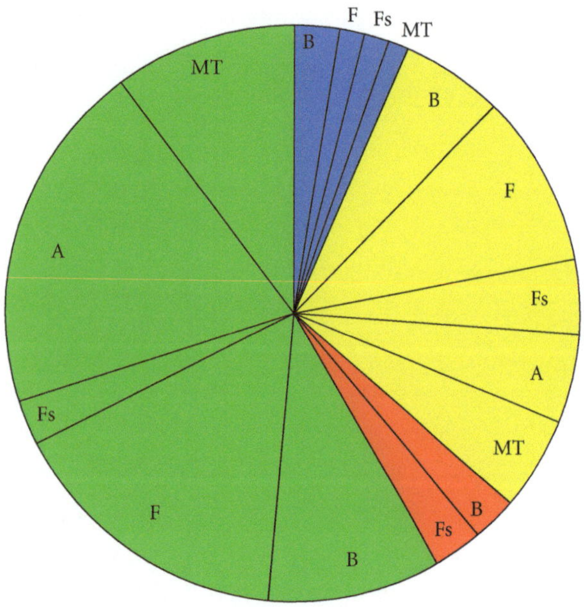

Figure 6: Pie chart representing the proportion of impacts for each ecosystem component for Impact scenario i5, floating mooring tidal turbine projects, categorized by whether the effect is positive (blue), minor negative (green), moderate negative (yellow), or major negative (red). B = Benthic habitat and resources; F = Fish; Fs = Fisheries; A = Avian species; MT = Marine mammals and sea turtles.

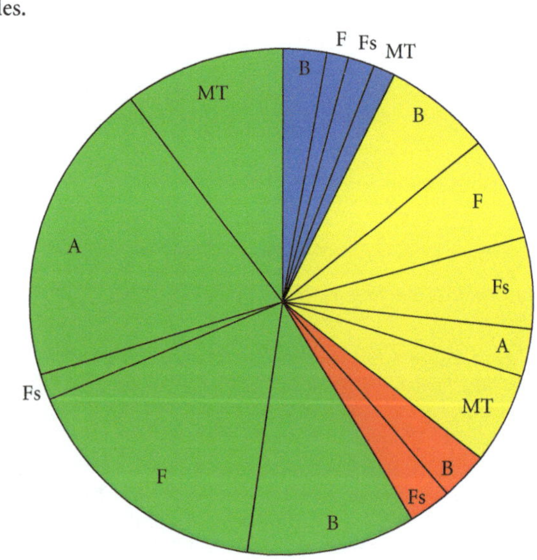

Figure 7: Pie chart representing the proportion of impacts for each ecosystem component for Impact scenario i6, wave energy project, categorized by whether the effect is positive (blue), minor negative (green), moderate negative (yellow), or major negative (red). B = Benthic habitat and resources; F = Fish; Fs = Fisheries; A = Avian species; MT = Marine mammals and sea turtles.

incorporated into monitoring efforts. As described above, we assign reference directions to indicators when we have insufficient data to establish a quantitative threshold. Protocols should therefore be flexible to incorporate new and existing data for better estimates of particular reference points. Where there is an ongoing environmental monitoring program in the project area, and if the methods in use are sufficient to

Table 11: Component decision tree for impacts on avian species. The key is followed to obtain a recommended suite of monitoring protocols.

Determine which impacts to avian species need to be monitored:	
The energy resource is	
Wind	Go to A
Waves	Go to B
Tidal	Go to C
(A) The stage of the wind energy project is	
Construction	Go to D
Operation	Go to D, Go to E, V11
Decommissioning	Go to D
(B) The stage of the tidal project is	
Construction	Go to D
Operation	Go to D, V12
Decommissioning	Go to D
(C) The stage of the wave energy project is	
Construction	Go to D
Operation	Go to D
Decommissioning	Go to D
(D) The target species are	
Easily disturbed, cryptic	V3, V4
Easily disturbed, noncryptic	V2, V3, V4
Not easily disturbed, cryptic	V1, V3, V4
Not easily disturbed, noncryptic	V1, V2, V3, V4
(E) The target species are	
Diurnal	V5, V6
Nocturnal	V5, V7

Recommended protocols:
V1: ship-based visual surveys.
V2: aerial surveys using human observers.
V3: aerial surveys using high-definition videography.
V4: aerial surveys using digital still photography.
V5: radar surveys.
V6: visual surveys.
V7: flight call surveys.
V11: remote detection system.
V12: sonar and video technology.

detect a change due to development, data collection should continue using the same methodology for comparable data. Examples of existing monitoring programs with appropriate data for incorporating into ORED studies may include species monitored under the Endangered Species or Marine Mammal Protection Acts for which monitoring may be occurring as part of a stock assessment program. Monitoring data collected for threatened, endangered, or other protected species could be compared directly to existing reference levels. For these species in particular it will be important to know if negative impacts are caused by ORED of any kind because such impacts will trigger immediate federal regulatory response/mitigation measures. In many cases, site-specific monitoring will still be desirable to analyze change related directly to the ORED.

TABLE 12: Component decision tree for impacts on marine mammals. The key is followed to obtain a recommended suite of monitoring protocols.

Determine which impacts to marine mammals need to be monitored:	
The energy resource is	
Wind	Go to A
Tidal	Go to B
Waves	Go to C
(A) The stage of the wind energy project is	
Construction	W1, W2, W3, W4, W5
Operation	W1, W2, W3, W4, W5
Decommissioning	W1, W2, W3, W4, W5
(B) The stage of the tidal energy project is	
Construction	W1, W2, W3, W4, W5
Operation	W1, W2, W3, W4, W5
Decommissioning	W1, W2, W3, W4, W5
(C) The stage of the wave energy project is	
Construction	W1, W2, W3, W4, W5
Operation	W1, W2, W3, W4, W5
Decommissioning	W1, W2, W3, W4, W5

Recommended protocols:
W1: visual surveys.
W2: passive acoustic monitoring.
W3: marine mammal observers.
W4: stranding response networks.
W5: tagging.
W6/7: underwater photography.
W8: SCUBA surveys.
W9: ROV surveys.

The time scales of monitoring protocols should be long enough to observe short-term or immediate impacts caused by an ORED, include enough data to limit some of the effects of natural variability on the analysis, and last long enough to observe whether conditions return to a preconstruction state. Developer-led monitoring will probably not be conducted on time scales of a length sufficient to observe very long-term effects from ORED (i.e., decades). Supplementary monitoring should be conducted for a decade or more in order to understand long-term effects. For example, five years of monitoring may be enough time to observe effects on some species but may not be sufficient to identify stock- or population-level effects, particularly on slow-growing or long-lived species such as elasmobranchs. Additionally, some have speculated that if offshore renewable energy devices result in reef effects, this could create secondary effects such as larval spillover if spawning is occurring around the devices. These sorts of secondary effects may not be observable during the time scales of developer-led monitoring. Thus we recommend that, where feasible, monitoring and supplementary studies take place well beyond the minimum time frames required by federal permitting agencies.

5.2. Demonstration-Scale Projects as Opportunities. Demonstration-scale projects provide an opportunity for research to reduce some of the existing uncertainty around the potential environmental effects of offshore renewable energy projects, assisting regulators in prioritizing monitoring needs and making better decisions. Due to their size, individual demonstration-scale projects should be considered separately from commercial-scale projects in the extent to which monitoring should be required. Demonstration-scale projects are not expected to result in environmental impacts of the same magnitudes as commercial projects for any of the renewable energy device types. These projects, or those testing new technologies, should however be subject to more extensive monitoring relative to the scale of the potential impact, at least in the early phases of these technologies, due to the high levels of uncertainty surrounding impacts. Greater monitoring effort at these early stages may later reduce monitoring requirements at commercial-scale facilities, as impacts are better understood. We recommend that the monitoring requirements for demonstration-scale projects be adaptive; more studies that respond to the major and moderate impacts of a commercial-scale project should also be conducted initially as these projects are deployed, particularly at a stage where there are few or no commercial-scale facilities available for monitoring. As impacts are better characterized and methodologies are made more efficient, individual monitoring activities could be phased down in order to maximize the suite of activities at each development. Overall, we recommend that as many monitoring protocols be implemented as is feasible for the early stages of ORED development in the U.S.

5.3. Multiple Projects and Cumulative Impacts. Without a more complete understanding of the direct impacts of ORED on the various environment components discussed here, it is infeasible to develop a monitoring framework to address multiple projects or cumulative impacts. While the likelihood of an effect at the stock or population scale may increase with multiple ORED projects in a given area, not enough conclusive evidence exists at this point to indicate whether those effects are additive or increase in a nonlinear fashion. At the stage in which there are multiple projects in an area, monitoring may need to occur on a regional scale to understand the magnitude of an impact and may need to occur over a longer time series, and data collected from separate projects may need to be analyzed together to more completely understand what is happening. Meta-type analyses could be considered to attempt to maximize the utility of data collected from multiple projects.

As monitoring data are collected at single projects and analyzed, the potential effects at the individual project level will become better understood, and some of these impacts

TABLE 13: Component decision tree for impacts on sea turtles. The key is followed to obtain a recommended suite of monitoring protocols.

Determine which impacts to sea turtles need to be monitored:	
The energy resource is	
Wind	Go to A
Tidal	Go to B
Waves	Go to C
(A) The wind turbine foundation is	
Monopile OR tripod OR floating mooring	Go to A1
Lattice OR gravity	Go to A2
(A1) The stage of the monopile or tripod or floating mooring wind turbine project is	
Construction	W1, W3, W4, W5, W8
Operation	W1, W3, W4, W5
Decommissioning	W1, W3, W4, W5, W8
(A2) The stage of the lattice structure or gravity foundation wind project is	
Construction	W1, W3, W4, W5, W8
Operation	W1, W3, W4, W5
Decommissioning	W1, W3, W4, W5, W6, W7, W8
(B) The tidal turbine type is	
Open OR shrouded bottom-mounted	Go to B1
Open OR shrouded floating mooring	Go to B2
(B1) The stage of the bottommounted tidal turbine project is	
Construction	W1, W3, W4, W5, W8
Operation	W1, W3, W4, W5
Decommissioning	W1, W3, W4, W5, W6, W7, W8
(B2) The stage of the floating mooring tidal turbine project is	
Construction	W1, W3, W4, W5, W8
Operation	W1, W3, W4, W5
Decommissioning	W1, W3 W4, W5, W8
(C) The stage of the wave energy device is	
Construction	W1, W3, W4, W5, W8
Operation	W1, W3, W4, W5
Decommissioning	W1, W3, W4, W5, W8

Recommended protocols:
W1: visual surveys.
W2: passive acoustic monitoring.
W3: marine mammal observers.
W4: stranding response networks.
W5: tagging.
W6: underwater photography.
W7: SCUBA surveys.
W8: ROV surveys.

may be found to be negligible. However, in some cases an impact could be negligible when there is a single project but more severe when combined from multiple projects. It is not known *a priori* which situation is applicable to a particular project-species interaction.

5.4. Effects and Indicators. Currently there is a great deal of uncertainty surrounding the environmental effects of offshore renewable energy technologies. Through implementing a comprehensive monitoring program at each new ORED and by comparing and aggregating data, much of this uncertainty will be reduced and negative impacts will be better understood. Additional studies, including both *in situ* and laboratory-based research, are needed to better understand ORED drivers of ecosystem change and impacts. In many cases this is beyond the scope of what can be determined by a straightforward monitoring study and may be unreasonable to require of a developer. Regardless, it would be best if studies on these unknowns are ongoing and occur alongside other monitoring efforts. Paradoxically, without a better understanding of these impacts, it may not be feasible to design a protocol to be applied on a widespread scale.

Our work has provided a starting point for identifying these types of research priorities. Those effects

TABLE 14: Monitoring protocols relevant to addressing the potential impacts identified from the Impact decision tree and component decision trees for the case study of a scale 2 (~200 devices) jacketed wind turbine farm involving pile driving. Protocols in bold face represent those that uniquely address the Impact scenario and local concerns and should thus be prioritized for implementation.

benthic habitat and resources:
(Z1) Seabed scour and/or deposition
(Z2) Changes in benthic community composition
(Z3) Increase in hard bottom habitat
Fish and fisheries resources
(X1a) Trawl surveys
(X1b) Ventless trap surveys
(X2) Habitat alteration/community composition: micro-scale changes to abundance and distribution—finfish
(X3) Reef effects
(X5) Spatial use of fishing activity
Avian species
(V3) Aerial surveys using high-definition videography
(V4) Aerial surveys using digital still photography
(V5) Radar surveys
(V6) Visual surveys of flight ecology
(V11) Remote detection system
Marine mammals/sea turtles
(W1) Visual surveys
(W2) Passive acoustic monitoring
(W3) Marine mammal observers
(W4) Stranding response networks
(W5) Tagging
(W8) ROV surveys

found to have a low level of certainty are those for which additional research should be conducted. We recommend research funding and effort be directed toward understanding the level of risk of the following potential effects.

Benthic habitat and resources:

(i) changes to currents or wave regimes,

(ii) increase in sediment temperature around cable,

(iii) reef effects on marine hydrokinetic (MHK) devices,

(iv) noise effects from construction, operation, and decommissioning,

(v) EMF effects from the power cables.

Fish species and fisheries:

(i) effect of pressure and velocity gradients around a rotor, and rotor wake, for tidal devices,

(ii) effects resulting from chemical discharge, including leaking, spills, or flaking of marine coating,

(iii) noise effects, including preconstruction noise from seismic surveying, construction noise, vessel noise, and operational noise,

(iv) EMF effects from the power cables,

(v) changes to community composition from reef effects or disturbance,

(vi) changes in species distribution.

Avian species:

(i) collision with rotating turbine blades from tidal devices,

(ii) pressure and velocity gradients around a rotor for both tidal and wind turbines,

(iii) EMF effects from the power cables,

(iv) changes to foraging due to changes in turbulent dissipation/boundary layers for MHK devices,

(v) changes to foraging due to changes in the wave energy regime, for all device types.

Marine mammals and sea turtles:

(i) reef effects from devices,

(ii) entanglement with mooring lines or cables,

(iii) potential for effects from diffusion or flaking of marine coating,

(iv) effects of operational noise, especially from tidal and wave energy devices,

(v) EMF effects from the power cables.

The potential for effects from EMF on all species has emerged as a particular issue of concern; there is considerable uncertainty around what the effects of EMF might be on each of the topic areas considered in this paper. When possible, we recommend site-based EMF studies be conducted alongside other monitoring projects.

6. Conclusions

The tools developed for this project represent an important first step in standardizing monitoring for offshore renewable energy projects in the United States. The data collected through a standardized monitoring program will provide a means for refining our understanding of the potential effects of offshore renewable energy projects and will feed back into better siting decisions. It is our hope that these results will prove useful to scientists, developers, managers, and regulators in understanding environmental changes resulting from offshore renewable energy development.

Table 15: An example monitoring protocol from the benthic habitat and resources category for seabed scour and/or deposition at a commercial-scale installation.

	High cost	Low cost
Indicator(s) of the impact	Scour: increase in median grain size, decrease in organic content, decrease in seabed volume Deposition: decrease in median grain size, increase in organic content, increase in seabed volume	
Methodology or technique to collect data	Particle size analysis, multibeam/interferometric bathymetry	
Description of methodology or technique(s) for collecting data	Seasonal surveys, 5 years Grain size: *5-sample transect at 3 devices out to 200 m Bathymetry: overlapping transects for 100% coverage (at least 0.5 m pixels) 1 km radius at 3 devices	Annual surveys, 3 years Grain size: *3-sample transect at 3 devices out to 200 m Bathymetry: overlapping transects for 100% coverage (at least 0.5 m pixels) 500 m radius 3 devices
Methodology for analyzing data	ANOVA on median grain size, volume change estimate using mosaicked bathymetry models	
Frequency and duration	1 preconstruction survey, seasonal operation, 1 postconstruction survey	1 preconstruction survey, annual operation, 1 postconstruction survey
Spatial scale	200 m–1 km radius around 3 devices	500 m radius around 3 devices
How well does this methodology account for environmental variability?	Seasonal and interannual variability	Interannual variability
Cost	$25 k/year, 5 years	$15 k/year, 3 years
Other considerations (e.g., advantages or disadvantages)	Can be combined with benthic community composition monitoring protocol	

Type of data output required: time series values for median grain size and standard deviations, time series on volume at each turbine and standard deviation.

Acknowledgments

This study was funded by the Bureau of Ocean Energy Management, U.S. Department of the Interior, Washington D.C., under Contract no. M10PC00097. Although the research described was funded by the Bureau of Ocean Energy Management, it does not necessarily reflect the views of the Agency. The authors wish to acknowledge the work and input of the members of the project advisory committee and individual topic area experts including Andrew Gill, David Beutel, Adrian Judd, Thomas Noji, Anna Malek, and Jeremy Collie.

References

[1] G. W. Boehlert and A. B. Gill, "Environmental and ecological effects of ocean renewable energy development," *Oceanography*, vol. 23, no. 2, pp. 68–81, 2010.

[2] R. Inger, M. J. Attrill, S. Bearhop et al., "Marine renewable energy: potential benefits to biodiversity? An urgent call for research," *Journal of Applied Ecology*, vol. 46, no. 6, pp. 1145–1153, 2009.

[3] A. B. Gill, "Offshore renewable energy: ecological implications of generating electricity in the coastal zone," *Journal of Applied Ecology*, vol. 42, no. 4, pp. 605–615, 2005.

[4] ICES (International Council for the Exploration of the Sea), "Report of the workshop on effects of offshore windfarms on marine benthos—facilitating a closer international collaboration throughout the North Atlantic region (WKEOMB)," Bremerhaven, Germany, March 2012.

[5] BOEM (U.S. Bureau of Ocean Energy Management), *BOEM Environmental Studies Program—Studies Development Plan FY, 2013-2015*, U.S. Department of the Interior, Washington, DC, USA, 2012.

[6] V. Stelzenmüller, P. Breen, T. Stamford et al., "Monitoring and evaluation of spatially managed areas: a generic framework for implementation of ecosystem based marine management and its application," *Marine Policy*, vol. 37, pp. 149–164, 2012.

[7] D. Wilhelmsson, T. Malm, R. Thompson et al., *Greening Blue Energy: Identifying and Managing the Biodiversity Risks and Opportunities of Offshore Renewable Energy*, IUCN, Gland, Switzerland, 2010.

[8] OSPAR, *Guidelines on Quality Assurance for Biological Monitoring in the OSPAR Area*, OSPAR Commission Agreement, Amsterdam, The Netherlands, 2002.

[9] J. S. Link, "Translating ecosystem indicators into decision criteria," *ICES Journal of Marine Science*, vol. 62, no. 3, pp. 569–576, 2005.

[10] J. F. Samhouri, P. S. Levin, C. A. James, J. Kershner, and G. Williams, "Using existing scientific capacity to set targets for ecosystem-based management: a puget sound case study," *Marine Policy*, vol. 35, no. 4, pp. 508–518, 2011.

[11] J. M. Carey and M. J. Keough, "The variability of estimates of variance, and its effect on power analysis in monitoring design," *Environmental Monitoring and Assessment*, vol. 74, no. 3, pp. 225–241, 2002.

[12] H. M. Leslie and K. L. McLeod, "Confronting the challenges of implementing marine ecosystem-based management," *Frontiers in Ecology and the Environment*, vol. 5, no. 10, pp. 540–548, 2007.

[13] I. Muxika, Á. Borja, and J. Bald, "Using historical data, expert judgement and multivariate analysis in assessing reference conditions and benthic ecological status, according to the European Water Framework Directive," *Marine Pollution Bulletin*, vol. 55, no. 1–6, pp. 16–29, 2007.

[14] H. L. Rees, J. L. Hyland, K. Hylland, C. S. L. M. Clarke, J. C. Roff, and S. Ware, "Environmental indicators: utility in meeting regulatory needs. An overview," *ICES Journal of Marine Science*, vol. 65, no. 8, pp. 1381–1386, 2008.

[15] B. P. Lapeña, K. M. Wijnberg, S. J. M. H. Hulscher, and A. Stein, "Environmental impact assessment of offshore wind farms: a simulation-based approach," *Journal of Applied Ecology*, vol. 47, no. 5, pp. 1110–1118, 2010.

[16] B. P. Lapeña, K. M. Wijnberg, S. J. M. H. Hulscher, and A. Stein, "Simulation-based impact assessment of offshore wind farms on seabirds," *Journal of Sustainability Science and Management*, vol. 6, no. 1, pp. 107–117, 2011.

[17] B. P. Lapeña, K. M. Wijnberg, A. Stein, and S. J. M. H. Hulscher, "Spatial factors affecting statistical power in testing marine fauna displacement," *Ecological Applications*, vol. 21, pp. 2756–2769, 2011.

[18] MMS (Minerals Management Service), *Alternative Energy Programmatic EIS: Potential Impacts of Alternative Energy Development on the OCS and Analysis of Potential Mitigation Measures*, U.S. Department of the Interior, Washington, DC, USA, 2007.

[19] J. Rice, C. Arvanitidis, A. Borja et al., "Indicators for sea-floor integrity under the European Marine Strategy Framework Directive," *Ecological Indicators*, vol. 12, pp. 174–184, 2011.

9

Numerical Investigation of Heat Transfer and Friction Factor Characteristics in a Circular Tube Fitted with V-Cut Twisted Tape Inserts

Sami D. Salman,[1,2] Abdul Amir H. Kadhum,[1] Mohd S. Takriff,[1] and Abu Bakar Mohamad[1]

[1] *Department of Chemical and Process Engineering, Faculty of Engineering and Built Environment, Universiti Kebangsaan Malaysia, 43600 Bangi, Selangor, Malaysia*
[2] *Biochemical Engineering Department, Al-Khwarizmi College of Engineering, University of Baghdad, Baghdad 47024, Iraq*

Correspondence should be addressed to Sami D. Salman; sami.albayati@gmail.com

Academic Editors: N. H. Afgan and A. Greco

Numerical investigation of the heat transfer and friction factor characteristics of a circular fitted with V-cut twisted tape (VCT) insert with twist ratio (y = 2.93) and different cut depths (w = 0.5, 1, and 1.5 cm) were studied for laminar flow using CFD package (FLUENT-6.3.26). The data obtained from plain tube were verified with the literature correlation to ensure the validation of simulation results. Classical twisted tape (CTT) with different twist ratios (y = 2.93, 3.91, 4.89) were also studied for comparison. The results show that the enhancement of heat transfer rate induced by the classical and V-cut twisted tape inserts increases with the Reynolds number and decreases with twist ratio. The results also revealed that the V-cut twisted tape with twist ratio y = 2.93 and cut depth w = 0.5 cm offered higher heat transfer rate with significant increases in friction factor than other tapes. In addition the results of V-cut twist tape compared with experimental and simulated data of right-left helical tape inserts (RLT), it is found that the V-cut twist tape offered better thermal contact between the surface and the fluid which ultimately leads to a high heat transfer coefficient. Consequently, 107% of maximum heat transfer was obtained by using this configuration.

1. Introduction

Heat exchangers with the convective heat transfer are widely used in many engineering applications. Enhancement of heat transfer in all types of thermotechnical apparatus is of great significance for the industry. Besides the savings of primary energy, it also leads to a reduction in size and weight. In general, heat transfer enhancement techniques can be divided into two categories: (1) active techniques which need external power source and (2) passive techniques which do not need external power source. Several experimental studies on heat augmentation techniques using twisted tape as passive technique have been reported in the literature [1–15]. Thereafter and due to advances in computer hardware and software and consequent increase in calculation speed, the CFD modeling technique was developed as a powerful and effective tool for better understanding of the complex hydrodynamics in many industrial processes. Kumar et al. [16] studied the hydrodynamics and heat transfer characteristics of tube in the form of the helical heat exchanger at a pilot plant scale fitted with semicircular plate in annuals area using a commercial CFD package to predict the flow and thermal properties in a tube of the helical heat exchanger. They found that the results of the simulated data of the Nusselt number and friction factor in the inner and outer tubes coincide with experimental data. Sivashanmugam et al. [17] presented the modeling of heat transfer augmentation in a circular tube fitted with a helical twist insert in a laminar and turbulent flow using CFD. Jayakumar et al. [18] offered a comparison study of CFD simulations to experiment on convective heat transfer in double pipe helical heat exchanger and developed an empirical correlation for estimation of the inner heat transfer coefficient of a helical coil. Lei et al. [19] conducted the hydrodynamics and heat transfer study on heat exchanger with a single-segment baffle, single-layer helical baffle, and two-layer helical baffle experimentally as well as numerically using the CFD method. They found that the configuration of helical baffles gives a higher heat transfer coefficient than a single-layer baffle at the same pressure drop and the configuration of a two-layer helical baffle has better heat performance than that of the single-layer baffle. Kharat et al. [20] developed a new correlation of heat transfer coefficient between concentric helical coils of the heat exchanger which depended on experimental work and CFD simulation. FLUENT 6.3.26 has been used to improve the heat transfer coefficient correlation for the flue gas side to optimize the gap between concentric coils. Nagarajan and Sivashanmugam [21] presented a simulation of heat transfer augmentation and friction factor characteristics of a circular tube fitted with a right-left helical twist insert with a 100 mm spacer using CFD. The results of a simulated Nusselt number, friction factor for a given flow rate, and twist ratio were compared with the experimental data and showed a good agreement. Shabanian et al. [22] conducted an experiment and CFD modeling studies on heat transfer, friction factor, and thermal performance of an air-cooled heat exchanger equipped with three types of tube insert including butterfly, classic, and jagged twisted tapes. They found that the predicted results in terms of turbulence intensity have a good agreement with measured values of Nu number and friction factor. Wang et al. [23] explicated the optimum configuration of regularly spaced short-length twisted tape in a circular tube by using computational fluid dynamics (CFD) modeling. The air was used as the test fluid with a turbulent flow with the configuration parameters including the free space ratio (s), twist ratio (y), and rotated angle (a). The predicted results were in a good agreement with experimental data. The results showed that the larger rotated angle yields a higher heat transfer value and greater flow resistance, whereas the smaller twist ratio resulted in better heat transfer performance except for a larger rotated angle at a high Reynolds number. Pathipakka and Sivashanmugam [12] proposed CFD simulation of the heat transfer characteristics of Al_2O_3 nanofluid in a circular tube fitted with helical twist inserts under constant heat flux using FLUENT version 6.3.26 in a laminar flow. The Al_2O_3 nanoparticles in water at different concentrations (0.5%, 1.0%, and 1.5%) and helical twist inserts with different twist ratios (y = 2.93, 3.91, and 4.89) were used for the simulation. The data obtained by simulation were compared with the literature value of water for plain tube helical tape inserts. Salman et al. [24] report an application of a mathematical model of the heat transfer enhancement and friction factor characteristics of water in constant heat-fluxed tube fitted with elliptical cut twisted tape inserts with twist ratios (y = 2.93, 3.91, and 4.89) and different cut depths (w = 0.5, 1, and 1.5 cm) under laminar flow using FLUENT version 6.3.26. The results elaborated that the enhancement of heat transfer rate and the friction factor induced by elliptical cut twisted tape inserts increases with the Reynolds number and decreases with twist ratio. In addition the results show that the elliptical cut twisted tape with twist ratio y = 2.93 and cut depth w = 0.5 cm offered higher heat transfer rate with significant increases in friction factor. In the present work a numerical investigation of heat transfer enhancement in a tube induced by V-cut twist tape inserts is reported using CFD simulation based on experimental work mentioned in [12]. The result obtained by this configuration enhanced the heat transfer rate than those obtained by classical and helical twist inserts. This study can be used as guideline for experimental work of heat transfer augmentation.

FIGURE 1: V-cut twisted tape insert.

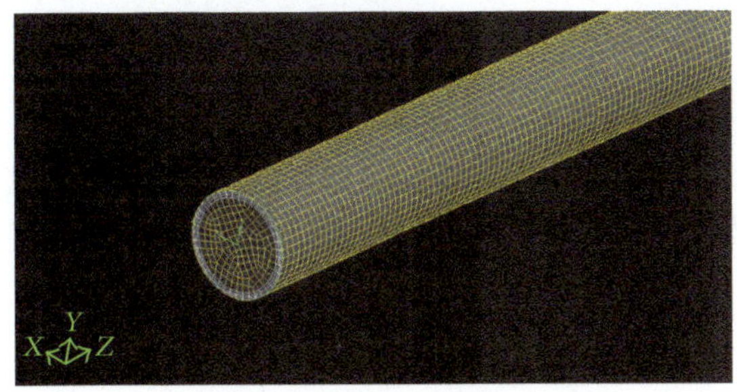

FIGURE 2: Grid for the plain tube.

2. Technical Details

The geometrical configuration of V-cut twisted tape (VCT) inserts with a thickness (t) 0.08 cm, width (W) of 2.45 cm, length (L) of 180 cm with relative twisted ratios (y = 2.93, 3.91, and 4.89), and cut depth (w = 0.5, 1 and 1.5 cm) is used for simulation. This configuration is illustrated in Figure 1.

3. Physical Model

The GAMBIT program is used for generation of the geometry and grids whereas FLUENT version 6.3.26 is used for module preprocessing. The geometry and the grid of the plain tube and V-cut twisted tape inserts are as shown in Figures 2 and 3, and these configurations were created in GAMBIT and imported into FLUENT for simulation. The geometry of classical and V-cut twisted tape inserts was made by winding a uniform strip of 25.54 mm width using the twist option in the sweeping of faces. The twist angle of 360° with a length of 75, 100, and 125 mm for various twist ratios (y = 2.93, 3.91, and 4) was generated by using a perpendicular type of sweeping for the entire length of 1800 mm. The volume required for simulation was created by the subtraction of the twisted tape insert from the plain tube geometry. The edged meshes were applied to each edge by using the particular interval count, whereas the front circular face was mashed by using tetrahedral and pave type meshing. The mesh face was swept over the entire volume using tetrahedral/hybrid elements and a T grid type. Boundary conditions for the mesh volume were inlet, outlet, wall, and type of fluid defined. Subsequently, the mesh file was exported to FLUENT for simulation. The following equations used to calculate

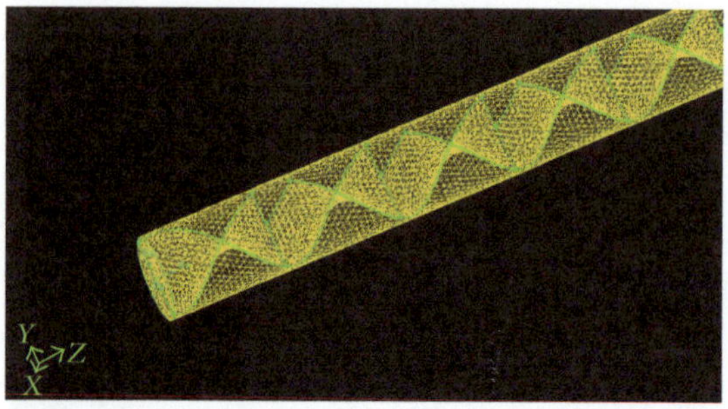

FIGURE 3: Grid for the plain tube fitted with V-cut twisted (VCT) tape insert.

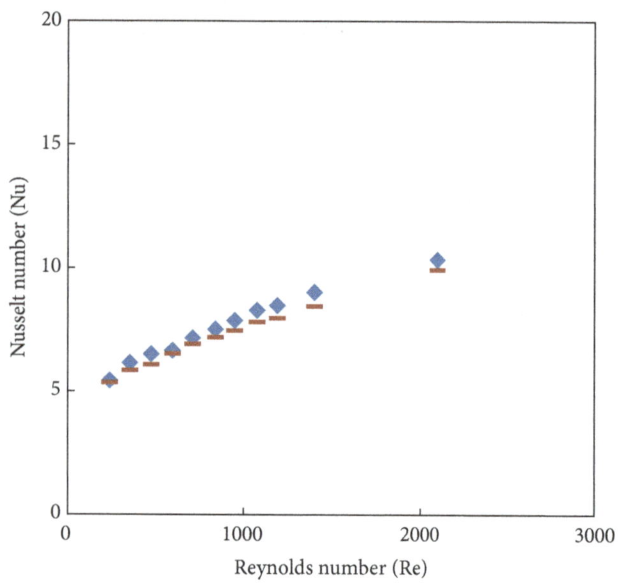

FIGURE 4: Plain tube simulated Nusselt number versus the literature data.

the Nsselt number (Nu) and the skin friction factor (f) (see Figure 4).

$$\begin{aligned} \mathrm{Nu} &= \frac{hD}{K}, \\ h &= \frac{q^{\bullet}}{T_w - T_b}, \\ f &= \frac{16}{\mathrm{Re}}, \\ \mathrm{Re} &= \frac{\rho u D}{\mu}, \end{aligned} \quad (1)$$

where D is the tube diameter, h is the heat transfer coefficient, k is the conductivity of water, $q^{\bullet}$ is the heat flux on the tube, T_w is the tube wall temperature, and T_b is the bulk temperature of water $T_b = (T_o + T_i)/2$. ρ is the density, μ is dynamic viscosity, and u is the water velocity.

TABLE 1: Thermo-physical properties of materials.

Materials	Density (Kg/m^3)	Specific heat (J/kg·K)	Thermal conductivity (W/m·K)	Viscosity (Pa·s)
Water	998.2	4182	0.6	0.001003
Steel	8030	502.48	16.27	—
Aluminium	2719	871	202.4	—

TABLE 2: Numerical values of the parameters used for simulations [12].

Mass flow rate (Kg/s)	Heat flux (W/m^2)
0.003	240.03
0.005	340.77
0.006	459.15
0.008	563.38
0.01	479.85
0.0116	1001.57
0.0133	1363.26
0.015	1512.68
0.0166	1893.56
0.02	2130.05
0.03	2445.25

4. Numerical Simulations

The problem investigated is a three-dimensional steady state laminar flow through a plain tube fitted with classical and V-cut twisted tape inserts under constant heat-fluxed tube using the following governing equations.

4.1. Continuity Equation for an Incompressible Fluid

$$\frac{\partial p}{\partial t} + \nabla \cdot (\rho \vec{v}) = S_m. \quad (2)$$

4.2. Conservation of Momentum

$$\frac{\partial v}{\partial t} + \rho (\vec{v} \cdot \nabla) \vec{v} = -\nabla p + \rho \overline{g} + \nabla \cdot \tau_{ij} + \vec{F}. \quad (3)$$

4.3. Conservation of Energy

$$\begin{aligned} &\rho \frac{\partial}{\partial t} (\rho E) + \nabla \cdot \{\vec{v} (\rho E + \rho)\} \\ &= \nabla \cdot \left\{K_{\mathrm{eff}} \nabla T - \sum h_i (\vec{\tau}_{\mathrm{eff}} \cdot \vec{v})\right\} + S_h. \end{aligned} \quad (4)$$

5. Modeling Parameters

The thermophysical properties of the fluid, materials, and the numerical values of the mass flow rate and heat flux which were used in a number of the simulations are given in Tables 1 and 2.

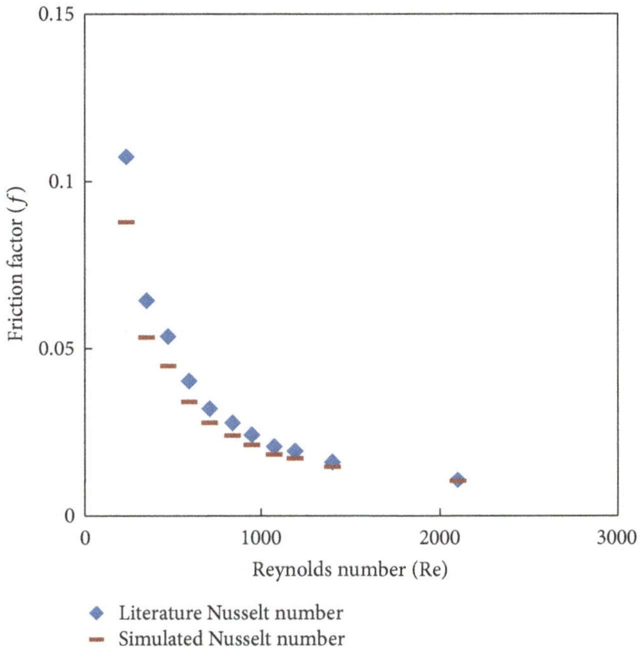

FIGURE 5: Plain tube simulated friction factor versus the literature data.

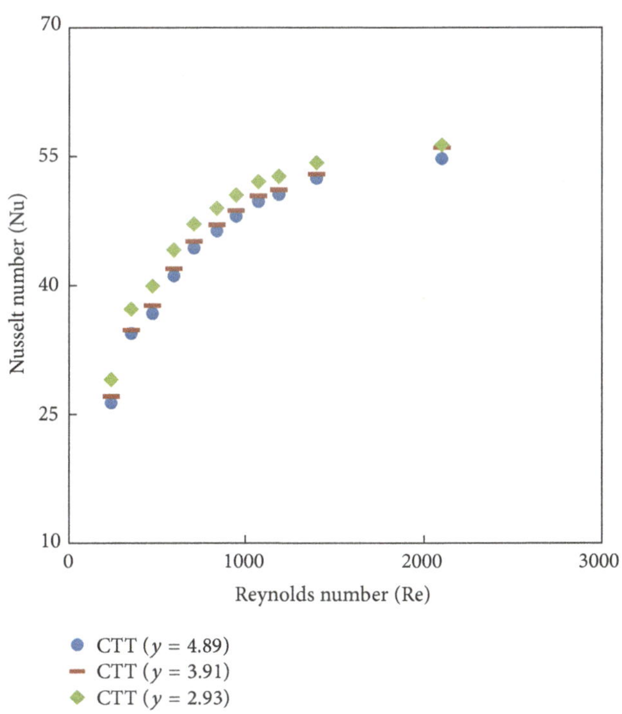

FIGURE 6: Simulated Nusselt friction factor of twisted tape with different twist ratios (y).

6. Numerical Method

The commercial package of CFD (FLUENT 6.3.26) was chosen as the CFD tool for this study to solve the above-mentioned governing equations accompanied with boundary conditions. Solution sequential algorithm (segregated

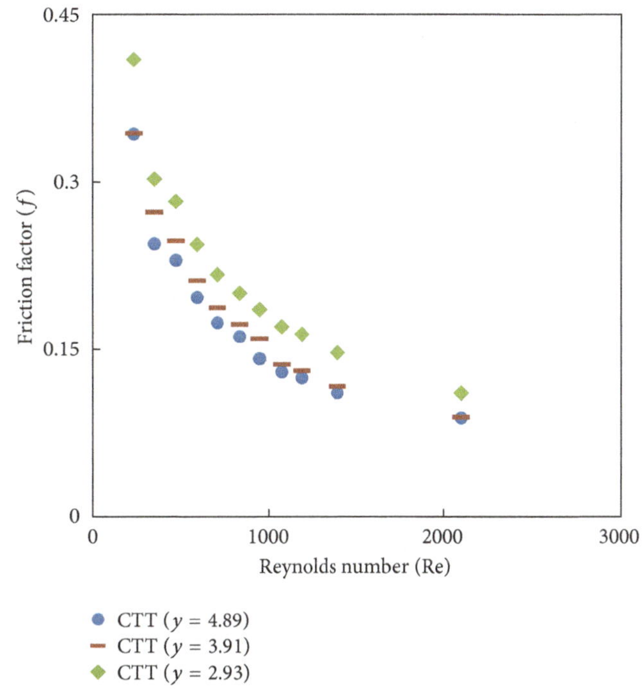

FIGURE 7: Simulated friction factor of twisted tape with different twist ratios (y).

solver algorithm) with settings including implicit formulation, steady (time-independent) calculation, viscous laminar model and energy equation, SIMPLE as the pressure-velocity coupling method, and first-order upwind scheme for energy and momentum equations was selected for simulation.

7. Results and Discussion

7.1. Grid Testing and Code Validation. Testing of the grid independence was carried out to evaluate the effects of grid sizes on the accuracy of the simulated results. In this study, five mesh volumes were considered, 200132, 233228, 292655, 373825, and 390191 at laminar flow Re = 2000. It is observed that the accuracy of the result increases with increase in mesh number (nodes) and the accuracy became constant around 292655 cells for plain tube and tube fitted with classical twisted tape inserts, and hence final CFD simulations were done at this cell. The results of the Nusselt number and friction factor for plain tube simulation are validated with Sieder and Tate [25] correlations as shown in Figures 5 and 6. Apparently, present results reasonably agree well with these correlations with maximum discrepancy of ±8% in the Nusselt number and ±10% for friction factor.

7.2. Effect of Twist Ratio on Heat Transfer and Friction Factor. The simulated data of the Nusselt number and friction factor and their variation with a Reynolds number of twisted tape inserts with twist ratios y = 2.93, 3.91, and 4.89 are shown in Figures 6 and 7. Figure 6 indicates that the Nusselt number increases with increasing Reynolds number and

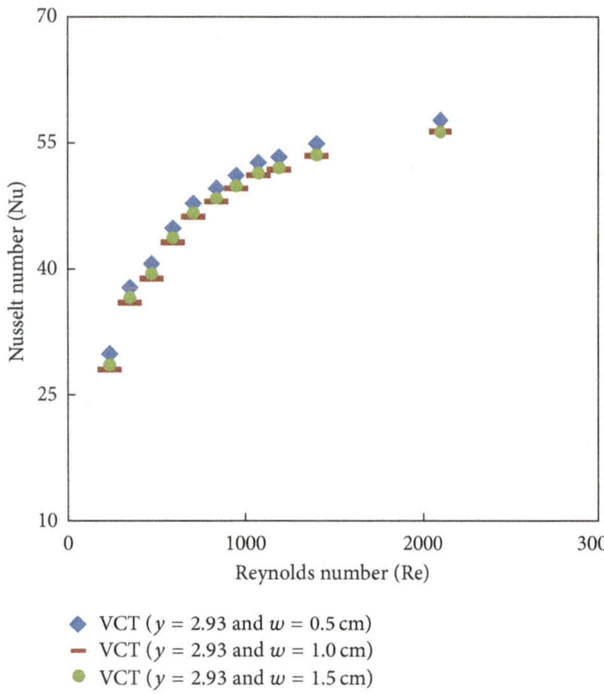

FIGURE 8: Simulated Nusselt number of V-cut twisted tape with different twist ratios (y) and different cut depths.

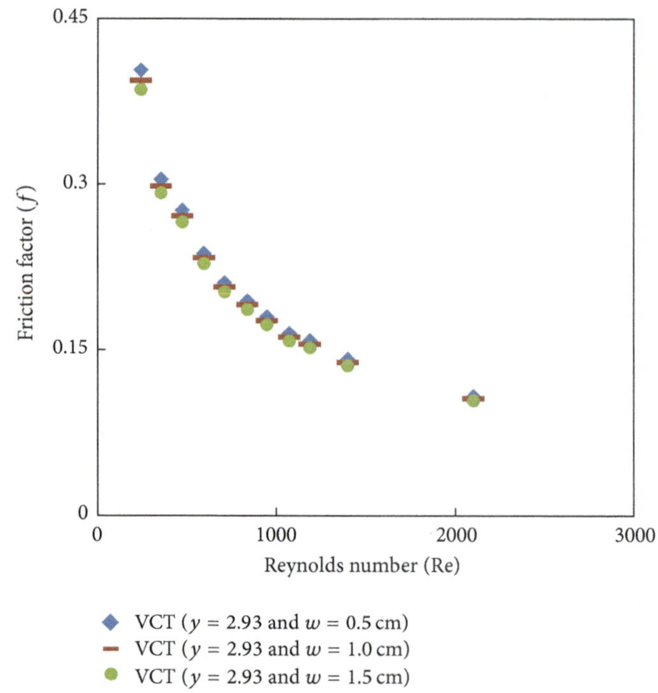

FIGURE 9: Simulated friction factor of V-cut twisted tape with different twist ratios (y) and different cut depths.

the heat transfer rate is higher for the twist tape set than the plain tube, because of strong swirl flow in the presence of the twist tape. It is found that the heat transfer rate with the twist ratio $y = 2.93$ is higher than those from other ratios ($y = 3.91$ and 4.89), and this means that the turbulent intensity obtained from the lower twist ratio is higher than those of higher ratios (y). Figure 7 shows the variation of friction factor with a Reynolds number for different twist ratios ($y = 2.93$, 3.91, and 4.89). The friction factor obtained from the tube with twisted tape insert is significantly higher than that from the plain tube. Moreover, the use of smaller twist ratio leads to higher tangential contact between the swirling flow and the tube surface. Therefore, the twisted tape with twist ratio ($y = 2.93$) has a maximum friction factor.

7.3. Effect of Cut Depth on Heat Transfer and Friction Factor. Simulated data for the Nusselt number and friction factor and their variation with Reynolds number for V-cut twisted tape inserts with twist ratio $y = 2.93$ and cut depths ($w = 0.5$, 1, and 1.5 cm) is shown in Figures 8, 9, 10, and 11. For a given Reynolds number, Nusselt number and friction factor are increased with decreasing cut depth, and this is mainly due to the combined effects of common swirling flow by the twisted tape and turbulence generated by the alternative cuts along the edge of the twisted tape which leads to the destruction of the thermal boundary layer and creating better flow mixing between the fluids at the core and heating wall surface. Furthermore the simulated results of the Nusselt number and friction factor of V-cut twisted tapes with twist ratio $y = 2.93$ and cut depth $w = 0.5$ cm are compared

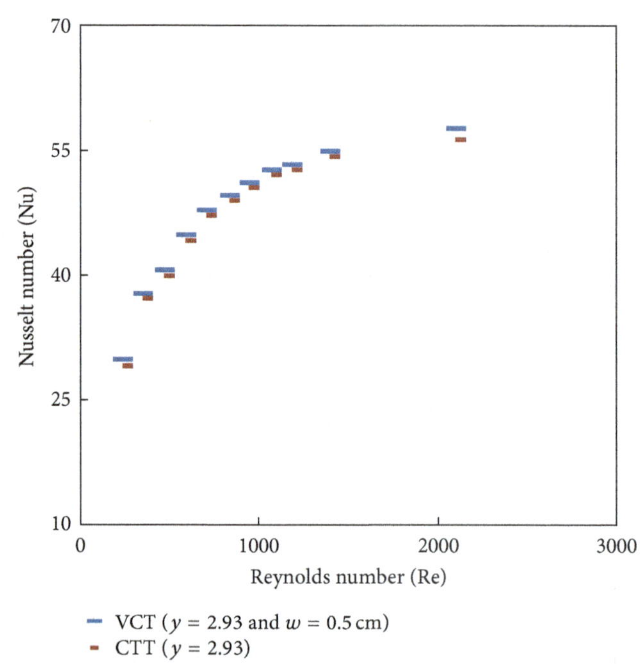

FIGURE 10: Simulated Nusselt number of classical and V-cut twisted tape inserts ($y = 2.93$) with cut depth $w = 0.5$ cm.

with experimental simulated data on helical twist tape [21] as shown in Figures 12 and 13. It is found that the VCT offered an additional heat transfer enhancement with less friction factor.

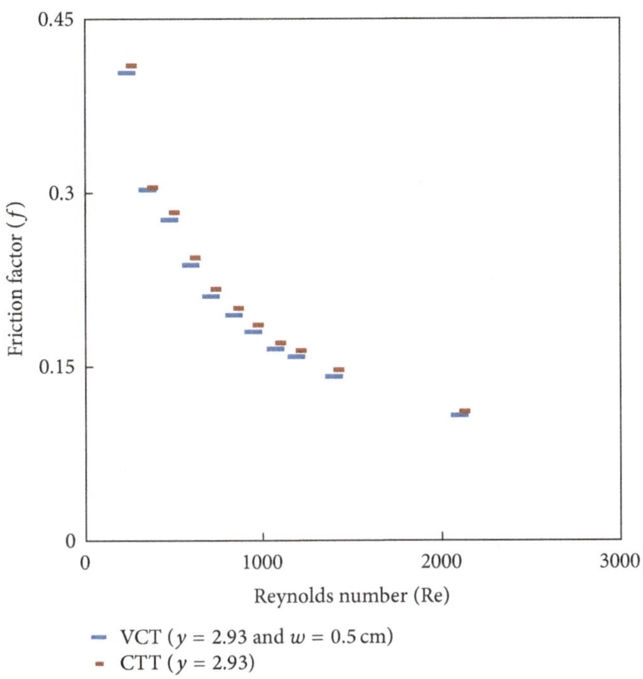

FIGURE 11: Simulated friction factor of classical and V-cut twisted tape inserts (y = 2.93) with cut depth w = 0.5 cm.

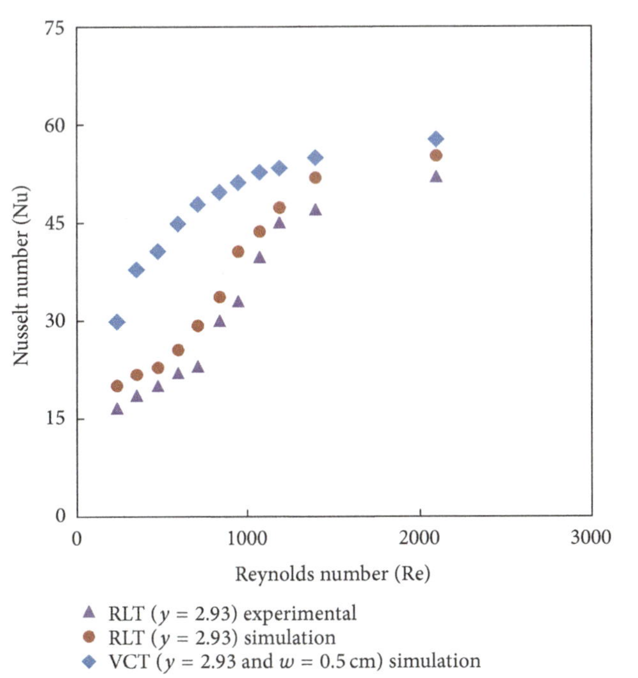

FIGURE 12: Simulated Nusselt number of V-cut twisted tape and RLT inserts.

8. Conclusion

CFD simulation for the heat transfer augmentation transfer and friction factor in a circular tube fitted with classical and V-cut twisted tape inserts with y = 2.93, 3.91, and 4.89 and cut depth w = 0.5, 1, and 1.5 cm in laminar flow conditions

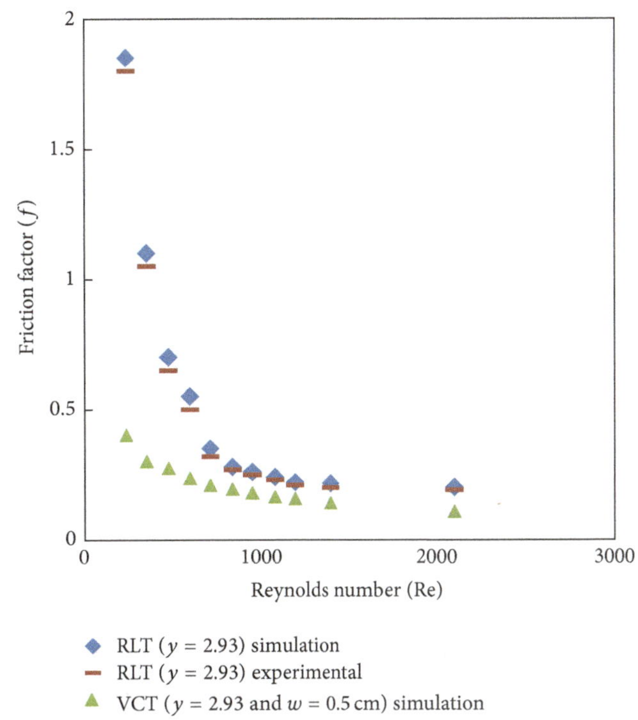

FIGURE 13: Simulated friction factor of V-cut twisted tape and RLT inserts.

has been reported using FLUENT version 6.3.26. The results are concluded as follows.

(i) The values of the Nusselt number and friction factor for the tube with V-cut twisted tape are noticeably higher than the values for the plain tube and tube equipped with classical twisted tapes as well.

(ii) Over the range of the Reynolds numbers considered, the tube equipped with the V-cut twisted tape y = 2.93 and w = 0.5 cm offered supreme heat transfer enhancement than the classical twisted tapes and other V-cut twisted tapes used.

(iii) The simulated data for V-cut twisted tape y = 2.93 and w = 0.5 cm is compared with theoretical and experimental work of Right-left helical tape inserts (RLT) with twist ratio y = 2.93 at the same conditions. The results show that the proposed model of V-cut twisted tape offered 107% heat transfer enhancement with less friction factor. This configuration can be used for heat transfer augmentation.

Nomenclature

E: Energy component in energy equation
F: Force component in momentum equation, N
f: Fanning friction factor
g: Acceleration due to gravity, m/s^2
k_{eff}: Thermal conductivity in energy equation, W/m K

m: mass flow rate of fluid, kg/s
Re: Reynolds number based on internal diameter of the tube, dimensionless
Nu: Nusselt number, dimensionless
p: Pressure component in momentum equation, N/m^2
S_m: Accumulation of mass, Kg
S_h: Accumulation of energy, J
T: Temperature, °C.
v: Velocity component in momentum equation, m/s
y: Twist ratio (length of one twist (360°)/diameter of the twist), dimensionless.

Greek Symbols

ρ: Density component in governing equations
$\vec{\tau}_{\text{eff}}$: Stress component in momentum equation, N/m^2.

Acknowledgments

The authors wish to especially acknowledge the financial assistance (FRGS/1/2013/TK07/UKM/01/1) offered by the University Kebangsaan Malaysia and MOSTI for carrying of this investigation

References

[1] P. K. Sarma, P. S. Kishore, V. D. Rao, and T. Subrahmanyam, "A combined approach to predict friction coefficients and convective heat transfer characteristics in A tube with twisted tape inserts for a wide range of Re and Pr," *International Journal of Thermal Sciences*, vol. 44, no. 4, pp. 393–398, 2005.

[2] S. Eiamsa-ard, C. Thianpong, and P. Promvonge, "Experimental investigation of heat transfer and flow friction in a circular tube fitted with regularly spaced twisted tape elements," *International Communications in Heat and Mass Transfer*, vol. 33, no. 10, pp. 1225–1233, 2006.

[3] S. Eiamsa-ard, "Study on thermal and fluid flow characteristics in turbulent channel flows with multiple twisted tape vortex generators," *International Communications in Heat and Mass Transfer*, vol. 37, no. 6, pp. 644–651, 2010.

[4] S. Eiamsa-ard and P. Promvonge, "Heat transfer characteristics in a tube fitted with helical screw-tape with/without core-rod inserts," *International Communications in Heat and Mass Transfer*, vol. 34, no. 2, pp. 176–185, 2007.

[5] L. S. Sundar, K. V. Sharma, and S. Ramanathan, "Experimental investigation of heat transfer enhancements with Al_2O_2 nanofluid and twisted tape insert in a circular tube," *International Journal of Nanotechnology and Applications*, vol. 1, pp. 21–28, 2007.

[6] P. Sivashanmugam and S. Suresh, "Experimental studies on heat transfer and friction factor characteristics of turbulent flow through a circular tube fitted with regularly spaced helical screw-tape inserts," *Applied Thermal Engineering*, vol. 27, no. 8-9, pp. 1311–1319, 2007.

[7] S. Eiamsa-ard, S. Pethkool, C. Thianpong, and P. Promvonge, "Turbulent flow heat transfer and pressure loss in a double pipe heat exchanger with louvered strip inserts," *International Communications in Heat and Mass Transfer*, vol. 35, no. 2, pp. 120–129, 2008.

[8] A. Kumar and B. N. Prasad, "Enhancement in solar water heater performance using twisted tape inserts," *National Journal of the Institution of Engineers*, vol. 90, pp. 6–9, 2009.

[9] M. Rahimi, S. R. Shabanian, and A. A. Alsairafi, "Experimental and CFD studies on heat transfer and friction factor characteristics of a tube equipped with modified twisted tape inserts," *Chemical Engineering and Processing*, vol. 48, no. 3, pp. 762–770, 2009.

[10] S. Jaisankar, T. K. Radhakrishnan, and K. N. Sheeba, "Experimental studies on heat transfer and friction factor characteristics of thermosyphon solar water heater system fitted with spacer at the trailing edge of twisted tapes," *Applied Thermal Engineering*, vol. 29, no. 5-6, pp. 1224–1231, 2009.

[11] S. Eiamsa-ard, K. Wongcharee, P. Eiamsa-ard, and C. Thianpong, "Thermohydraulic investigation of turbulent flow through a round tube equipped with twisted tapes consisting of centre wings and alternate-axes," *Experimental Thermal and Fluid Science*, vol. 34, no. 8, pp. 1151–1161, 2010.

[12] G. Pathipakka and P. Sivashanmugam, "Heat transfer behaviour of nanofluids in a uniformly heated circular tube fitted with helical inserts in laminar flow," *Superlattices and Microstructures*, vol. 47, no. 2, pp. 349–360, 2010.

[13] J. Guo, A. Fan, X. Zhang, and W. Liu, "A numerical study on heat transfer and friction factor characteristics of laminar flow in a circular tube fitted with center-cleared twisted tape," *International Journal of Thermal Sciences*, vol. 50, no. 7, pp. 1263–1270, 2011.

[14] E. Z. Ibrahim, "Augmentation of laminar flow and heat transfer in flat tubes by means of helical screw-tape inserts," *Energy Conversion and Management*, vol. 52, no. 1, pp. 250–257, 2011.

[15] A. V. N. Kapatkar, A. S. Padalkar, and C. Sanjay Kasbe, "Experimental investigation on heat transfer enhancement in laminar flow in circular tube equipped with different inserts," *AMAE International Journal on Manufacturing and Material Science*, vol. 1, no. 1, 2011.

[16] V. Kumar, S. Saini, M. Sharma, and K. D. P. Nigam, "Pressure drop and heat transfer study in tube-in-tube helical heat exchanger," *Chemical Engineering Science*, vol. 61, no. 13, pp. 4403–4416, 2006.

[17] P. Sivashanmugam, P. K. Nagarajan, and S. Suresh, "Experimental studies on heat transfer and friction factor characteristics of turbulent flow through a circular tube fitted with right and left helical screw-tape insert," *Chemical Engineering Communications*, vol. 195, no. 8, pp. 977–987, 2008.

[18] J. S. Jayakumar, S. M. Mahajani, J. C. Mandal, P. K. Vijayan, and R. Bhoi, "Experimental and CFD estimation of heat transfer in helically coiled heat exchangers," *Chemical Engineering Research and Design*, vol. 86, no. 3, pp. 221–232, 2008.

[19] Y.-G. Lei, Y.-L. He, P. Chu, and R. Li, "Design and optimization of heat exchangers with helical baffles," *Chemical Engineering Science*, vol. 63, no. 17, pp. 4386–4395, 2008.

[20] R. Kharat, N. Bhardwaj, and R. S. Jha, "Development of heat transfer coefficient correlation for concentric helical coil heat exchanger," *International Journal of Thermal Sciences*, vol. 48, no. 12, pp. 2300–2308, 2009.

[21] P. K. Nagarajan and P. Sivashanmugam, "CFD Simulation of heat transfer augmentation in circular tube filled with right-left helical inserts with spacer," *International Journal of Chemical Engineering Research*, vol. 6, no. 1, pp. 1–11, 2009.

[22] S. R. Shabanian, M. Rahimi, M. Shahhosseini, and A. A. Alsairafi, "CFD and experimental studies on heat transfer enhancement in an air cooler equipped with different tube inserts,"

International Communications in Heat and Mass Transfer, vol. 38, no. 3, pp. 383–390, 2011.

[23] Y. Wang, M. Hou, X. Deng et al., "Configuration optimization of regularly spaced short-length twisted tape in a circular tube to enhance turbulent heat transfer using CFD modeling," *Applied Thermal Engineering*, vol. 31, no. 6-7, pp. 1141–1149, 2011.

[24] S. D. Salman, A. A. H. Kadhum, M. S. Takriff, and A. B. Mohamad, "CFD simulation of heat transfer and friction factor augmentation in a circular tube fitted with elliptic-cut twisted tape inserts," *Mathematical Problems in Engineering*, vol. 2013, Article ID 163839, 7 pages, 2013.

[25] E. N. Sieder and G. E. Tate, "Heat transfer and pressure drop of liquids in tubes," *Industrial Engineering and Chemistry Research*, vol. 28, pp. 1429–1435, 1936.

10

Determinants of Electricity Consumption Intensity in China: Analysis of Cities at Subprovince and Prefecture Levels in 2009

X. H. Xia[1] and Yi Hu[2]

[1] *Institute of China's Economic Reform and Development, Renmin University of China, Beijing 100872, China*
[2] *Academy of Mathematics and Systems Science, Chinese Academy of Sciences, Beijing 100190, China*

Correspondence should be addressed to X. H. Xia, xia.email@yahoo.com.cn

Academic Editors: C. Bin and C. Zhan-Ming

China has experienced great social and economic vicissitudes that caused the vast complexity and uncertainty for electricity consumption. This paper attempts to identify the main determinants of the electricity consumption intensity by using the data from Chinese cities at subprovince and prefecture levels in 2009. The key category factors, including urban morphology, industrial structure, regulation context, urbanization degree, price, natural condition, and resource endowment, are abstracted and the influence of these determinants is evaluated by adopting the finite mixture models. The variation of each determinant across regions, the comparative weights of all the factors, and the detailed classifications of the cities are reported for facilitating the understanding of electricity consumption in China. The corresponding policies for electricity administration are addressed as well.

1. Introduction

Industrialization and urbanization are two distinguishing features in the present social and economic development of China. The rural-to-urban migration and changes in industrial and economic structure are regarded as two major factors which determine the electricity consumption and economic output. The cities at subprovince and prefecture levels especially districts under the jurisdiction of cities which are, in general, cores of the region, play the leading roles and have the dominant positions in the processes of social development.

In the hierarchy of the administrative divisions of China mainland, cities at subprovince and prefecture levels (hereinafter briefly called the cities) cover 47.859% of China's total land area, 87.260% of total population, and 93.903% of Gross Regional Product (GRP), while 52.847% of GRP, 49.843% of total investment in fixed assets, and 51.695% of local government revenue are concentrated in the districts under the jurisdiction of the cities with only 12.476% of population and 27.654% of land area of the cities in 2009 [1, 2]. Meanwhile, the electricity consumption of the districts of the cities amounts to 1639.546 billion kwh which is about 44.273% of the total electricity consumption in China [2].

Prefecture level divisions (including subprovincial level, similarly hereinafter) are the second level of the administrative structure in the Constitution of China. The government at this level undertakes multiple responsibilities in terms of the administration of energy, economy, and environment issues.

The progress of urbanization and industrialization is always highly promoted by local governments, as it could bring about economic growth and better official performances which are two of the most important measures of GRP. Under the system of intergovernmental fiscal decentralization, about 52.400% of fiscal revenues are extracted by central government in 2009 [1]. The fiscal revenue of local government is another incentive to contest for output increase between different administrative divisions such as counties. As a general rule, the government at higher level is entitled to balance the development of the jurisdictional area. Hence the government at prefecture level is not only the implementer and promoter of urbanization and industrialization, but also the coordinator and participator of regional competition.

Table 1: Selected categorical indicators determined electricity consumption in the literature.

Reference	Regions	Contents	Period	Related variables	Categorical indicator
Larivière and Lafrance [23]	Canadian cities	Electricity consumption	1991	Urban density	Urban morphology
Halicioglu [29]	Turkey	Residential electricity consumption	1968–2005	Annual degree-days below 18°C Real residential electricity price Urbanization rate	Natural condition Price Urbanization
Bessec and Fouquau [34]	15 European countries	Electricity consumption	1985–2000	Temperature	Natural condition
Lai et al. [35]	Macao	Electricity consumption	2000–2006	Monthly mean temperature	Natural condition
Permana et al. [24]	Bandung city, Indonesia	Household consumption of electricity and liquefied petroleum gas	2007	Urban development forms	Urban morphology
Ziramba [32]	South Africa	Per capita residential electricity consumption	1978–2005	Real residential electricity price	Price
Liu [30]	China	Energy (including electricity) consumption	2009	Urbanization level	Urbanization
Aasen et al. [26]	Norway	Electricity consumption of enterprise	—	Electricity disclosure scheme	Regulation context
Bianco et al. [33]	Romania	Nonresidential electricity consumption	1975–2008	Electricity price	Price
Burke [36]	133 countries	Electricity ladder	1960–2003	Resource reserves	Resource endowment
Wiesmann et al. [31]	Portugal	Residential electricity consumption	2001	Heating /cooling degree-days Urbanization level	Natural condition Urbanization
Chen [27]	China	Energy (mainly electricity) conservation	2004–2011	Differential electricity pricing policy	Regulation context
Wang et al. [25]	China	Change in industrial electricity consumption	1998–2007	Change in industrial structure	Industrial structure

As a basic decision-making unit, the government at prefecture level is responsible for designing the layout of industry, especially that of small- and medium-sized enterprises (SMEs). For instance, the industrial park, being regarded as the incubator of local development, is often planned by government at this level. Meanwhile they also own the right to select industrial projects. In fact, the local government could reject the industrial settlement even programmed by the higher authorities for environmental or other reasons, or strive for it in order to earn more fiscal revenues when the projects are not settled. Thus the spatial arrangement of urban morphology and economy is mainly influenced and managed by authority at this level.

It is also a crucial node for the execution of energy policies including the regulation of electricity consumption in consideration of the more direct contact with local enterprises compared with higher authorities. The Chinese energy systems are intensively studied from the perspectives of international contexts [3, 4], provinces and regions [5, 6], macro systems [7–11], historical evolution [12, 13], and economic sectors [14, 15]. Further study focusing on the analysis of divisions at the prefectural level could provide detailed information on the enforcement of the energy legislation, the motivation for the behavior of local government, and the heterogeneity and exogenous variables determining the electricity consumption.

As an efficient method allowing introducing unobservable heterogeneity, which exists largely in Chinese regions due to the unbalance of development in terms of electricity consumption, the finite mixture model has been widely used for capturing the different effects among variables in regressions and segregating samples for the convenience of analysis. The application of finite mixture models includes Beard et al. [16] for the identification of the technical diffusion and changing among US banks, Wedel and Desarbo [17] for the clustering of market segment on the satisfaction study in Europe, Yau et al. [18] for the estimation of the effects of pertinent factors which influenced the hospital neonatal length of stay in Austria, and Imai and Tingley [19] for the measuring the performance of rival theories. Considering the possibility of inaccurateness caused by statistical mistakes of migration population and the consistency with China's statistical system, the electricity intensity defined as the ratio of electricity to GRP is used as the indicator of electricity consumption in different areas. This paper aims to abstract the key factors determining the electricity consumption intensity of the cities at prefecture level in China, evaluate the influence of the processes of urbanization and industrialization on the electricity consumption intensity, compare the difference of these determinants' impact, and understand the channels through which each variable functions for policy makers.

2. Literature Review

The determinants of electricity consumption are extensively studied in much of the contemporary literature. Overall, seven factors which attracted great attention considering the influences of economic, social, and geographic contexts can be classified, as shown in, Table 1.

Urban morphology refers to the characteristics of the spatial structure and the patterns of the regional economy. It is associated with many environmental topics as well, such as air pollution [20], land use [21], and urban conservation [22]. As for the electricity consumption, Larivière and Lafrance [23] demonstrated that raising the urban density which is defined as inhabitant per km^2 could decrease the annual city electricity consumption per inhabitant in Canada. Permana et al. [24] confirmed the differences of energy consumption per unit income in three kinds of urban forms in Bandung City which is selected as a typical sample of developing country cities.

Industrial structure refers to the number and size distribution of enterprises. It causes significant impacts on the distribution of electricity consumption. Wang et al. [25] decomposed the determinant factors influencing the change trend of industrial electricity consumption in China and characterized these elements. Besides, the economic output, as well as the social and environmental function of different sectors, should be taken into consideration from the systematic perspectives [10]. For instance, Xia et al. [14] introduced the employment variable to evaluate the Chinese industrial sectors.

Regulation context is related to the institution backgrounds of environmental legislation, administration structure, and other settings. Aasen et al. [26] analyzed the influences of the electricity disclosure scheme required by the EU Electricity Directive on Norwegian enterprises and found out that the regulation policy may be inefficient because of the malfunction of program implementation. Chen [27] described the reform process of electricity pricing system proposed by China's central authorities and presented three difficulties of the current promotion for the conservation of energy consumption. The administration powers from the higher authorities are the mainspring of promoting energy efficiency and conservation in China. The regulation of energy consumption relies heavily on mandatory plans or other compulsory measures from above due to lacking of market mechanisms [28].

Urbanization rate is measured by the ratio of urban population to the total population. Since the urbanized area does not only include megalopolises, metropolises, but also small cities and towns, urbanization is different from the concentration of population which gives more attention to the central point of the city. For instance, a person who is living in a small town which is far from the central city and undertaking the production of mechanical instruments should be regarded as the urbanization population but not as the people concentrated in the district of the city. Halicioglu [29] found the long-run causality relationship between residential electricity consumption and urbanization in Turkey. Liu [30] studied the linkage between urbanization and energy consumption based on econometric methods including ARDL testing and ECM methods and proposed that urbanization acceleration may harmonize the sustainable development in China. Wiesmann et al. [31] also use the urbanization as an explanation variable to model the residential electricity consumption.

Price is bound up with the electricity utilization in terms of the demand and supply of electricity. As the instrument of market adjustment, price scheme together with other factors such as urbanization rate [29] impacts the selection of energy type of the individual and enterprises and eventually determines the amount of electricity consumption. Ziramba [32] estimated the price and income elasticities of residential electricity consumption, showing the price insensitivity of electricity demand in South Africa. Bianco et al. [33] forecasted the nonresidential electricity consumption based on the price analysis using trigonometric grey model.

Natural condition is a set of factors which could affect the producing and living conditions and influences the demand function of electricity consumption as the independent variable. For example, the necessity of heating, ventilation, and air condition system in the building absolutely depends on the temperature and humidity. Bessec and Fouquau [34] found the nonlinear relationship between temperature and electricity consumption and distinguished different patterns of this link in European countries by virtue of panel threshold model. Lai et al. [35] modeled the electricity consumption in Macao using mathematic methods and identified the quadratic influence of temperature on electricity consumption. The significance of temperature impacts on electricity was also reported by Larivière and Lafrance [23] and Wiesmann et al. [31].

Resource endowment is associated with the reserves of mineral, water, and other resources in the region, exerting great influence on the selection and development of industry and indirectly determining the electricity consumption. Burke [36] found that the electricity ladder which is defined by the transition from traditional electricity to renewable power relied heavily on the energy endowments after the investigation of 133 countries.

3. Methodology and Data Descriptions

3.1. Finite Mixture Model and PCA. The finite mixture model would be employed to identify the complexity of the electricity consumption in China and scrutinize the influence of the key factors mentioned before. The integration of factor analysis and clustering with category indicators are provided for implementing the policies of electricity administration. Let y_i be the electricity consumption intensity (electricity consumption/GRP) observed for the city i, $i = 1, \dots, N$. The sample is assumed to be drawn from a finite mixture model with K components. The probability density function for observations is given by

$$f\left(y \mid X_p, X_c, \Pi, \Theta, \alpha\right) = \sum_{l=1}^{K} \Pi_l(X_c, \alpha) f_l\left(y \mid \Theta\left(X_p\right)\right). \quad (1)$$

Here, X_p denotes the predictor vector, X_c is the concomitant variable vector, and f $(l = 1, \dots, K)$ indicates the component density function with the parameter vector Θ. Π_l represents the component weight and is defined in the $(K-1)$-dimensional simplex as

$$\sum_{l=1}^{K} \Pi_l(X_c, \alpha) = 1, \quad 0 < \Pi_l(X_c, \alpha) < 1, \forall l, \quad (2)$$

where $\alpha = (\alpha_l')'_{l=1,\dots,K}$ is the parameter vector determining the probability of components. Suppose further that the model is the mixture of Gaussian linear regression, and (1) could be constructed as

$$\begin{aligned} &\sum_{l=1}^{K} \Pi_l(X_c, \alpha) N\left(y \mid \mu_l\left(X_p\right), \sigma_l^2\right) \\ &\quad = \sum_{l=1}^{K} \Pi_l(X_c, \alpha) N\left(y \mid X_p' \beta_l, \sigma_l^2\right). \end{aligned} \quad (3)$$

Here $\mu_l(\cdot)$ is the mean with linear structure $\mu_l(\cdot) = X_p'\beta_l$ and σ_l^2 is the variance for the lth component normal density $N(y \mid \mu_l(X_p), \sigma_l^2)$. The coefficients of mixture regression β_l could be *fixed* across all the components, *nested* with some coefficients of components equally, or *varied* between the K components. Maximum likelihood (ML) estimation could be used to fit the multivariate Gaussian linear mixture models through the Expectation-Maximization (EM) algorithm [37–40]. In computation practices the component weight Π_l could be assumed to follow a multinomial logit model [41] as

$$\Pi_l(X_c, \alpha) = \frac{\exp(X_c'\alpha_l)}{\sum_t \exp(X_c'\alpha_t)}, \quad \forall l \quad (4)$$

with the constraint $\alpha_1 = 0$. Information criteria including Akaike Information Criterion (AIC) [42], Bayes Information Criterion (BIC) [43] and Hannan-Quinn Criterion (HQ) [44], and classification criteria including Normalized Entropy Criterion (NEC) [45], Classification Likelihood Criterion (CLC) [46] and Integrated Completed Likelihood-BIC (ICL-BIC) [47] are available for choosing the optimal structure of finite mixture models.

Principal component analysis, a linear dimension reduction method, is applied to combine the explanatory variables for the finite mixture model. This treatment was inspired by Bai and Ng [48] and Ng and Bai [49] and the analogous technique was also used in Xia et al. [14].

3.2. Data Descriptions

3.2.1. Data Definitions. As urban sprawl is generally developed from a specific focal point [50], the district concentration is a main feature of urban morphology. The district under the jurisdiction of the city is regarded as the focal point of the city, especially in China this district is normally the political, economical, and cultural center of the city. The concentrations of GRP (X_{11}), population (X_{12}), industrial value added (X_{13}), fiscal revenue (X_{14}), and investment in fixed assets (X_{15}) are selected for measuring *urban morphology* as defined in Table 2.

Table 2: Description of explanatory variables and category factors.

Variable	Description
Urban morphology[a]	
GRP concentration	The ratio of Gross Regional Product (GRP) of the district to that of the city
Population concentration	The ratio of population of the district to that of the city
Industrial concentration	The ratio of industrial value added of the district to that of the city
Fiscal concentration	The ratio of government revenue by the district to that by the city
Investment concentration	The ratio of total investment in fixed assets of the district to that of the city
Industrial structure	
Share of industrial output	The ratio of industrial value added to GRP in the city
Share of industrial employment	The ratio of the number of employed person in industry to the total of employed person in the city
Share of industrial electricity consumption	The ratio of electricity consumption in local industry to the total of energy consumption in the city
Regulation context	
Decease rate of energy intensity	The decease rate of energy consumption per unit of GRP in local province in the corresponding period
Decease rate of energy intensity of industry	The decease rate of energy consumption per unit of industrial value added in local province in the corresponding period
Decease rate of electricity consumption intensity	The decease rate of electricity consumption per unit of GRP in local province in the corresponding period
Urbanization rate	
Urbanization degree of the province	The proportion of urban population in the local province
Share of nonagricultural population	The proportion of nonagricultural population in the city
Urbanization degree of the city	The proportion of urban population in the city
Electricity price	
Average retail price of electricity	The weighted average of industry, commerce, and residence prices
Natural condition	
Annual average temperature	The mean of annual temperature according to urban temperature record in long term
Resource endowment[b]	
Ensured reserves of iron ore	The amount of iron ore that can be utilized under current conditions in the local province

[a]District refers to the districts under the jurisdiction of the city at subprovince and prefecture levels.
[b]The effects of ensured reserves of coal which had been also taken into account are not significant. The resource endowment of the city may be disturbed by the coal transportation, while the iron ore is often utilized locally in China.

The shares of industrial output (X_{21}), industrial employment (X_{22}), and industrial electricity consumption (X_{23}) are evaluated by the ratio of industrial value added to GRP, the ratio of industrial employment to total labor population and the weight of industrial electricity consumption in the city respectively. These indicators embody the aspects of economic, social, and environmental impacts and could give a comprehensive description of *industrial structure* (X_2).

The energy intensity and electricity consumption intensity are the main indicators to be supervised by Chinese energy regulatory system and official statistical bureau. The former one becomes a constraint target in national development programs such as the 12th Five-Year Plan; the latter would be easily monitored in the view of government authorities and also treated as a relatively accurate indicator to reflect the actual conditions of energy consumption due to fewer incentives to manipulate the data on electricity consumption in comparison with those on energy consumption. The decease rates of energy intensity (X_{31}), energy intensity of industry (X_{32}), and electricity consumption intensity (X_{33}) in the province are taken into consideration comprehensively to evaluate the regulation pressure which is exerted by provincial governments on the city authorities for the reflection of *regulation context* (X_3).

The urbanization rate of the province (X_{41}) is used to capture the provincial heterogeneity in the development due to uneven economic development in China. Since the urbanization process is often closely associated with the transfer of rural surplus labor to urban sectors, the share of nonagricultural population (X_{42}) is chosen to measure industrial dimension of urbanization. The combination of the urbanization rate of the city (X_{43}) and the other two variables could overcome the possible statistical bias coming from the incorrect estimated flowing population and gives a comprehensive evaluation of the categorical factor X_4, that is, *urbanization* (X_4).

Average retail price of electricity which is the synthesis of all kinds of price including residential price, commercial

Table 3: Descriptive statistics and units of original data.

Variable	Unit	Mean	Standard deviation	Minimum	Maximum
Output concentration (X_{11})	%	45.533	23.585	8.038	100.000
Population concentration (X_{12})	%	33.224	22.877	4.368	100.000
Industrial concentration (X_{13})	%	48.899	25.140	5.447	100.000
Fiscal concentration (X_{14})	%	52.600	25.490	5.931	100.000
Investment concentration (X_{15})	%	45.047	23.766	7.300	100.000
Share of industrial output (X_{21})	%	50.378	11.814	9.740	82.390
Share of industrial employment (X_{22})	%	45.419	15.178	6.550	79.450
Share of industrial electricity consumption (X_{23})	%	65.964	18.926	7.243	98.839
Decease rate of energy intensity (X_{31})	%	4.908	2.295	0.830	9.730
Decease rate of energy intensity of industry (X_{32})	%	8.984	3.057	0.030	15.100
Decease rate of electricity consumption intensity (X_{33})	%	5.330	0.812	2.810	6.970
Urbanization degree of the province (X_{41})	%	46.903	8.743	29.890	63.400
Share of nonagricultural population (X_{42})	%	61.423	25.211	13.480	100.000
Urbanization degree of the city (X_{43})	%	44.321	14.790	13.840	100.000
Average retail price of electricity (X_5)	Yuan/MWH	512.013	85.014	298.630	699.400
Annual average temperature (X_6)	°C	14.653	5.294	−1.880	25.500
Ensured reserves of iron ore (X_7)	100 million tons	9.916	16.698	0.000	70.200

Note. Some maximum values of concentration variables are 100% because of the fulfillment of urbanization in some region.

price, and industrial price in China is chosen as an integrative indicator to measure the factor *price* (X_5).

Since temperature is one of the most important factors affecting the electricity consumption as presented in literature, on account of the availability of data, annual average temperature which is recorded in climatological data is selected to measure the characteristics of *natural condition* (X_6).

In view of the dependence of manufacture industry in China's economic growth and the high growth in demand for mineral resource, ensured reserves of iron ore could be proxy for *resource endowment* (X_7).

3.2.2. Data Sources. The main data sources are China Statistical Yearbook and China City Statistical Yearbook, published by National Bureau of Statistics of China (NBSC). The variables associated with provincial characteristics are taken from NBSC [1] and those associated with city characteristics are abstracted from NBSC [2]. After removing the data with missing explanation variables, a final sample set of 277 cities including 15 cities at subprovince level and 262 cities at prefecture level in 2009 is obtained. The annual average temperature of each city is collected by the authors from public data sources such as the official website of local government. The average retail prices of electricity in the cities are drawn from the report of State Electricity Regulatory Commission [51].

3.2.3. Sample Description. The descriptive statistics and units of original data are given in Table 3. The great disparity between regions in China exists in consideration of concentration degree. The minimum values of concentration variables are below 10% while the maximum values of these reach to 100%, showing the spatial differences in Chinese urban development. The mean shares of industrial output, employment, and electricity consumption in industry are 50.378%, 45.419%, and 65.964%, respectively, indicating the high weight of electricity consumption in industry even if the output and employment contributions are accounted. Seeing that the average decease rate of industrial energy intensity (8.984%) is smaller than that of total energy intensity (4.908%) at the provincial level during the same period, the energy consumption of industry presents more elasticity than other sectors. The lowest retail price of electricity (298.630 Yuan/MWH) is only about 42.698% of the highest one (699.400 Yuan/MWH) in 2009. Despite the fact that the electricity price is regulated by Chinese government, price mechanism may play certain roles in resource allocations.

4. Results and Discussion

4.1. Dimensionality Reduction. Dimensionality reduction technique could provide crucial variables for understanding the key phenomena of interest without losing the information conveyed by the original data. The results of combined factors via PCA method could be given by

$$X_1 = 0.977X_{11} + 0.924X_{12} + 0.935X_{13} + 0.905X_{14} + 0.955X_{15},$$
$$X_2 = 0.895X_{21} + 0.847X_{22} + 0.776X_{23},$$
$$X_3 = 0.493X_{31} + 0.841X_{32} + 0.927X_{33},$$
$$X_4 = 0.747X_{41} + 0.842X_{42} + 0.786X_{43}, \quad (5)$$

The selected principle components of urban morphology (X_1), industrial structure (X_2), regulation context (X_3) and urbanization rate (X_4) extracted 88.242%, 70.724%, 60.356%, and 62.840% of squared loading sums, respectively.

Table 4: Estimation results and variable structure for the finite mixture model[a].

Variable	Structure	Coefficient estimates			
		Component 1	Component 2	Component 3	Component 4
		Predictor			
Intercept	Varied	5461.407*** (919.115)	379.656*** (87.482)	1774.662*** (158.055)	982.128*** (61.198)
X_1	Varied	113.915*** (40.749)	23.923*** (2.821)	59.830*** (6.378)	40.738*** (3.021)
X_2	Varied	66.189 (92.219)	22.124*** (6.895)	48.875*** (12.326)	32.844*** (4.232)
X_3	Fixed	−9.677** (4.896)	−9.677** (4.896)	−9.677** (4.896)	−9.677** (4.896)
X_4	Nested	19.305*** (7.366)	19.305*** (7.366)	17.365** (6.991)	17.365** (6.991)
X_5	Varied	−7.509*** (1.315)	−0.287 (0.185)	−2.910*** (0.414)	−0.759*** (0.130)
X_6	Varied	−52.716 (35.108)	6.815** (2.750)	25.498*** (7.160)	−7.531*** (1.774)
		Concomitant variable			
Intercept	Varied	—	2.187*** (0.441)	1.976*** (0.473)	2.285*** (0.516)
X_7	Varied	—	−0.032*** (0.012)	−0.062** (0.024)	−0.186** (0.092)

[a]The standard errors of coefficient estimates are in parentheses. ** and *** denote significance at 5% and 1% levels, respectively. All the explanatory variables are standardized.

4.2. Estimation Results. Considering the commonness of BIC [14, 49] and the robustness of ICL-BIC [47], the two criteria are used for appraising the fitting of mixture models. Table 4 shows the estimation results for the finite mixture model with best performance of fitting.

It is notable that the acting structure of factors that impacted on the electricity intensity is distinguishable across the components which are corresponding to the city clusters as shown in Table 5. The factors of urban morphology (X_1), industrial structure (X_2), price (X_5), natural condition (X_6), and resource endowment (X_7) present variance between components, while the influence of regulation pressure (X_3) is fixed and the impact of urbanization degree (X_4) shows a nest structure. Meanwhile, the factor of resource endowment (X_7) is modeled in the finite mixture model as a concomitant variable, indicating this factor influences the electricity intensity indirectly, whereas the other factors as predictors make more direct impacts on the dependent variable.

The estimations of each factor would be analyzed next. The positive coefficients on urban morphology (X_1) which is related to the urban concentration degree suggest that promoting the urban concentration would increase the electricity intensity. The moderate concentration of region would bring about economies of scale in the sense of electricity consumption, while the urban sprawling around a central point immoderately would vanish the scale merits and cause diseconomies of scale. The estimated results manifest the excess centralization of urban expansion in China. This centralized trend is most significant in the component 1 with the largest estimated coefficient which is evaluated *cluster 1* of cities.

The extent of industrial structure (X_2) impacts on electricity intensity is in keeping with that of urban morphology (X_1). That is, as for the electricity intensity, the higher marginal contribution of urban concentration means the greater marginal influence of the rise of industrial weights as viewed from the sizes of estimated coefficients between the categorical factors. It is the evidence that the highly centered development in regions is partially attributed to the significant dependence on industrial output. The combined function of these factors leads to the increasing of electricity intensity and the boost of electricity demand.

The estimated coefficients of −9.677 on regulation context (X_3) suggest that the administrative pressure from the higher level authorities on electricity consumption is the powerful measure to induce decreasing the electricity consumption intensity. Furthermore, as the estimation values are equal between the components, the validity of regulation pressure is identical across the cities owing to the homogeneity of administration system in China. The administration power from top to bottom is still a useful but expedient approach in the context of lacking long-term mechanisms for energy savings.

Regarding the urbanization rate (X_4), the impacts on electricity consumption intensity presented two levels as the coefficients of 19.305 and 17.365 illustrated. The variables are standardized before estimation, so the impact extents could be compared. The signs of estimates betoken the possibility of long-term increasing of energy consumption intensity due

Table 5: Clustering membership of the finite mixture model.

Cluster 1
Wuhai (Inner Mongolia), Anshan (Liaoning), Fushun (Liaoning), Benxi (Liaoning), Yingkou (Liaoning), Fuxin (Liaoning), Liaoyang (Liaoning), Huludao (Liaoning), Laiwu (Shandong), Jiaozuo (Henan), Panzhihua (Sichuan), Qujing (Guizhou), Tongchuan (Shaanxi), Jiayuguan (Gansu), Baiyin (Gansu), Shizuishan (Ningxia), Zhongwei (Ningxia)
Cluster 2
Shijiazhuang (Hebei), Handan (Hebei), Baoding (Hebei), Cangzhou (Hebei), Langfang (Hebei), Hengshui (Hebei), Jincheng (Shanxi), Shuozhou (Shanxi), Luliang (Shanxi), Hohhot (Inner Mongolia), Erdos (Inner Mongolia), Hulunbuir (Inner Mongolia), Bayannur (Inner Mongolia), Ulanqab (Inner Mongolia), Shenyang (Liaoning), Dalian (Liaoning), Dandong (Liaoning), Jinzhou (Liaoning), Panjin (Liaoning), Tieling (Liaoning), Chaoyang (Liaoning), Changchun (Jilin), Liaoyuan (Jilin), Baishan (Jilin), Songyuan (Jilin), Baicheng (Jilin), Harbin (Heilongjiang), Jiamusi (Heilongjiang), Wuxi (Jiangsu), Suzhou (Jiangsu), Nantong (Jiangsu), Lianyungang (Jiangsu), Yancheng (Jiangsu), Taizhou (Jiangsu), Suqian (Jiangsu), Zhoushan (Zhejiang), Hefei (Anhui), Wuhu (Anhui), Huangshan (Anhui), Chuzhou (Anhui), Bozhou (Anhui), Xuancheng (Anhui), Fuzhou (Fujian), Xiamen (Fujian), Putian (Fujian), Quanzhou (Fujian), Longyan (Fujian), Ningde (Fujian), Nanchang (Jiangxi), Jingdezhen (Jiangxi), Fuzhou (Jiangxi), Jinan (Shandong), Qingdao (Shandong), Zaozhuang (Shandong), Dongying (Shandong), Yantai (Shandong), Weifang (Shandong), Taian (Shandong), Weihai (Shandong), Dezhou (Shandong), Liaocheng (Shandong), Xuchang (Henan), Luohe (Henan), Zhoukou (Henan), Xiangfan (Hubei), Xiaogan (Hubei), Changsha (Hunan), Changde (Hunan), Yiyang (Hunan), Guangzhou (Guangdong), Heyuan (Guangdong), Yangjiang (Guangdong), Jieyang (Guangdong), Beihai (Guangxi), Fangchenggang (Guangxi), Yulin (Guangxi), Chengdu (Sichuan), Zigong (Sichuan), Luzhou (Sichuan), Deyang (Sichuan), Mianyang (Sichuan), Suining (Sichuan), Neijiang (Sichuan), Nanchong (Sichuan), Meishan (Sichuan), Yibin (Sichuan), Guangan (Sichuan), Dazhou (Sichuan), Bazhong (Sichuan), Ziyang (Sichuan), Yuxi (Guizhou), Baoshan (Guizhou), Lincang (Guizhou), Xi'an (Shaanxi), Baoji (Shaanxi), Xianyang (Shaanxi), Weinan (Shaanxi), Yan'an (Shaanxi), Hanzhong (Shaanxi), Yulin (Shaanxi), Shangluo (Shaanxi), Jiuquan (Gansu), Qingyang (Gansu)
Cluster 3
Tangshan (Hebei), Qinhuangdao (Hebei), Xingtai (Hebei), Zhangjiakou (Hebei), Chengde (Hebei), Yangquan (Shanxi), Yuncheng (Shanxi), Hegang (Heilongjiang), Daqing (Heilongjiang), Quzhou (Zhejiang), Maanshan (Anhui), Tongling (Anhui), Xinyu (Jiangxi), Zibo (Shandong), Rizhao (Shandong), Zhengzhou (Henan), Luoyang (Henan), Anyang (Henan), Shangqiu (Henan), Huangshi (Hubei), Ezhou (Hubei), Xiangtan (Hunan), Zhangjiajie (Hunan), Huaihua (Hunan), Shaoguan (Guangdong), Shantou (Guangdong), Dongguan (Guangdong), ZhongShan (Guangdong), Guigang (Guangxi), Baise (Guangxi), Hezhou (Guangxi), Laibin (Guangxi), Guangyuan (Sichuan), Leshan (Sichuan), Yaan (Sichuan), Guiyang (Guizhou), Zunyi (Guizhou), Anshun (Guizhou), Lanzhou (Gansu), Zhangye (Gansu), Pingliang (Gansu), Xining (Qinghai), Wuzhong (Ningxia)
Cluster 4
Taiyuan (Shanxi), Datong (Shanxi), Changzhi (Shanxi), Jinzhong (Shanxi), Xinzhou (Shanxi), Linfen (Shanxi), Baotou (Inner Mongolia), Chifeng (Inner Mongolia), Tongliao (Inner Mongolia), Jilin (Jilin), Siping (Jilin), Tonghua (Jilin), Qiqihar (Heilongjiang), Jixi (Heilongjiang), Shuangyashan (Heilongjiang), Yichun (Heilongjiang), Qitaihe (Heilongjiang), Mudanjiang (Heilongjiang), Heihe (Heilongjiang), Suihua (Heilongjiang), Nanjing (Jiangsu), Xuzhou (Jiangsu), Changzhou (Jiangsu), Huaian (Jiangsu), Yangzhou (Jiangsu), Zhenjiang (Jiangsu), Hangzhou (Zhejiang), Ningbo (Zhejiang), Wenzhou (Zhejiang), Jiaxing (Zhejiang), Huzhou (Zhejiang), Shaoxing (Zhejiang), Jinhua (Zhejiang), Taizhou (Zhejiang), Lishui (Zhejiang), Bengbu (Anhui), Huainan (Anhui), Huaibei (Anhui), Anqing (Anhui), Fuyang (Anhui), Suzhou (Anhui), Chaohu (Anhui), Liuan (Anhui), Chizhou (Anhui), Sanming (Fujian), Zhangzhou (Fujian), Nanping (Fujian), Pingxiang (Jiangxi), Jiujiang (Jiangxi), Yingtan (Jiangxi), Ganzhou (Jiangxi), Jian (Jiangxi), Yichun (Jiangxi), Shangrao (Jiangxi), Jining (Shandong), Linyi (Shandong), Binzhou (Shandong), Heze (Shandong), Kaifeng (Henan), Pingdingshan (Henan), Hebi (Henan), Xinxiang (Henan), Puyang (Henan), Sanmenxia (Henan), Nanyang (Henan), Xinyang (Henan), Zhumadian (Henan), Wuhan (Hubei), Shiyan (Hubei), Yichang (Hubei), Jingmen (Hubei), Jingzhou (Hubei), Huanggang (Hubei), Xianning (Hubei), Suizhou (Hubei), Zhuzhou (Hunan), Hengyang (Hunan), Shaoyang (Hunan), Yueyang (Hunan), Chenzhou (Hunan), Yongzhou (Hunan), Loudi (Hunan), Shenzhen (Guangdong), Zhuhai (Guangdong), Foshan (Guangdong), Jiangmen (Guangdong), Zhanjiang (Guangdong), Maoming (Guangdong), Zhaoqing (Guangdong), Huizhou (Guangdong), Shanwei (Guangdong), Qingyuan (Guangdong), Chaozhou (Guangdong), Nanning (Guangxi), Liuzhou (Guangxi), Guilin (Guangxi), Wuzhou (Guangxi), Qinzhou (Guangxi), Hechi (Guangxi), Chongzuo (Guangxi), Haikou (Hainan), Sanya (Hainan), Liupanshui (Guizhou), Kunming (Guizhou), Zhaotong (Guizhou), Lijiang (Guizhou), Puer (Guizhou), Ankang (Shaanxi), Tianshui (Gansu), Wuwei (Gansu), Dingxi (Gansu), Longnan (Gansu), Yinchuan (Ningxia), Guyuan (Ningxia)

to the urbanization trends in China, but the influences of the factor are inferior to those of urban morphology (X_1) and industrial structure (X_2).

The electricity consumption intensity is sensitive to the price of electricity (X_5) in most regions in spite of the fact that the electricity price is regulated and controlled by the government administration. The coefficient of -7.509 on the electricity price (X_5) in component 1 indicates that the enhancement of the electricity price may largely reduce the electricity consumption intensity of the regions in cluster 1. The detailed examination of cluster 1 in Table 5 reveals that most of these cities abound with mineral deposits and rely on resource intensive industry for economic development. This is the evidence of the distortion of electricity price which may induce the discordances of industrial structure such as the overdevelopment of electricity intensive industry.

As the key variable indicated the natural condition the annual average temperature (X_6) affects the utilization of electricity significantly in the region of clusters 2–4. The values of −0.032, −0.062, and −0.186 on ensured reserves of iron ore (X_7) give the detailed description of the cluster probability which is determined by the concomitant variable. It is revealed by the coefficients that the regions in cluster 1 are most related with the endowment in electricity consumption, followed by the regions in cluster 2, cluster 3, and cluster 4 in order. The industrial structure may be the intermediate link which could explain the indirection of the connection between electricity consumption intensity and the resource endowment. These findings show that the industry distribution near the ore sources is one of the characteristics in China and the influence degrees of the determinants of electricity consumption intensity may be varied with the resource endowment.

5. Conclusions and Policy Implications

This study extracts the key factors which determine the electricity consumption intensity of the cities at subprovince and prefecture levels and reveals the acting structure and extent of each determinant which reflects the main characteristics of Chinese social and economic development. The findings of this study are crucial for electricity administration and the specific policies are suggested as follows.

(1) Scrutinizing the underlying influences of potential actions before decision making and formulating the comprehensive policy in consideration of regional characteristics may cause better results. The unitary policy may distort the aims of the policymakers and even upset the initial targets because of the vast spread of development degrees and various effects of different factors.

(2) Optimizing the spatial layout of regional development would improve the efficiency of electricity consumption and decrease the electricity intensity. Since the GPR, population, industry, public finance, and investment are highly concentrated around the core areas of the cities in China, decentralization of regional planning should be taken into account to upgrade the performance of electricity utilization. It is necessary for the government to balance the economic and social development within the regions.

(3) Diversifying the industrial structure, decentralizing economic activities, and establishing multiple cores for regional development would be helpful to achieve lower growth of electricity intensity in China. The integrated consideration of industrial distribution and regional cooperation is needed for the sustainability of economic growth. These immediate actions should be taken especially for the regions in cluster 1 which is most related to the mineral dependent cities.

(4) Besides the regulation powers from the administrative systems, the synthetic system with multiple policy instruments should be established. More attention should be paid to the market oriented mechanism. Also, the price system should be given full scope to adjust the relationship between electricity consumption and economic activities.

(5) Promoting the urbanization in decentralization way and increasing the employment capacities of industry could relieve the pressure on electricity caused by urbanization in China. The influences of the regional concentrating and industry intensifying are greater than that of the urbanization in development. More attention should be paid on the former factors in order to manage the electricity consumption.

(6) Deregulating the electricity price and accelerating the market-oriented reform in energy field could phase down the distortion possibility of energy price system. Extending the industrial chains could lower the electricity consumption in the regions where the development is heavily relied on energy intensive industries. Meanwhile, the unitary energy policy within the country with a vast territory and various degrees of development as China may lead to unrealized results. Energy policies should be synthesized and the multiple factors should be considered adequately before implementation.

Appendix

See Table 5.

Acknowledgments

This study has been supported by the Fundamental Research Funds for the Central Universities, and the Research Funds of Renmin University of China (Grant no. 12XNF017).

References

[1] NBSC (National Bureau of Statistics of China), *China Statistical Yearbook 2010*, China Statistical Press, Beijing, China, 2011.

[2] NBSC (National Bureau of Statistics of China), *China City Statistical Yearbook 2010*, China Statistical Press, Beijing, China, 2011.

[3] Z. Zhang, "The U.S. proposed carbon tariffs, WTO scrutiny and China's responses," *International Economics and Economic Policy*, vol. 7, no. 2, pp. 203–225, 2010.

[4] J. Guo, L. L. Zou, and Y. M. Wei, "Impact of inter-sectoral trade on national and global CO_2 emissions: an empirical analysis of China and US," *Energy Policy*, vol. 38, no. 3, pp. 1389–1397, 2010.

[5] L. X. Zhang, Z. F. Yang, and G. Q. Chen, "Emergy analysis of cropping-grazing system in Inner Mongolia Autonomous Region, China," *Energy Policy*, vol. 35, no. 7, pp. 3843–3855, 2007.

[6] L. Zhang, Z. Yang, B. Chen, and G. Chen, "Rural energy in China: pattern and policy," *Renewable Energy*, vol. 34, no. 12, pp. 2813–2823, 2009.

[7] G. Q. Chen and B. Chen, "Extended-exergy analysis of the Chinese society," *Energy*, vol. 34, no. 9, pp. 1127–1144, 2009.

[8] B. Zhang and G. Q. Chen, "Physical sustainability assessment for the China society: exergy-based systems account for resources use and environmental emissions," *Renewable and*

Sustainable Energy Reviews, vol. 14, no. 6, pp. 1527–1545, 2010.

[9] Z. F. Yang, M. M. Jiang, B. Chen, J. B. Zhou, G. Q. Chen, and S. C. Li, "Solar emergy evaluation for Chinese economy," *Energy Policy*, vol. 38, no. 2, pp. 875–886, 2010.

[10] Z. M. Chen, G. Q. Chen, J. B. Zhou, M. M. Jiang, and B. Chen, "Ecological input-output modeling for embodied resources and emissions in Chinese economy 2005," *Communications in Nonlinear Science and Numerical Simulation*, vol. 15, no. 7, pp. 1942–1965, 2010.

[11] H. Liao and Y. M. Wei, "China's energy consumption: a perspective from Divisia aggregation approach," *Energy*, vol. 35, no. 1, pp. 28–34, 2010.

[12] J. B. Zhou, M. M. Jiang, and G. Q. Chen, "Estimation of methane and nitrous oxide emission from livestock and poultry in China during 1949–2003," *Energy Policy*, vol. 35, no. 7, pp. 3759–3767, 2007.

[13] X. Ji and G. Q. Chen, "Unified account of gas pollutants and greenhouse gas emissions: Chinese transportation 1978–2004," *Communications in Nonlinear Science and Numerical Simulation*, vol. 15, no. 9, pp. 2710–2722, 2010.

[14] X. H. Xia, G. T. Huang, G. Q. Chen, B. Zhang, Z. M. Chen, and Q. Yang, "Energy security, efficiency and carbon emission of Chinese industry," *Energy Policy*, vol. 39, no. 6, pp. 3520–3528, 2011.

[15] X. H. Xia and G. Q. Chen, "Energy abatement in Chinese industry: cost evaluation of regulation strategies and allocation alternatives," *Energy Policy*, vol. 45, pp. 449–458, 2012.

[16] T. R. Beard, S. B. Caudill, and D. M. Gropper, "The diffusion of production processes in the U.S. Banking industry: a finite mixture approach," *Journal of Banking and Finance*, vol. 21, no. 5, pp. 721–740, 1997.

[17] M. Wedel and W. S. Desarbo, "Market segment derivation and profiling via a finite mixture model framework," *Marketing Letters*, vol. 13, no. 1, pp. 17–25, 2002.

[18] K. K. W. Yau, A. H. Lee, and A. S. K. Ng, "Finite mixture regression model with random effects: application to neonatal hospital length of stay," *Computational Statistics and Data Analysis*, vol. 41, no. 3-4, pp. 359–366, 2003.

[19] K. Imai and D. Tingley, "A statistical method for empirical testing of competing theories," *American Journal of Political Science*, vol. 56, no. 1, pp. 218–236, 2012.

[20] P. Edussuriya, A. Chan, and A. Ye, "Urban morphology and air quality in dense residential environments in Hong Kong. Part I: district-level analysis," *Atmospheric Environment*, vol. 45, no. 27, pp. 4789–4803, 2011.

[21] T. Van de Voorde, W. Jacquet, and F. Canters, "Mapping form and function in urban areas: an approach based on urban metrics and continuous impervious surface data," *Landscape and Urban Planning*, vol. 102, no. 3, pp. 143–155, 2011.

[22] J. W. R. Whitehand, K. Gu, S. M. Whitehand, and J. Zhang, "Urban morphology and conservation in China," *Cities*, vol. 28, no. 2, pp. 171–185, 2011.

[23] I. Larivière and G. Lafrance, "Modelling the electricity consumption of cities: effect of urban density," *Energy Economics*, vol. 21, no. 1, pp. 53–66, 1999.

[24] A. S. Permana, R. Perera, and S. Kumar, "Understanding energy consumption pattern of households in different urban development forms: a comparative study in Bandung City, Indonesia," *Energy Policy*, vol. 36, no. 11, pp. 4287–4297, 2008.

[25] W. Wang, H. Mu, X. Kang, R. Song, and Y. Ning, "Changes in industrial electricity consumption in china from 1998 to 2007," *Energy Policy*, vol. 38, no. 7, pp. 3684–3690, 2010.

[26] M. Aasen, H. Westskog, H. Wilhite, and M. Lindberg, "The EU electricity disclosure from the business perspective-a study from Norway," *Energy Policy*, vol. 38, no. 12, pp. 7921–7928, 2010.

[27] J. Chen, "China's experiment on the differential electricity pricing policy and the struggle for energy conservation," *Energy Policy*, vol. 39, no. 9, pp. 5076–5085, 2011.

[28] N. Zhou, M. D. Levine, and L. Price, "Overview of current energy-efficiency policies in China," *Energy Policy*, vol. 38, no. 11, pp. 6439–6452, 2010.

[29] F. Halicioglu, "Residential electricity demand dynamics in Turkey," *Energy Economics*, vol. 29, no. 2, pp. 199–210, 2007.

[30] Y. Liu, "Exploring the relationship between urbanization and energy consumption in China using ARDL (autoregressive distributed lag) and FDM (factor decomposition model)," *Energy*, vol. 34, no. 11, pp. 1846–1854, 2009.

[31] D. Wiesmann, I. Lima Azevedo, P. Ferrão, and J. E. Fernández, "Residential electricity consumption in Portugal: findings from top-down and bottom-up models," *Energy Policy*, vol. 39, no. 5, pp. 2772–2779, 2011.

[32] E. Ziramba, "The demand for residential electricity in South Africa," *Energy Policy*, vol. 36, no. 9, pp. 3460–3466, 2008.

[33] V. Bianco, O. Manca, S. Nardini, and A. A. Minea, "Analysis and forecasting of nonresidential electricity consumption in Romania," *Applied Energy*, vol. 87, no. 11, pp. 3584–3590, 2010.

[34] M. Bessec and J. Fouquau, "The non-linear link between electricity consumption and temperature in Europe: a threshold panel approach," *Energy Economics*, vol. 30, no. 5, pp. 2705–2721, 2008.

[35] T. M. Lai, W. M. To, W. C. Lo, and Y. S. Choy, "Modeling of electricity consumption in the Asian gaming and tourism center-Macao SAR, People's Republic of China," *Energy*, vol. 33, no. 5, pp. 679–688, 2008.

[36] P. J. Burke, "Income, resources, and electricity mix," *Energy Economics*, vol. 32, no. 3, pp. 616–626, 2010.

[37] A. P. Dempster, N. Laird, and D. B. Rubin, "Maximum likelihood estimation from incomplete data using the EM algorithm (with discussion)," *Journal of the Royal Statistical Society B*, vol. 39, pp. 1–38, 1977.

[38] G. Mclachlan and T. Krishnan, *The EM Algorithm and Extensions*, John Wiley & Sons, New York, NY, USA, 1997.

[39] G. McLachlan and D. Peel, *Finite Mixture Models*, Wiley Series in Probability and Statistics, New York, NY, USA, 2000.

[40] S. Frühwirth-Schnatter, *Finite Mixture and Markov Switching Models*, Springer, New York, NY, USA, 2007.

[41] B. Grün and F. Leisch, "Flexmix version 2: finite mixtures with concomitant variables and varying and constant parameters," *Journal of Statistical Software*, vol. 28, no. 4, pp. 1–35, 2008.

[42] H. Akaike, "Information theory and an extension of the maximum likelihood principle," in *Proceedings of the 2nd International Symposium on Information Theory*, B. N. Petrov and F. Csáki, Eds., pp. 267–281, Académiai Kiadó, Budapest, Hungary, 1973.

[43] G. Schwarz, "Estimating the dimension of a model," *The Annals of Statistics*, vol. 6, pp. 461–464, 1978.

[44] E. J. Hannan and B. G. Quinn, "The determination of the order of an autoregression," *Journal of the Royal Statistical Society B*, vol. 41, pp. 190–195, 1979.

[45] G. Celeux and G. Soromenho, "An entropy criterion for assessing the number of clusters in a mixture model," *Journal of Classification*, vol. 13, no. 2, pp. 195–212, 1996.

[46] C. Biernacki and G. Govaert, "Using the classification likelihood to choose the number of clusters," *Computing Science and Statistics*, vol. 29, pp. 451–457, 1997.

[47] C. Biernacki, G. Celeux, and G. Govaert, "Assessing a mixture model for clustering with the integrated completed likelihood," *IEEE Transactions on Pattern Analysis and Machine Intelligence*, vol. 22, no. 7, pp. 719–725, 2000.

[48] J. Bai and S. Ng, "Confidence intervals for diffusion index forecasts and inference for factor-augmented regressions," *Econometrica*, vol. 74, no. 4, pp. 1133–1150, 2006.

[49] S. Ng and J. Bai, "Selecting instrumental variables in a data rich environment," *Journal of Time Series Econometrics*, vol. 1, no. 1, article 4, 2009.

[50] E. L. Glaeser and M. Kahn, "Sprawl and Urban Growth," in *The Handbook of Regional and Urban Economics*, V. Henderson and J. Thisse, Eds., North Holland, Amsterdam, The Netherlands, 2004.

[51] SERC (State Electricity Regulatory Commission), "The regulatory report on the execution of electricity price and the settlement of electricity charges in 2009," http://www.serc.gov.cn/zwgk/jggg/201009/W020100913435310821659.doc.

11

Double-Grating Displacement Structure for Improving the Light Extraction Efficiency of LEDs

Zhibin Wang,[1] Yang Hao,[1] Zhongdong Wang,[2] Xian Liu,[1] Qian Zhang,[1] and Dandan Zhu[1]

[1] *College of Electrical Engineering, Yanshan University, Qinhuangdao 066004, China*
[2] *Mechanical Science and Engineering Institute, Northeast Petroleum University, Daqing 163318, China*

Correspondence should be addressed to Zhibin Wang, wzb_ysu@ysu.edu.cn

Academic Editors: B. Chen, T. Han-Song, and C. Zhan-Ming

To improve the light extraction efficiency of light-emitting diodes (LEDs), grating patterns were etched on GaN and silver film surfaces. The grating-patterned surface etching enabled the establishment of an LED model with a double-grating displacement structure that is based on the surface plasmon resonance principle. A numerical simulation was conducted using the finite difference time domain method. The influence of different grating periods for GaN surface and silver film thickness on light extraction efficiency was analyzed. The light extraction efficiency of LEDs was highest when the grating period satisfied grating coupling conditions. The wavelength of the highest value was also close to the light wavelength of the medium. The plasmon resonance frequencies on both sides of the silver film were affected by silver film thickness. With increasing film thickness, plasmon resonance frequency tended toward the same value and light extraction efficiency reached its maximum. When the grating period for the GaN surface was 365 nm and the silver film thickness was 390 nm, light extraction efficiency reached a maximum of 55%.

1. Introduction

Low-carbon economy and green economy have recently become major themes of global development. As a type of green lighting energy, light-emitting diodes (LEDs) have attracted considerable research attention [1]. LEDs present the advantages of energy conservation, environment friendliness, fast response, and long lifetimes. Thus, LEDs lighting systems have been considered to reduce greenhouse gas emissions by replacing fuel-based lighting in the developing world [2]. Not only that, these light sources are extensively used as backlight sources of liquid crystal displays, advertisements, night lights, traffic signals, outdoor lighting, full color displays [3], and biometric devices [4], in addition, the array of LEDs has been used to inactivate bacteria in liquid suspension and on exposed surfaces [5], among other applications. Despite these advantages, the development of LED core technology is constrained by the photons produced in the lighting devices. Such a constraint prevents photons from being effectively radiated and converted into available optical power. This problem stems from the inherent structure of solid-state lighting devices: the refractive index of solid semiconductor luminophore is higher than that of the surrounding medium [6]. Thus, the light emitted by the lighting devices easily causes total internal reflection in the interface of semiconductor and air, after which the light retraces the optical path of the luminophore and is converted into heat [7]. This phenomenon makes the LEDs working in high-temperature environments for a long time; it not only wastes energy, but also shortens LED lifetimes. So the improvement of LEDs light extraction efficiency can not only extend lifetime, but also make important contribution to the development of the new green energy.

Foreign and domestic experts have begun improving the structure of LED chips, thereby significantly enhancing light extraction efficiency. These improvements primarily include changing the geometric shape of the chips, reducing the duration of total internal reflection in the chips, and decreasing light energy losses. The methods generally applied in realizing such enhancements are using the inverted pyramid structure [8, 9], performing surface roughening [10], constructing a mirror with Bragg grating reflection [11], and creating a two-dimensional photonic crystal structure on the surface of GaN [12–15]. Using the characteristic of surface plasmons (SPs) to enhance the light extraction efficiency of LEDs has recently become a research focus [16]. Compare with a single grating structure of the LED model described in [17]. In the current work, a double-grating displacement structure was constructed for an LED model to improve the light extraction efficiency of LEDs. The structure was realized by etching grating patterns on GaN and metallic silver film surfaces.

Under certain conditions, the free electrons of metal surfaces produce collective coherent oscillation, which is stimulated by light; the electromagnetic wave at the metal/air interface is called SP [18, 19]. The internal quantum efficiency and external quantum efficiency of LEDs can be improved by using SPs. The improvement in the internal quantum efficiency of LEDs by applying SPs is based on a principle that is related to the spontaneous radiative rate of excitons and state density [20]. When the light center is located in the micro cavity of the wavelength scale, the state density of photons changes, which, in turn, causes variations in the spontaneous radiation rate of excitons. The principle governing the use of SPs to improve the external quantum efficiency of LEDs is based on the observation that light can excite SPs. The light reflection angle is greater than the total internal reflection angle, but the former prevents light from radiating outward. A reasonable metal structure can excite SPs, which are then radiated outward in the form of light. Part of the SPs can be coupled to the electromagnetic waves on the surface of metal, and radiated outward in the form of light by etching a grating-coupled structure on the surface. Part of the incident light that causes total internal reflection can be diminished by grating-patterned etching of the GaN surface. Reducing total internal reflection can also be realized by using a dielectric material with a high coefficient. Using this material produces an evanescent field in total internal reflection. The evanescent field is used as the excitation source that stimulates the SP resonance. It can effectively extract a limited amount of light and reduce the junction temperature of LED chips. Compared with the traditional LED structure, the grating etched onto the GaN surface is relatively complex when used without system optimization. In this work, therefore, the grating period for GaN and metallic silver film thickness was optimized. The influence of these parameters on light extraction efficiency was also simulated and analyzed.

2. Theories and Methods

2.1. Introduction to SP Resonance. When the wave frequency for the excitation of SPs is larger than the plasmon resonance frequency (ω_p), the electromagnetic field is radiatively propagated far from the space between the metal/dielectric material interface, a phenomenon described as the radiative mode of SPs [20, 21]. Only when the horizontal direction wave vector $\mathbf{k}_x$ is very small (i.e., resonance frequency $\approx \omega_p$) can the life cycle of the radiant mode of SPs on a metal surface be defined. The gradual increase in $\mathbf{k}_x$ results in a resonance mode life cycle that is too short to have any practical significance. The SP resonance in the range of visible frequency

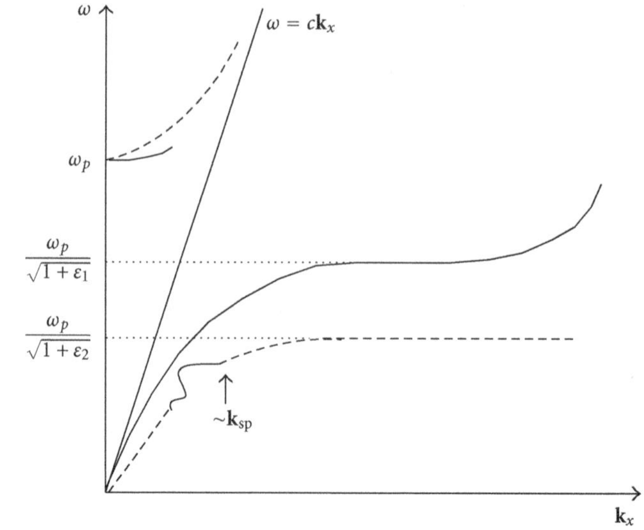

ω: Incident light frequency.
c: Light velocity.
$\varepsilon_1 \varepsilon_2$: Material dielectric constant.

FIGURE 1: Dispersion relationship curve of SPs.

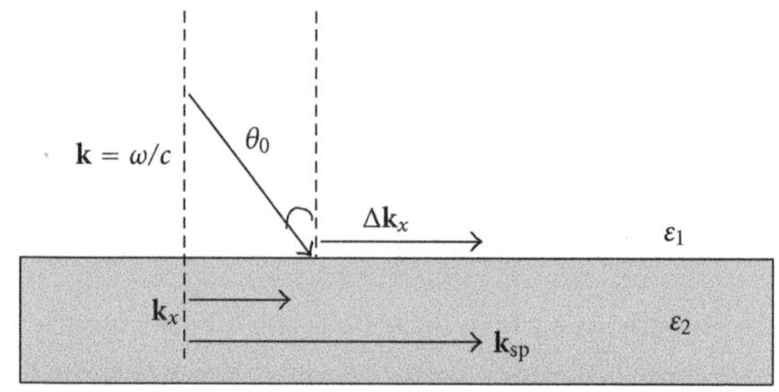

$\mathbf{k}$: Incident light vector
θ_0: Considered angle of incident light and normal line

FIGURE 2: SPs excited by light.

is the nonradiative SP mode. The electromagnetic field produced by this mode is limited to an area near the metal surface. This phenomenon is called electromagnetic wave dispersion.

The SP dispersion relationship is depicted in Figure 1. The dispersion curve of the nonradiative SPs produced on a metal surface lies completely on the right side of the incident electromagnetic wave line of the dielectric material [20]. This result indicates that the SPs have a longer horizontal component of wave vector $\mathbf{k}_{sp}$ than that of the incident electromagnetic wave vector $\mathbf{k}_x$ of the same energy. Thus, the general electromagnetic waves incident from the dielectric material cannot stimulate nonradiative SPs and reach the resonance mode. To match the excitation conditions for the SP resonance mode, a coupled mechanism should be applied. This mechanism increases wave vector $\Delta\mathbf{k}_x$ and causes the incident electromagnetic wave to achieve a higher $\mathbf{k}_x$ (Figure 2).

Two coupled mechanisms are commonly used [20]. One involves producing a small-raster periodic structure on a metal surface as a coupled medium. Under an external electromagnetic field, the free electrons of the metal surface in these cycle structures cause polarized electron oscillation for a specific wavelength. The generated electromagnetic field can provide an additional value $\Delta \mathbf{k}_x$ to the incident electromagnetic wave. Subsequently, SP resonance is stimulated. The other coupled mechanism uses the total internal reflection dissipation field produced in a material with a high dielectric constant as the excitation source to stimulate SP resonance.

The electromagnetic wave vector can be written as

$$\mathbf{k}^2 = \frac{\omega^2 \varepsilon \mu}{c^2}, \tag{1}$$

where ε and μ are the relative permittivity and relative permeability of the material, respectively. When the electromagnetic wave passes through the prism of high permittivity, the wave vector is greater than that observed in dielectric material ε_1. When the incident light produces total internal reflection in the prism/dielectric material interface, part of the evanescent field is tunneled into the dielectric material near the total internal reflection interface. The wave vector $\mathbf{k}_x$ of the evanescent field and the total internal reflection wave vector have the same size when the distance between the prism and the metal surface is sufficiently small. At the same time, when the incident light wave vector horizontal component of total internal reflection satisfies the conditions of SPs, the SPs in the dielectric material/metal interface are stimulated.

2.2. Finite Difference Time Domain Method. The finite difference time domain (FDTD) method [22], an extension of the finite difference method, is a numerical analysis approach that directly performs computer simulations by using Maxwell equations for electromagnetic fields. Using the Maxwell curl equations in accordance with the field quantities that have special configurations in the Yee grid yields [23, 24]

$$\begin{aligned}
&\frac{\varepsilon(i+1/2,j,k)\left[\mathbf{E}_x^{n+1}(i+1/2,j,k) - \mathbf{E}_x^{n}(i+1/2,j,k)\right]}{\Delta t} \\
&\quad + \frac{\sigma(i+1/2,j,k)\left[\mathbf{E}_x^{n+1}(i+1/2,j,k) + \mathbf{E}_x^{n}(i+1/2,j,k)\right]}{2} \\
&\quad = \frac{\left[\mathbf{H}_z^{n+1/2}(i+1/2,j+1/2,k) - \mathbf{H}_z^{n+1/2}(i+1/2,j-1/2,k)\right]}{\Delta y} \\
&\quad - \frac{\left[\mathbf{H}_y^{n+1/2}(i+1/2,j,k+1/2) - \mathbf{H}_y^{n+1/2}(i+1/2,j,k-1/2)\right]}{\Delta z}.
\end{aligned} \tag{2}$$

Let

$$\begin{gathered}
p = \left(i+\frac{1}{2}, j, k\right), \qquad q^{+} = \left(i+\frac{1}{2}, j+\frac{1}{2}, k\right), \\
q^{-} = \left(i+\frac{1}{2}, j-\frac{1}{2}, k\right), \\
r^{+} = \left(i+\frac{1}{2}, j, k+\frac{1}{2}\right), \qquad r^{-} = \left(i+\frac{1}{2}, j, k-\frac{1}{2}\right).
\end{gathered} \tag{3}$$

Thus,

$$\begin{aligned}
(\mathbf{E}_x)_p^{n+1} &= \frac{\left[1 - \Delta t \sigma_p / 2\varepsilon_p\right](\mathbf{E}_x)_p^{n}}{\left[1 + \Delta t \sigma_p / 2\varepsilon_p\right]} \\
&\quad + \frac{\left(\Delta t/\varepsilon_p\right)\left\{\begin{array}{l}\left[(\mathbf{H}_z)_{q^+}^{n+1/2} - (\mathbf{H}_z)_{q^-}^{n+1/2}\right]/\Delta y - \\ \left[(\mathbf{H}_y)_{r^+}^{n+1/2} - (\mathbf{H}_y)_{r^-}^{n+1/2}\right]/\Delta z\end{array}\right\}}{\left[1 + \Delta t \sigma_p / 2\varepsilon_p\right]},
\end{aligned} \tag{4}$$

where the superscripts $n + 1/2$ and n are the time step-numbers; a group of changes (i,j,k) and Δt are the time steps; Δx, Δy, Δz are the distances of the adjacent lattice point in the x, y, z directions, respectively. Other field quantities can be treated in the same manner.

To perform a stable numerical simulation, the variables should satisfy

$$v_{\max} \Delta t \le \frac{1}{\sqrt{(1/\Delta x)^2 + (1/\Delta y)^2 + (1/\Delta z)^2}}, \tag{5}$$

where $v_{\max}$ is the maximum phase velocity, Δt denotes the time step, and Δx, Δy, Δz are the distance steps in the x, y, z directions, respectively. When $\Delta x = \Delta y = \Delta z = \Delta$, (5) can be simplified as $v_{\max}\Delta t < \Delta/\sqrt{3}$. To reduce the error caused by numerical dispersion effects, (5) should also satisfy the conditions $\Delta/\lambda_{\min} < 1/10$, where $\lambda_{\min}$ is considered the shortest wavelength of the electromagnetic wave considered.

The FDTD method was used to solve the numerical analysis problems presented by grating-patterned etching on the metallic silver film. This approach also enables the excitation of SPs and application of the total internal reflection dissipation field for exciting the SPs. The changes in time step Δt reflect the variations in energy flow on the receiving surface. This reflection is realized by placing a receiving surface over the structure. The changes in energy flow facilitate the analysis of the effects of structural parameter variations on the light extraction efficiency of LEDs.

3. Physical LED Model

The physical model of the LED with a double-grating displacement structure was constructed by grating-patterned etching on the GaN and metallic silver film surfaces (Figure 3). The model comprises the silver film layer, P-GaN layer, active layer, N-GaN layer, and SiC substrate. During the experiment, the calculated area dimensions of 4200 nm × 420 nm were realized; the thicknesses of the P-GaN and N-GaN layers were 200 and 400 nm, respectively. Compare with the physical LED model as described in [17], a grating-patterned etching was added to the GaN surface, which had two reasons. (1) Grating-patterned etching can diminish part of the total internal reflection produced by incident light. (2) A portion of the light causes the evanescent field of total internal reflection to excite SPs; The study in [7] has discussed that grating-patterned etching on the top surface can effectively enhance the light extraction efficiency of LEDs; thus, this phenomenon results in a combined effect with

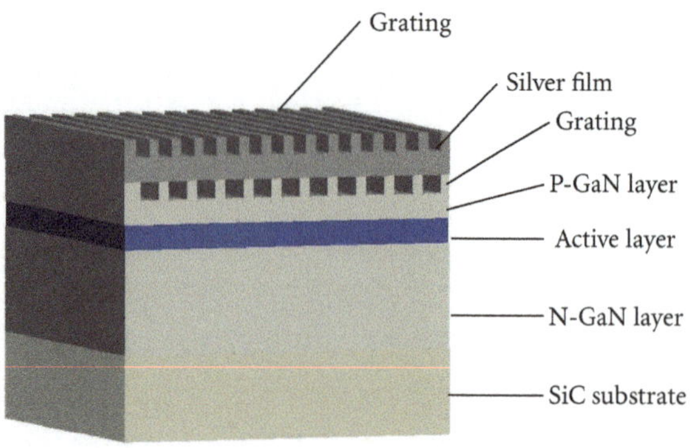

FIGURE 3: Physical model of an LED with a double-grating displacement structure.

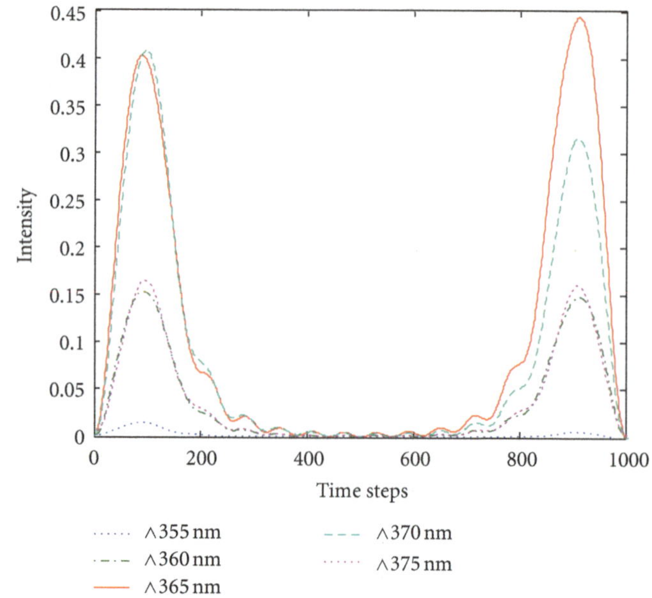

FIGURE 4: Energy flow of GaN at different grating periods.

grating-patterned etching on the metallic silver film surface, thereby enhancing optical transmission and improving the light extraction efficiency of LEDs.

4. Numerical Simulation and Analysis

A numerical simulation was conducted using the FDTD method. The wavelength of light was 500 nm, the refractive index of air was 1.0, and the refractive index of GaN was 2.4. The silver film was used in the Lorentz dispersion model. In analyzing the light absorption process, the silver film is disregarded. The light emitted by the active layer is replaced by the total field-scattered field source, an approach that realizes incident plane waves from all directions. The intensity was 1 and the grating period of the metallic silver film was 375 nm. A receiving surface, which is capable of reflecting light intensity, was placed above the model to keep track of energy flow.

4.1. Effect of Grating Period on the Light Extraction Efficiency of the GaN Surface. Figure 4 presents the energy flow of the GaN with a double-grating displacement structure at different grating periods. The maximum flow intensity value was 0.44, which can be obtained at a GaN surface grating period of 365 nm. Figure 5 shows the relationship between the surface grating period of GaN and light extraction efficiency when the silver film thickness changes from 300 nm to 400 nm. For light extraction efficiency with oscillatory changes in grating period, the maximum value acquired constantly ranged from 360 nm to 380 nm. This result confirms that grating period influences the light extraction efficiency of LEDs. At a silver film thickness of 300 nm to 400 nm, the relationship curves of grating period and light extraction efficiency are similar and the maximum light extraction efficiency occurred at a 365 nm grating period. When the grating period continued to increase, the light extraction efficiency gradually declined. This result is attributed to two factors. First, the surface grating period is related to the wave vector that causes plasmon resonance. Thus, only when the grating period satisfies the excitation conditions can the SPs be excited. Consequently, visible light is extracted and transmissible light is enhanced. Second, the ideal size of diffraction grating satisfies the wavelength of light on a metal surface when the period is considerably smaller than the wavelength of light. Otherwise, conductivity to light diffraction is low and the effect of changing light directions caused by light diffraction decreases. On the other hand, when the period is far from the wavelength of light, the plane of the grating section is excessively large, thereby causing stronger total internal reflection, which slows down light emission.

4.2. Effect of Silver Film Thickness on Light Extraction Efficiency. The grating period was set at 365 nm on the basis of the analysis in Section 4.1. The thickness of the metallic silver film was set at 300 nm to 400 nm. Figures 6 and 7 show the energy flow of the silver film with a double-grating displacement structure and that of the silver film with a single grating, respectively. The light extraction efficiency of LEDs with a double-grating displacement structure improved by 43% over that of the silver film surface with a single grating. The graph shows that the silver film thickness should satisfy the grating coupling conditions to enhance light extraction efficiency. A silver film that is too thick or too thin cannot induce coupling. The plasmon resonance mode existed on both sides of the metal/dielectric material interface. Thus, when the film thickness was at the nanometer scale, the evanescent field formed by the SPs was strong enough to pass through the other side of the metal. Consequently, the SP electromagnetic fields on both sides of the metal film interacted with each other and formed a group of coupled SPs, as illustrated by the energy flow curve of silver thickness at 390 nm in Figure 6.

The plasmon resonance frequency reached the same value and energy flow intensity reached a maximum of 0.55.

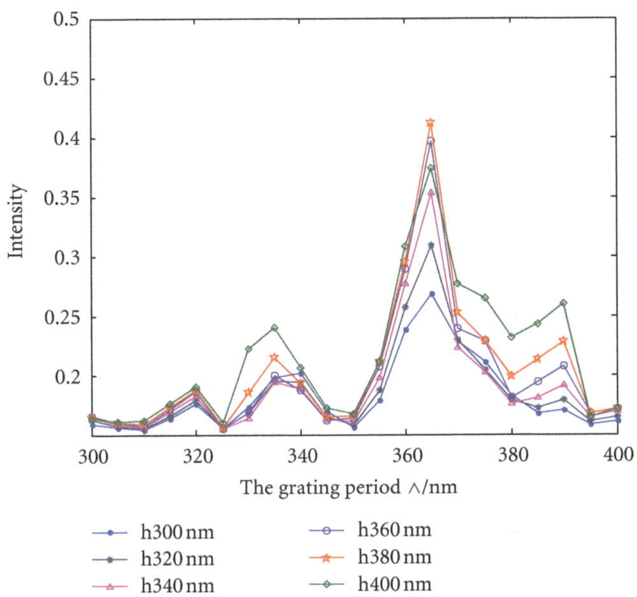

FIGURE 5: Relationship between the grating period of GaN and light extraction efficiency with silver film thickness ranging from 300 nm to 400 nm.

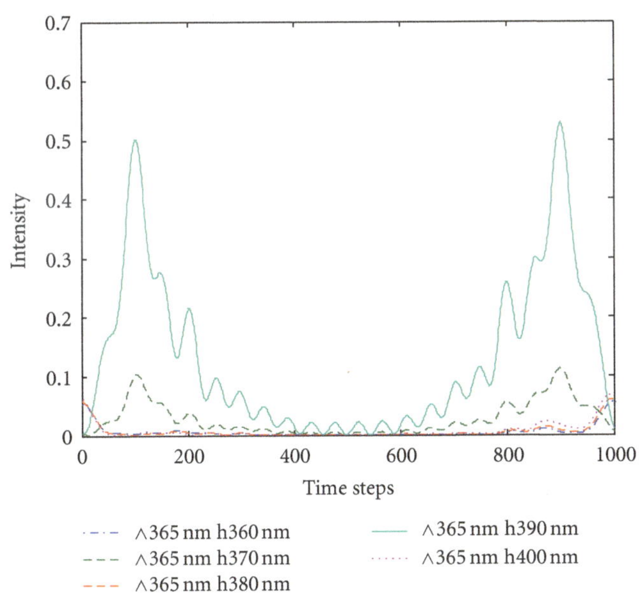

FIGURE 6: Energy flow of silver film (of different thicknesses) with a double-grating structure.

However, when the silver film thickness was smaller than 390 nm, the SP resonance frequency remained at the degenerate state, in which the energy flow intensity exhibited no significant enhancement. The results indicate the absence of a coupled phenomenon. With increasing silver film thickness, the energy flux rate gradually decreased, indicating the continuous division of the SP resonance frequency. Consequently, coupling disappeared when the silver film thickness tended toward infinity.

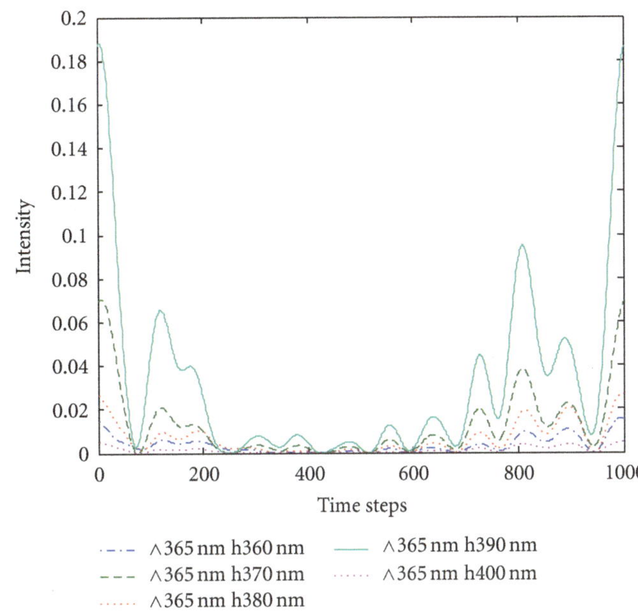

FIGURE 7: Energy flow of silver film (of different thicknesses) with a single grating structure.

5. Conclusion

Grating-patterned etching was performed on GaN and silver film surfaces to establish an LED model with a double-grating displacement structure. The FDTD method was used in conducting a numerical simulation. A grating etched on the GaN surface effectively extracts a limited amount of light and enhances the light extraction efficiency of LEDs. At a GaN grating period of 365 nm and a silver film thickness of 390 nm, light extraction efficiency reaches a maximum of 55%.

Acknowledgments

This work was supported by the National Natural Science Foundation of China under Grant no. 61107039 and no. 61004050 and the Natural Science Foundation of Heilongjiang Province under Grant no. F201006.

References

[1] X. Gu, T. Qiu, W. Zhang, and P. K. Chu, "Light-emitting diodes enhanced by localized surface plasmon resonance," *Nanoscale Research Letters*, vol. 6, no. 1, article 199, 2011.

[2] E. Mills and A. Jacobson, "From carbon to light: a new framework for estimating greenhouse gas emissions reductions from replacing fuel-based lighting with LED systems," *Energy Efficiency*, vol. 4, no. 4, pp. 523–546, 2011.

[3] H. Lin, S. Liu, X. S. Zhang, B. L. Liu, and X. C. Ren, "Enhanced external quantum efficiency of light emitting diodes by fabricating two-dimensional photonic crystal sapphire substrate with holographic technique," *Acta Physica Sinica*, vol. 58, no. 2, pp. 959–963, 2009.

[4] N. Kourkoumelis and M. Tzaphlidou, "Eye safety related to near infrared radiation exposure to biometric devices," *TheScientificWorldJournal*, vol. 11, pp. 520–528, 2011.

[5] L. E. Murdoch, M. MacLean, E. Endarko, S. J. MacGregor, and J. G. Anderson, "Bactericidal effects of 405nm light exposure demonstrated by inactivation of escherichia, salmonella, Shigella, Listeria, and mycobacterium species in liquid suspensions and on exposed surfaces," *The Scientific World Journal*, vol. 2012, Article ID 137805, 8 pages, 2012.

[6] D. H. Kim, C. O. Cho, Y. G. Roh et al., "Enhanced light extraction from GaN-based light-emitting diodes with holographically generated two-dimensional photonic crystal patterns," *Applied Physics Letters*, vol. 87, no. 20, Article ID 203508, 3 pages, 2005.

[7] S. Trieu, X. Jin, A. Ellaboudy et al., "Top transmission grating GaN LED simulations for light extraction improvement," in *Physics and Simulation of Optoelectronic Devices XIX*, vol. 7933 of *Proceedings of SPIE*, San Francisco, Calif, USA, January 2011.

[8] C. M. Tsai, J. K. Sheu, W. C. Lai et al., "Enhanced output power in GaN-based LEDs with naturally textured surface grown by MOCVD," *IEEE Electron Device Letters*, vol. 26, no. 7, pp. 464–466, 2005.

[9] W. Zhou, G. Min, Z. Song, J. Zhang, Y. Liu, and J. Zhang, "Enhanced efficiency of light emitting diodes with a nano-patterned gallium nitride surface realized by soft UV nanoimprint lithography," *Nanotechnology*, vol. 21, no. 20, Article ID 205304, 2010.

[10] K. S. Kim, S. M. Kim, H. Jeong, M. S. Jeong, and G. Y. Jung, "Enhancement of light extraction through the wave-guiding effect of ZnO sub-microrods in InGaN blue light-emitting diodes," *Advanced Functional Materials*, vol. 20, no. 7, pp. 1076–1082, 2010.

[11] C. M. Lim and G. Hugh Song, "Design of superperiodic photonic-crystal light-emitting plates with highly directive luminance characteristics," *Journal of the Optical Society of America B*, vol. 26, no. 2, pp. 328–336, 2009.

[12] A. David, H. Benisty, and C. Weisbuch, "Optimization of light-diffracting photonic-crystals for high extraction efficiency LEDs," *IEEE/OSA Journal of Display Technology*, vol. 3, no. 2, pp. 133–148, 2007.

[13] D. H. Long, I. K. Hwang, and S. W. Ryu, "Design optimization of photonic crystal structure for improved light extractionof GaN LED," *IEEE Journal on Selected Topics in Quantum Electronics*, vol. 15, no. 4, pp. 1257–1263, 2009.

[14] A. I. Zhmakin, "Enhancement of light extraction from light emitting diodes," *Physics Reports*, vol. 498, no. 4-5, pp. 189–241, 2011.

[15] T. A. Truong, L. M. Campos, E. Matioli et al., "Light extraction from GaN-based light emitting diode structures with a noninvasive two-dimensional photonic crystal," *Applied Physics Letters*, vol. 94, no. 2, Article ID 023101, 33 pages, 2009.

[16] V. K. Komarala, W. H. Guo, and M. Xiao, "Surface plasmon density of states at the metal-dielectric interface: dependence of metal layer thickness and dielectric material," *Journal of Applied Physics*, vol. 107, no. 1, Article ID 014309, 5 pages, 2010.

[17] Y. Z. Lin, K. Li, F. M. Kong, J. Zhao, L. G. Du, and H. Gao, "Comprehensive numeric study of gallium nitride light-emitting diodes adopting surface-plasmon-mediated light emission technique," *IEEE Journal on Selected Topics in Quantum Electronics*, vol. 17, no. 4, pp. 942–951, 2010.

[18] Z. Wu, J. W. Haus, Q. Zhan, and R. L. Nelson, "Plasmonic notch filter design based on long-range surface plasmon excitation along metal grating," *Plasmonics*, vol. 3, no. 2-3, pp. 103–108, 2008.

[19] S. H. Choi, S. J. Kim, and K. M. Byun, "Characteristics of light emission from surface plasmons based on rectangular silver gratings," *Optics Communications*, vol. 283, no. 14, pp. 2961–2966, 2010.

[20] H. Raether, *Surface Plasmons on Smooth and Rough Surface and on Gratings*, Springer, Berlin, Germany, 1998.

[21] S. Glasberg, A. Sharon, D. Rosenblatt, and A. A. Friesem, "Long-range surface plasmon resonances in grating-waveguide structures," *Applied Physics Letters*, vol. 70, no. 10, pp. 1210–1212, 1997.

[22] K. S. Yee and J. S. Chen, "The finite-difference time-domain (FDTD) and the finite-volume time-domain (FVTD) methods in solving Maxwell's equations," *IEEE Transactions on Antennas and Propagation*, vol. 45, no. 3, pp. 354–363, 1997.

[23] X. Zhou and W. Lin, "Numerical stability and numerical dispersion of compact conformal mapping 2D-FDTD method used for the dispersion analysis of waveguides," *International Journal of Infrared and Millimeter Waves*, vol. 20, no. 7, pp. 1403–1412, 1999.

[24] M. T. Bettencourt, "Flux limiting embedded boundary technique for electromagnetic FDTD," *Journal of Computational Physics*, vol. 227, no. 6, pp. 3141–3158, 2008.

12

Assessment for Fuel Consumption and Exhaust Emissions of China's Vehicles: Future Trends and Policy Implications

Yingying Wu, Peng Zhao, Hongwei Zhang, Yuan Wang, and Guozhu Mao

School of Environmental Science and Engineering, Tianjin University, Tianjin 300072, China

Correspondence should be addressed to Peng Zhao, zhp_tj@126.com

Academic Editors: B. Chen and Z.-M. Chen

In the recent years, China's auto industry develops rapidly, thus bringing a series of burdens to society and environment. This paper uses Logistic model to simulate the future trend of China's vehicle population and finds that China's auto industry would come into high speed development time during 2020–2050. Moreover, this paper predicts vehicles' fuel consumption and exhaust emissions (CO, HC, NO_x, and PM) and quantificationally evaluates related industry policies. It can be concluded that (1) by 2020, China should develop at least 47 million medium/heavy hybrid cars to prevent the growth of vehicle fuel consumption; (2) China should take the more stringent vehicle emission standard V over 2017–2021 to hold back the growth of exhaust emissions; (3) developing new energy vehicles is the most effective measure to ease the pressure brought by auto industry.

1. Introduction

During the past decade, China's automotive industry has experienced a dramatically growth. Since 2009 China had become the largest vehicle producer and consumer. Auto industry has become a pillar industry in China's national economy and played an increasingly important role in economic development, employment promotion and domestic demand stimulation.

In recent years, the auto industry in developed countries has tended saturated state, with a slow growth (e.g., America) even a negative growth (e.g., Japan and Germany) in vehicle population. However, compared with developed countries, the per capita car ownership in China is very low and auto industry is still in a rapid increasing period. By the end of 2008, China's vehicle population was 51 million, just ranked 3rd after America and Japan (Table 1), and by August 2011, it has reached 93.5 million, second only to America (about 240 million) in the world. On a global scale, per capita car ownership is 0.15, while China's is 0.06, only 40% of the world average level. Calculated by the average level of the world, vehicle population in China should be 207 million, and that should reach 1.24 billion according to the American average level of 0.82 cars per capita. Obviously, it would bring severe challenges for China, including resource demand and exhaust emissions.

In fact, the crude oil consumption in China had reached 383.845 million tons in 2009, with average annual growth rate of 6.5% from 1990s, which is 4 times greater than the world growth rate (1.5%) over the same period. One of the major drivers of this increase in China's oil consumption is the rapid growth of the transportation sector in general and motor vehicles in particular [1]. About 60% of China's oil consumption in 2009 was for transportation [2]. Similarly, vehicle exhaust emissions grew from 17.24 million tons in 1995 to 40.555 million tons in 2009, with an annual average growth rate of 6.3% [3]. Thus, while auto industry expands rapidly, these environmental problems will become more stand out, seriously impeding the sustainable development of China.

Concerning the rapid development of the auto industry in China, scholars in the recent years have begun to keen to study a series of environmental issues associated with auto industry, analyzing the status, and predicting the future development. For example, Yan and Crookes [4] set two scenarios, best case scenario and baseline scenario, to assess the reduction potentials of energy demand and greenhouse gas (GHG) emissions in China's road transport sector till

Table 1: Worldwide vehicle population and per capita car ownership in 2008.

Country	Vehicle population/10^4	Per capita cars
USA	25024	0.82
Japan	7553	0.59
China	5100	0.04
Germany	4400	0.54
Italy	4089	0.69
Russia	3826	0.27
France	3721	0.60
Britain	3562	0.58
Spain	2761	0.60
Brazil	2748	0.14
Mexico	2531	0.24
Canada	2052	0.62
Poland	1909	0.50
India	1851	0.02
Korea	1679	0.35
Australia	1468	0.67
Turkey	1019	0.15
Thailand	977	0.15
The Netherlands	891	0.52
Argentina	846	0.21
Indonesia	825	0.04
South Africa	749	0.15
Belgium	586	0.53
Sweden	480	0.53
Austria	468	0.59
Switzerland	436	0.62
Others	15752	0.07
Total	97306	0.15

Data resource: National Bureau of Statistics of China, Japan Automobile Manufacturers Association, German Automobile Industry Association, Ward's Auto, and ANFIA.

2030. Hao et al. [5] established a bottom-up model to deliver the future trends of fuel consumption and life cycle GHG emissions by China's on-road trucks.

Meanwhile, the study in this field is also hot around the world. Dargay and Gately [6] analyzed and predicted the worldwide vehicle shock from 1960 to 2015, on the view of per capita income. Bastani et al. [7] predicted the uncertainty on US transport-related GHG emissions and fuel consumption out to 2050 and got the view that developing new energy vehicles is the necessary choice for auto industry, but negative influences brought about should not be ignored. Kloess and Müller [8] assess the energy consumption and GHG emissions on the passenger car fleet in Austria, from the aspect of policy, energy prices, and technological progress, and then concluded that material cuts and appropriate taxation on fuels and cars are positive measures for the development of auto industry. In addition, scholars also did studies on different fields, such as fuel types, energy efficiency, and industry policy, and discussed the fuel consumption and GHG emission in Malaysia [9], Europe [10, 11], China [12], and other countries [13].

However, the current literature focus little on quantification of specific implementation effects of auto industry policies. Additionally, the concentration on vehicle emissions is much more on GHG [4, 7–13] while it is little on the prediction of air pollutants (such as CO, HC, NO_x, and PM). This paper aims to make up the defects in research, by predicting and evaluating the growth of auto industry in China during the next few decades, including vehicle population, fuel consumption, and vehicle exhaust emissions.

2. Prediction of Vehicle Population

2.1. Methodology. The change of vehicle population should be traced back to the combined actions of inner needs and external environmental restrictions, which could be properly described by Logistic equation. Logistic equation is a model to describe the trend of dependent variable over time. It can properly reflect the market expansion of new products [14] and shows the following characters: slow growth of the dependent variable at the initial stage, then experiencing a rapid increase stage, finally entering a market saturation stage.

As stated above, the total number of vehicles in China has been considerable. However, from the perspective of per capita car ownership, the auto industry in China is just at the primary phase, comparing with developed countries and even the world average level (Table 1). Thus, this paper assimilated China's auto industry to a market expansion process of new product and then forecasted the developing trend of China's vehicle population according to Logistic model.

Compared with other similar prediction models, for example, the Gompertz model [19, 20], the superiorities of Logistic model can be concluded to two aspects: few constraint conditions, and only one parameter estimation which the whole process involves. In fact, there have been some works in the literature applying Logistic model to the prediction of vehicle population in other countries. Button et al. [21] modeled vehicle ownership and use in five groups of low-income countries by quasilogistic model. Singh developed projections for future mobility in India, based on Logistic and Gompertz models, respectively [22].

The differential form of Logistic model can be expressed as [23]

$$\frac{dF_t}{dt} = aF_t(1 - F_t), \tag{1}$$

where $F_t = N_t/N_m$, N_t represents vehicle population at moment t, N_m is the maximum based on the market, and a is the instantaneous rate of increase, a constant. While $(1-F_t) > 0$, the vehicle population increases; while $(1-F_t) < 0$, it decreases; while $(1 - F_t) = 0$, it remains stable.

TABLE 2: Vehicle population from 1978 to 2008 in China.

Year	1978	1979	1980	1981	1982	1983	1984	1985
Population	142.92	156.57	168.10	187.30	205.32	222.71	243.37	288.71
Year	1986	1987	1988	1989	1990	1991	1992	1993
Population	357.45	412.29	477.64	527.47	583.59	611.41	701.47	817.58
Year	1994	1995	1996	1997	1998	1999	2000	2001
Population	941.95	1040.00	1100.08	1219.09	1319.30	1452.94	1608.91	1802.04
Year	2002	2003	2004	2005	2006	2007	2008	
Population	2053.17	2382.93	2693.71	3159.66	3697.35	4358.36	5099.61	

Unit: 10^4.

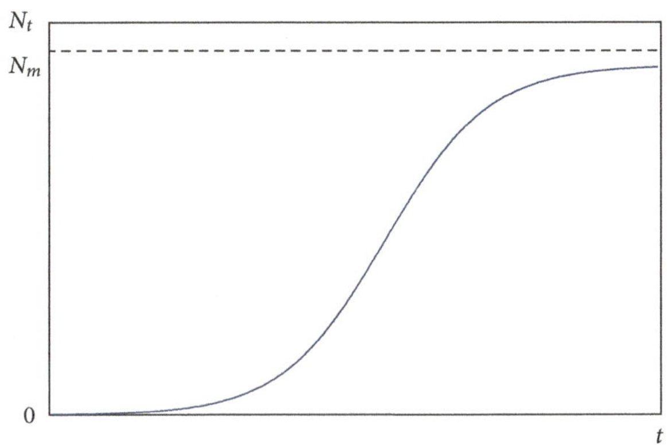

FIGURE 1: Logistic model curve.

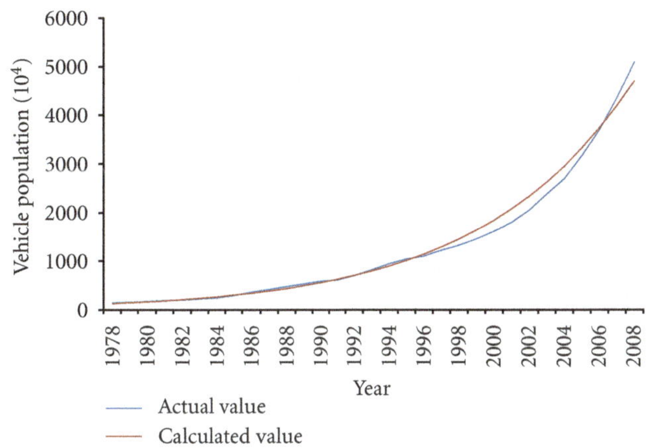

FIGURE 2: Comparison of calculated vehicle population and actual value.

Solving (1) by separating variables, we can see that

$$F_t = \frac{1}{1 + e^{-(b+at)}}, \quad (2)$$

where b is a constant. Equation (2) is shown in Figure 1, which indicates that the Logistic model is an S-shaped growth curve. A feedback regulation exists among three variables (F_t, $(1 - F_t)$, a), which makes vehicle population tend to the maximum N_m.

Take the logarithm of (2):

$$\ln \frac{F_t}{1 - F_t} = at + b. \quad (3)$$

It can be concluded from (3) that $\ln F_t/(1 - F_t)$, with t, constitutes a linear relationship. Hence, estimating a maximum population value N_m based on the actual situation of our country, then figuring out $\ln F_t/(1 - F_t)$, we can get the model parameters (a and b).

2.2. Model Calculation and Estimation Results. This paper consulted the per capita car ownership data (Figure 1) and population densities of developed countries to estimate N_m. In 2008, Americans owned 0.82 cars per capita, while this value ranged from 0.54 to 0.69 in some other developed countries, such as Japan, Germany, Italy, Britain, France, and Spain. The auto industries had developed to a mature period in these countries above, even showing a negative growth since 2007 in the US, Japan, and Germany. Moreover, while the population density in the US (32 people per km^2) was low, the value in Italy (196 people per km^2) and France (113 people per km^2) was closer to that in China (138 people per km^2). Thus, this paper assumed that the per capita car ownership in China would be 0.6 when reaching the maximum. Suppose that Chinese population would keep around 1.4 billion in the next few decades, so N_m was calculated to be 840 million.

Calculate the model parameters (a and b) by (3) and show the result in (4). The original data were the statistic vehicle population in China from 1978 to 2008 (Table 2):

$$\ln \frac{F_t}{1 - F_t} = 0.1217t - 6.5971. \quad (4)$$

The correlation coefficient R^2 of (4) was 0.9932, showing a good linear relationship. Draw the calculated curve by (4) and compare with the actual value in Figure 2. The fitting degree in Figure 2 was tested to be 98.76%, which could guarantee the validity of this model.

The forecast outcome of future vehicle population in China was shown in Figure 3. The vehicle population in China would reach 171.02 million in 2020 and then enter a high speed growth period. In 2035, it would achieve 515.23 million, and at the same time the growth rate would begin to slow down. Around 2050, the vehicle population would have been 762.55 million, and then it would turn into a smooth growth period. Thus, the vehicle population in China should

Table 3: Statuses and planning targets of domestic and foreign vehicle average fuel consumption per hundred kilometers per car.

	Current	2015	2020	2025
China	8.25[1]	7.5[2]	5.78[3]	—
Japan	6.13 [15]	—	4.93 [15]	—
EU	6.13 [16]	5.46 [15]	3.99 [15]	2.94 [16]
USA	7.50 [15]	—	—	4.31 [15]

Unit: L/100 km.
Tips: [1]8.25 = (1 + 0.1) × 7.50, according to the report that current fuel consumption in China was 10% higher than that of USA (available on http://www.find800.cn/news/bus/ye/110930/12778.html).
[2]7.5 = 7 × 0.75 + 9 × 0.25, according to [17, 18] and weights for passenger car and commercial car in China.
[3]5.78 = 5 × 0.75 + 8.1 × 0.25, the same as Tip 2.

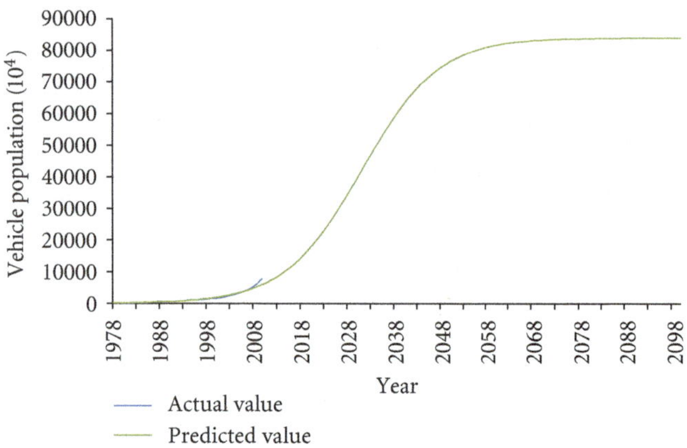

Figure 3: Prediction curve of future vehicle population in China.

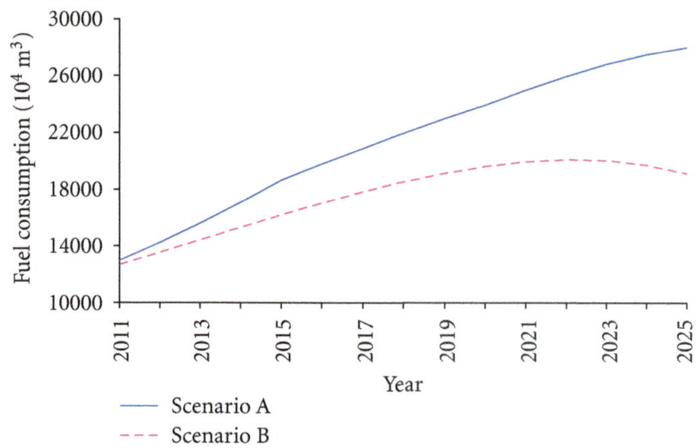

Figure 4: Forecast curve of China's vehicle fuel consumption from 2011 to 2025.

run through a quite rapid growth period inevitably over 2020–2050. Moreover, the enormous environmental pressure followed would bring China huge challenges, such as fuel consumption, exhaust emissions, and scrap car disposal.

3. Prediction of Fuel Consumption

According to the average vehicle fuel consumption per hundred kilometers and future vehicle population predicted above, this paper proposed a prediction of fuel consumption of future auto industry in China. The equation was as follows:

$$V_t = 10^{-5} \cdot v_t S N_t, \quad (5)$$

where V_t is the total vehicle fuel consumption of China in year t, $10^4\,m^3$; v_t is the average fuel consumption per hundred kilometers per car in year t, L/100 km; S is the annual average driving distance per car in China, km; N_t is the predictive value of vehicle population in year t, 10^4.

In addition, v_t could be calculated on the basis of current statuses and planning objectives both in China and overseas (Table 3). The value of S was 2.42×10^4 km, according to Table 3 and the total fuel consumption in 2010.

3.1. Scenario Prediction. Depending on different planning objectives for fuel consumption per hundred kilometers, it is proposed to set the following two scenarios.

Scenario A. In terms of average fuel consumption per hundred kilometers, Chinese auto industry develops under the current domestic set of development goals, which is described as the 1st line in Table 3.

Scenario B. The average fuel consumption of hundred kilometers in China will reach the world's most advanced level in 2025, which is equivalent to the EU fuel consumption level (the 3rd line in Table 3). That is to say, in 2018 the average level in China is assumed to achieve 5.46 L/100 km, the level of EU in 2015, and in 2025 it should achieve 2.94 L/100 km, equal to that of EU.

Calculate all the data in (5) and attain the forecast value of fuel consumption under the above two scenarios, shown in Figure 4.

In Scenario A, the fuel consumption in China will increase year by year in the next 15 years, with no trend to slow down. In 2022, the fuel consumption (239.07 million m^3) will be twice than that in 2011 (121.31 million m^3). In Scenario B, the fuel consumption will achieve the largest in 2022 (185.10 million m^3) and then begin to decline. The accumulative fuel consumptions from 2011 to 2025 will have been 2.3 billion tons and 1.9 billion tons, respectively. However, the Chinese oil reserve was only 2.95 billion tons in 2009. According to the developing trend in Scenario A, the oil

TABLE 4: Implementation schedule of various Motor Vehicles Emission Standard in China.

Year	2000–2003	2004–2006	2007–2010	2011	2012	2013
Emission standard	I	II	III		IV in batches	

TABLE 5: Annual emissions of 4 pollutants for various standards.

	I before	I	II	III	IV[1]
CO	1.4648	0.6421	0.2107	0.0715	0.0286
HC	0.1809	0.0662	0.0221	0.0110	0.0044
NO_x	0.2477	0.0977	0.0400	0.0200	0.0080
PM	0.0296	0.0099	0.0036	0.0010	0.0004
Total	1.9230	0.8159	0.2766	0.1035	0.0414

Unit: tons/car.
Data resource: *China Vehicle Emission Control Annual Report (2010)* [3].
Tip: [1]According to news report, the data for Standard IV was 60% lower than Standard III (Available on http://auto.qq.com/a/20050530/000043.htm).

reserve in China would be depleted out before 2030 without relying on oil import.

3.2. Fuel Consumption Policy Evaluation. In July 2012, China has officially released Energy Saving and New Energy Vehicles Industry Development Planning (2012–2020) (hereinafter referred to as the "*Planning*"), which detailed goals of energy saving and new energy vehicles in China. In *Planning*, there are two stage goals on the market share of new energy vehicles: (1) up to 2015, the number of pure electric vehicles (PEVs) and plug-in hybrid electric vehicles (PHEVs) should outnumber 0.5 million, and medium/heavy hybrid cars should reach one million; (2) up till 2020, the number of PEVs and PHEVs should reach 5 million, and medium/heavy hybrid cars should popularize on a large scale.

To assess the effect of *Planning*, this paper contrasts it with the two scenarios above, shown in Figure 5. From Figure 5, we can see that according to *Planning*, the fuel consumption in China would still keep in a rising trend, but since 2015 the growth will slow down significantly. The fuel consumption would be 1.18% lower than that in Scenario A by 2015, and 9.72% lower by 2020. Thus, we conclude that developing new energy vehicles is the absolutely necessary to reduce the fuel consumption in China. However, to achieve the world's most advanced level, the developing track in Scenario B, China would continue to reduce to 86% of the total fuel consumption by 2015, and 89% by 2020, on the base of *Planning*. That is, by 2020, under the premise of completing the target of 5 million PEVs and PHEVs, medium/heavy hybrid cars should be developed to 47 million at least, which is equivalent to 60% of current domestic vehicle population. Thus, China faces difficult challenges to reduce the vehicle fuel consumption.

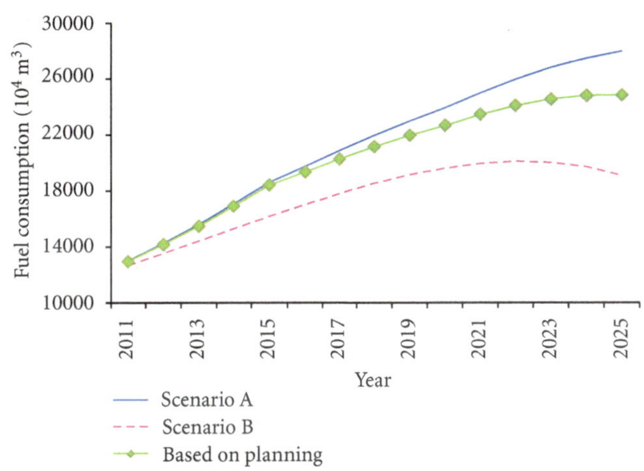

FIGURE 5: Forecast curve of China's vehicle fuel consumption under the implementation of *Planning*.

4. Prediction of Exhaust Emissions

Since 2000, when Motor Vehicles Emission Standard I implemented, China has implemented 4 emission standards, as shown in Table 4. According to *China Vehicle Emission Control Annual Report (2010)* [3], the reduction effects of stringent implement of motor vehicle emission standards emission is very significant. In 2009, vehicles produced before Standard I, which just accounted for 17.1% of the total, discharged four main pollutants (CO, HC, NO_x, and PM) in excess of 50%, while vehicles produced after Standard III, which made up 25.4% of the total, only discharged 4% of the four pollutants.

4.1. Scenario Prediction. To predict and assess changes of four major pollutants (CO, HC, NO_x, and PM) in vehicle emissions, the paper intends to figure out the annual vehicle exhaust emissions by:

$$M_{t,k} = \sum_{i} m_{t,k} N_{i,t}, \quad (6)$$

where $M_{t,k}$ is the emissions of pollutant k in year t, $10^4 t$; $m_{i,k}$ is the annual emissions of pollutant k per car performing Standard i (Table 5), t; $N_{i,t}$ is the predictive value of vehicle population performing Standard i in year t, 10^4; i presents Standard I before, Standard I, Standard II, Standard III, and

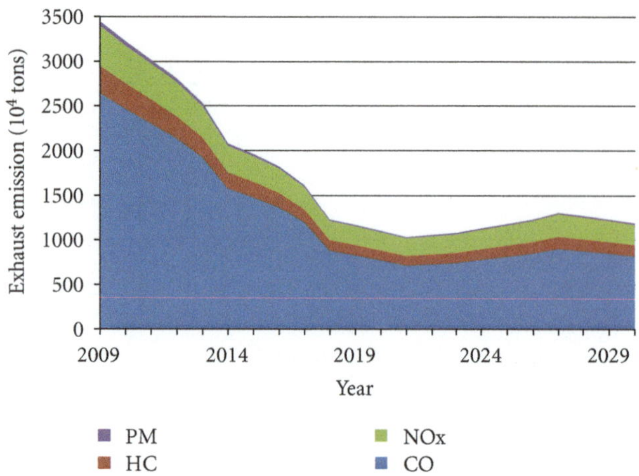

Figure 6: Forecast curve of China's exhaust emissions in Scenario A.

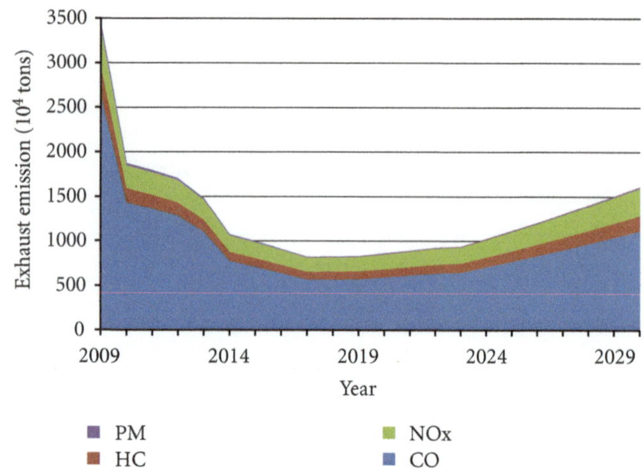

Figure 7: Forecast curve of China's exhaust emissions in Scenario B.

Standard IV; k presents the four main exhaust pollutants, that is, CO, HC, NO_x, and PM.

$N_{i,k}$ is calculated by the relation of vehicles scrapped and vehicle sales per year. Accounting for the different vehicle life cycles, it is proposed to set the following two scenarios.

Scenario A. The vehicle life cycle is 15 years, the maximum prescribed years in *Motor Vehicle Rejection Standard (1997)* [24]. That is, vehicles before Standard I would be fully scrapped till 2014, by the same token, vehicles performing Standards I, II, and III would be fully scrapped, respectively till 2018, 2021, and 2027, while new vehicles would perform Standard IV.

Scenario B. The vehicle life cycle is 11 years, which is calculated according to *Motor Vehicle Rejection Standard (1997)* [24] and the portion of different vehicle types. Thus, vehicles before Standard I would be fully scrapped till 2010, by the same token, vehicles performing Standards I, II, and III would be fully scrapped, respectively till 2014, 2017, and 2023, while new vehicles before 2013 would perform Standard III, then would perform Standard IV.

The prediction curves of China's four major pollutants in vehicle emissions under two scenarios are shown in Figures 6 and 7. It can be found that changes of four pollutants are roughly consistent under the two settings and show a trend that first decreases and then grows. At the lowest point, total emissions, respectively, amount to 10.28 million tons in 2021, and 8.22 million tons in 2017.

It can be concluded that prior to these two time points (2021 in Scenario A, and 2017 in Scenario B), the current implementation of emission standards would continue to play a role in the next 5–10 years, and then the increase in vehicle population would surpass the constraint function of Standard IV, resulting in the regrowth of pollutant emissions. Thus, to ensure the exhaust emissions growth, China should take the more stringent emission standard V at least in 2017–2021.

4.2. Exhaust Emission Policy Evaluation. Further analysing the two scenarios and evaluating the effectiveness of emission standard V, we make two assumptions: firstly, Standard V was 60% lower on the basis of Standard IV, referring the reduction effects of Standard I to Standard IV; secondly, Standard V would be implemented after the time points of lowest emissions (2021 in Scenario A, and 2017 in Scenario B). The prediction curves are shown in Figures 8 and 9. It can be seen that, due to the timely implementation of Standard V, China's vehicle pollutant emissions continue to decline, but the rate is slowing down. Under Scenario B, emissions of the four pollutants restart the upward trend after 2028. However, Table 5 shows that the annual emission of four pollutants per car performing Standard V is only 0.016t, about 1/100 of that of vehicles before Standard I. If only implementing emission standard VI, rather than taking other measures, it would achieve little success, comparing the rapid growth of vehicle population. Therefore, to further control vehicle exhaust emissions, China should actively develop new energy vehicles, on the basis of continuing to strengthen phased emission standards.

5. Conclusions

In this paper, we developed a prediction of vehicle population in China in the next few decades, based on Logistic model, by using a statistical data set over the period of 1978–2008. Given the per capita car ownership and population density of developed countries, we assumed that per capita would own 0.6 cars when entering the saturation state. We predicted that the total number of China's vehicles would be approximately 15 times higher in 2050 than in 2008, increasing to more than 750 million vehicles (Figure 3). It also can be concluded that China's auto industry would run through a high speed development period during 2020–2050, which would inevitably aggravate China's environmental burden.

Moreover, fuel consumption prediction over 2011 to 2025, based on the average vehicle fuel consumption per

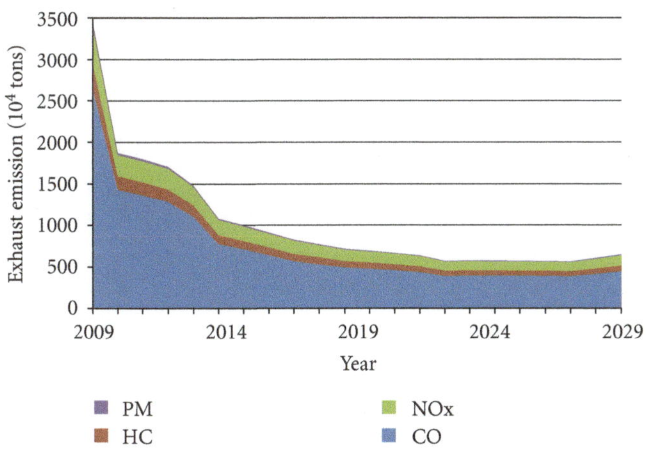

FIGURE 8: Forecast curve of China's exhaust emissions under the implementation of Standard V in Scenario A.

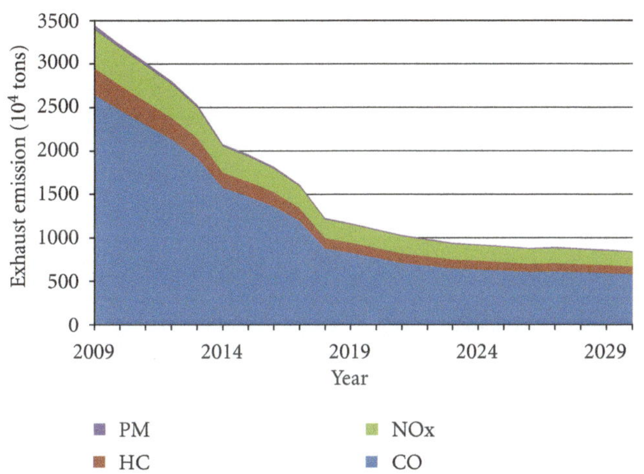

FIGURE 9: Forecast curve of China's exhaust emissions under the implementation of Standard V in Scenario B.

hundred kilometers, shows that under the current scenario, China's fuel consumption for vehicles would grow without trend to slow down, reaching 239.07 million m^3 in 2022 (Figure 4), which is the twice in 2011. However, under the world's most advanced scenario, it would peak in 2022 and then begin to decline. Considering the tension of fuel supply, the current development goals for China auto industry could not be suited to demand. Further, we evaluated *Planning*, China's latest policy for vehicle fuel consumption, and found that the growth trend of fuel consumption in China would meet a marked slowness since 2015 (Figure 5), on the basis of current development goals. Nevertheless, to achieve the world's most advanced level, China should develop medium/heavy hybrid cars to 47 million at least by 2020, under the existing premise of completing the target of 5 million PEVs and PHEVs.

Furthermore, allowing for different vehicle life cycles (the maximum and the average), exhaust emissions prediction for China's vehicles up to 2029 was developed in this paper, on the basis of vehicle emission standards for different phrases. We projected that the four pollutants (CO, HC, NO_x, and PM) would display roughly consistent changing trends under the two settings and reach the lowest points, respectively, in 2021 and in 2017 (Figures 6 and 7). Thus, in order to hold back the exhaust emission growth, China should continue to implement the more stringent emission standard V at least in 2017–2021. However, further study on the effect of emission standards showed that the Standard VI would achieve little, when it meets the rapid growth of vehicle population (Figures 8 and 9).

Finally, our results suggest that the future strong growth in China's vehicle population will bring huge burdens on fuel demand and exhaust emissions. China should actively develop new energy vehicles, which is the most effective measure to ease the pressure on fuel demand and exhaust emissions brought by auto industry.

References

[1] K. He, H. Huo, Q. Zhang et al., "Oil consumption and CO_2 emissions in China's road transport: current status, future trends, and policy implications," *Energy Policy*, vol. 33, no. 12, pp. 1499–1507, 2005.

[2] The International Council on Clean Transportation, "Overview of China's vehicle emission control program," 2010.

[3] Ministry of Environmental Protection of the People's Republic of China, China Vehicle Emission Control Annual Report (2010), 2011.

[4] X. Yan and R. J. Crookes, "Reduction potentials of energy demand and GHG emissions in China's road transport sector," *Energy Policy*, vol. 37, no. 2, pp. 658–668, 2009.

[5] H. Hao, H. Wang, and M. Ouyang, "Fuel consumption and life cycle GHG emissions by China's on-road trucks: future trends through 2050 and evaluation of mitigation measures," *Energy Policy*, vol. 43, pp. 244–251, 2012.

[6] J. Dargay and D. Gately, "Income's effect on car and vehicle ownership, worldwide: 1960–2015," *Transportation Research A*, vol. 33, no. 2, pp. 101–138, 1999.

[7] P. Bastani, J. B. Heywood, and C. Hope, "The effect of uncertainty on US transport-related GHG emissions and fuel consumption out to 2050," *Transportation Research A*, vol. 46, pp. 517–548, 2012.

[8] M. Kloess and A. Müller, "Simulating the impact of policy, energy prices and technological progress on the passenger car fleet in Austria—a model based analysis 2010–2050," *Energy Policy*, vol. 39, no. 9, pp. 5045–5062, 2011.

[9] T. M. I. Mahlia, S. Tohno, and T. Tezuka, "History and current status of the motor vehicle energy labeling and its implementation possibilities in Malaysia," *Renewable and Sustainable Energy Reviews*, vol. 16, pp. 1828–1844, 2012.

[10] T. Zachariadis, "On the baseline evolution of automobile fuel economy in Europe," *Energy Policy*, vol. 34, no. 14, pp. 1773–1785, 2006.

[11] M. F. Torchio and M. G. Santarelli, "Energy, environmental and economic comparison of different powertrain/fuel options using well-to-wheels assessment, energy and external costs—european market analysis," *Energy*, vol. 35, no. 10, pp. 4156–4171, 2010.

[12] X. Ou, X. Zhang, and S. Chang, "Alternative fuel buses currently in use in China: life-cycle fossil energy use, GHG emissions and policy recommendations," *Energy Policy*, vol. 38, no. 1, pp. 406–418, 2010.

[13] L. Kliucininkas, J. Matulevicius, and D. Martuzevicius, "The life cycle assessment of alternative fuel chains for urban buses and trolleybuses," *Journal of Environmental Management*, vol. 99, pp. 98–103, 2012.
[14] Z. Mohamed and P. Bodger, "A comparison of Logistic and Harvey models for electricity consumption in New Zealand," *Technological Forecasting and Social Change*, vol. 72, no. 8, pp. 1030–1043, 2005.
[15] L. Jiu, "Subsidy policy for energy saving cars is of far-reaching significance," *Automobile & Parts*, vol. 41, pp. 14–16, 2011 (Chinese).
[16] H.-H. Zhou, "Internal combustion engine is still the main force for automotives'energy conservation and emission reduction in the future," *Automobile & Parts*, vol. 32, p. 26, 2011 (Chinese).
[17] "Rules of average fuel consumption may come on brands in freedom confront again puzzledom," *Automobile Parts*, no. 8, p. 15, 2011.
[18] GB, 19578–2004, "Passenger car fuel consumption limit value".
[19] M. Keshavarzian, S. K. Anaraki, M. Zamani, and A. Erfanifard, "Projections of oil demand in road transportation sector on the basis of vehicle ownership projections, worldwide: 1972–2020," *Economic Modelling*, vol. 29, pp. 1979–1985, 2012.
[20] J. Dargay, D. Gately, and M. Sommer, "Vehicle ownership and income growth, worldwide: 1960–2030," *Energy Journal*, vol. 28, no. 4, pp. 143–170, 2007.
[21] K. Button, N. Ngoe, and J. Hine, "Modeling vehicle ownership and use in low-income countries," *Journal of Transport Economics and Policy*, vol. 27, pp. 51–67, 1993.
[22] S. K. Singh, "Future mobility in India: implications for energy demand and CO_2 emission," *Transport Policy*, vol. 13, no. 5, pp. 398–412, 2006.
[23] P. Zhao, H.-W. Zhang, and Z.-G. Niu, "Nonlinear dynamics of urban domestic water consumption," *Journal of Tianjin University Science and Technology*, vol. 39, no. 6, pp. 735–738, 2006.
[24] "Motor Vehicle Rejection Standard," 1997.

13

Effects of Thermal Mass, Window Size, and Night-Time Ventilation on Peak Indoor Air Temperature in the Warm-Humid Climate of Ghana

S. Amos-Abanyie,[1] F. O. Akuffo,[2] and V. Kutin-Sanwu[1]

[1] *College of Architecture and Planning, Kwame Nkrumah University of Science and Technology, Kumasi, Ghana*
[2] *College of Engineering, Kwame Nkrumah University of Science and Technology, Kumasi, Ghana*

Correspondence should be addressed to S. Amos-Abanyie; sumwelamos@yahoo.com

Academic Editors: A. J. N. Khalifa and A. Klimenko

Most office buildings in the warm-humid sub-Saharan countries experience high cooling load because of the predominant use of sandcrete blocks which are of low thermal mass in construction and extensive use of glazing. Relatively, low night-time temperatures are not harnessed in cooling buildings because office openings remain closed after work hours. An optimization was performed through a sensitivity analysis-based simulation, using the Energy Plus (E+) simulation software to assess the effects of thermal mass, window size, and night ventilation on peak indoor air temperature (PIAT). An experimental system was designed based on the features of the most promising simulation model, constructed and monitored, and the experimental data used to validate the simulation model. The results show that an optimization of thermal mass and window size coupled with activation of night-time ventilation provides a synergistic effect to obtain reduced peak indoor air temperature. An expression that predicts, indoor maximum temperature has been derived for models of various thermal masses.

1. Introduction

The predominant hot and humid tropical nature of the climate of Ghana makes climatic conditions predominantly fall outside the human comfort zone and thus require cooling for about 50 and 100% of the year for which air conditioning (AC) is used [1]. Figure 1 shows the Building Bioclimatic Chart for Kumasi, the second largest city in Ghana, showing the mean outdoor air temperature and relative humidity represented by the lines falling outside the comfort zone (shaded). The application of passive and low energy cooling techniques (PLECT) can reduce AC use to below 7% of the time of the year [1]; however, AC is adopted as a straightforward response to achieve thermal comfort without recourse to the potential of PLECT to reduce energy use in space cooling.

In recent years, there has been a movement towards universal building design approaches with extensive use of glazing that is poorly adapted to the local climatic conditions and also permits high penetration of solar radiation. Construction of building envelope in Ghana predominantly employs sandcrete blocks that are of low thermal mass. No empirical study has been carried out to assess its performance with that of other available materials. The relatively low night outdoor air temperature is not harnessed to cool the thermal mass of buildings. At night, there is reasonably very low air temperature; however, because the windows remained closed, as it is in normal practice in office buildings at night, the cool night air does not contribute to cooling the thermal mass at night. The cooling of the mass at night from the outside is hampered by the thermal resistance of the mass, and thus the inner layers are not cooled to provide a heat sink during the following day. The above leads to higher average peak indoor air temperatures with a corresponding higher cooling load.

There is lack of practical advice and empirical data on potential of PLECT to reduce peak air temperatures in figures readily appreciated by building designers and

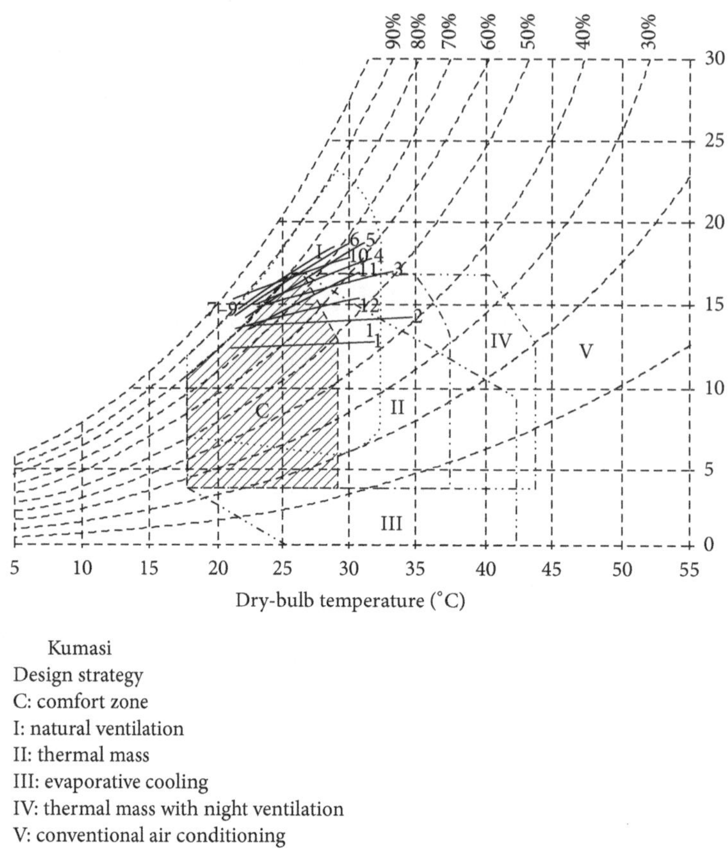

Figure 1: Building Bioclimatic Chart of Kumasi [1].

prospective builders. In many instances, the building design is not supported by a detailed analysis and evaluation of thermally relevant features and options related to orientation, envelope, glazing ratio, shading devices, and thermal mass. Thus, design decision making is not sufficiently informed by pertinent expertise pertaining to energy-efficient building design methods and technologies [2]. Selection of materials for a building in Ghana has often been based on affordability, aesthetics, appearance, durability, and in some cases availability, other than functional, thermal performance, and environmental considerations.

It has been projected that Ghana will need more than 7 times its 2007 electric power capacity by 2020 if it should succeed in developing its economy into a middle income one [3]. It is clear that Ghana will have to expand and diversify installed energy capacity. However, expansion is costly and takes time. In the meantime, any energy resources that can be saved through efficiency or conservation can be used for other purposes.

The aim of this paper is to investigate, in a systematic way, using both simulation and experimental approaches, the effects of thermal mass, window size, and night ventilation technique to reduce peak indoor air temperature in the mode of building design in Ghana to reduce cooling load in buildings.

2. Methodology

In investigations involving cooling potential of passive cooling techniques, energy simulation models of the building need to be validated with scale measured data from a building designed to run on the technique in question so that retrofit strategies can be targeted to solve the building's energy problems [4]. During the last two decades, several calibration methodologies have been developed [4–6]. Most of these methods employed site measurement and monitoring of the building's performance. However, while this approach is appropriate for retrofit projects with a limited budget [7], no building is designed in Ghana to run on night ventilation technique from which data can be collected for the validation.

In the above context, the procedure employed in this study included the following tasks to achieve the aim of the research (Figure 2): (1) data collection of the climatic data and building parameters; (2) development of a simulation model and prediction of indoor air temperature; (3) design of experimental setup, construction, and monitoring of indoor air temperature; and (4) validation of the predicted (simulated) results with the experimental results employing graphical and statistical tools. Simulation was adopted to normalize confounding factors such as differences in obtained and

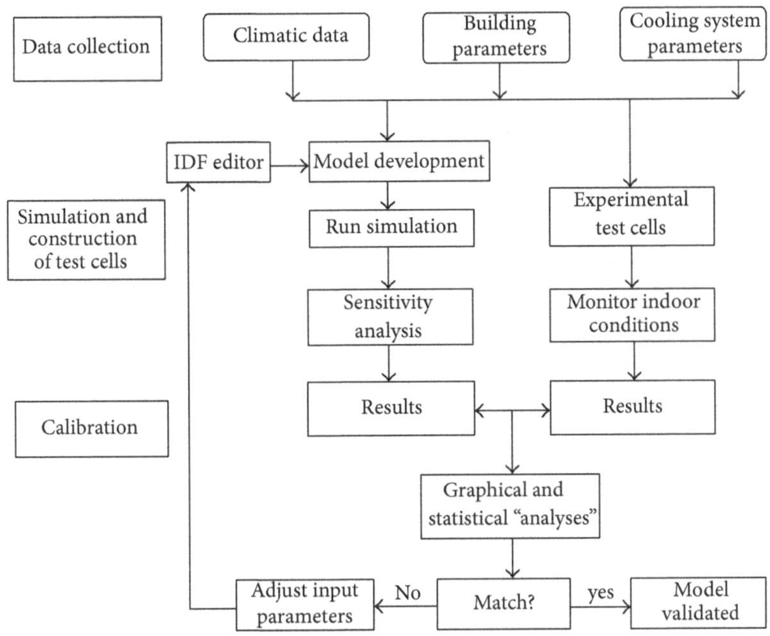

FIGURE 2: Flowchart of activities.

actual weather file and construction mixtures [6]. Version 4 of Energy Plus Building Simulation Program for the analysis [8] was used.

2.1. Data Collection. The data collected included climatic data and building parameters. Simulation requires a detailed and comprehensive meteorological input data [5]. For this study, a weather file in a timestep of 6 was generated via Meteotest 2008 file [2]. The weather comprised outdoor dry-bulb temperature, relative humidity, wind speed, solar radiation, and ground temperature. The building parameters included the thermophysical properties of the basic building materials used for the model. Building parameters were obtained from the specifications and architectural drawings of the conventional mode of building design as specified in the Ghana Building Regulations. Building parameters helped to correctly dimension the model and specify zone data.

2.2. Simulation Model Development. A series of comparative simulations were performed using the weather file of Kumasi. The period for the simulation was from the month of November to March inclusively, a total of 212 days. This period represents the warmest months in Ghana [9]. A control model simulating the conventional mode of building design and construction in Ghana was first developed. The control models had dimensions of 2.40 m high, 1.2 m wide, and 2.4 m deep. The floor was made of 150 mm thick mass concrete, and the roof was made of 150 mm thick reinforced concrete slab. The wall of the control model was made of 150 mm thick solid sandcrete block which in this research represents the low mass material. The pilot survey of selected buildings in the region of study established the average window-to-floor ratio to be 54%. This was used as the base case of the control model. A total window-to-floor ratio of 1.62 m^2 of glazing area was obtained with a floor area of 2.44 m^2 of the control model. The control model had windows closed for 24 hours, with no night-time ventilation.

To assess the effect of thermal mass, three models were used, one with each of the three different thermal masses: solid sandcrete blocks (SSB), baked bricks (BB), and concrete (CONC) all with window-to-floor ratio of 54% as in Table 1. Nine models were used to assess the effect of window size. There were three models made of each of the solid sandcrete blocks (SSB), baked bricks (BB), and concrete (CONC). Among the three models of given thermal masses, each had 54% (window area of 1.62 m^2), 27% (window area of 0.82 m^2), and 0% window-to-floor ratios. The models used in assessing the effects of thermal mass and window size were not subjected to night-time ventilation. In assessing effect of night-time ventilation, the nine models used in assessing the effects of window size were subjected to a night-time ventilation rate of 10 ACH. Table 2 shows the characteristics of models.

To assess the effect of varying night-time ventilation rate, the air change rates per hour were increased from 10 ACH to 20 ACH, then to 30 ACH. The test models were subjected to night-time ventilation from 19:00 hours (evening) till 7:00 hours (next morning). It is very common for night-time ventilation to be provided by natural ventilation. However, the site location of this research had irregular wind patterns and low wind speeds. This could not have provided effective natural ventilation rates for effective night-time ventilation. For this reason, mechanical ventilation was adopted with the use of an intake fan and an exhaust fan to provide night-time ventilation. During the daytime periods, the fans were provided with dampers to avoid internal heat gains by convection. Intake and exhaust ventilation fans were modelled for each model.

TABLE 1: Treatments made to the control model.

Thermal mass		SSB			BB			CONC	
Model no.	**1**	**2**	**3**	**4**	**5**	**6**	**7**	**8**	**9**
W-F-R	54%	27%	0	54%	27%	0	54%	27%	0
Effect of mass	√			√			√		
Effect of window size	√	√	√	√	√	√	√	√	√
Effect of NV	√	√	√	√	√	√	√	√	√

TABLE 2: Description of models.

Test model	Description
Model 1	Light-weight mass with 0.54% window-to-floor ratio
Model 2	Light-weight mass with 0.27% window-to-floor ratio
Model 3	Light-weight mass with no window
Model 4	Light-weight mass with 0.54% window-to-floor ratio
Model 5	Light-weight mass with 0.27% window-to-floor ratio
Model 6	Light-weight mass with no window
Model 7	Heavy-weight mass with 0.54% window-to-floor ratio
Model 8	Heavy-weight mass with 0.27% window-to-floor ratio
Model 9	Heavy-weight mass with no window

2.3. Performance Evaluation of Simulation Models. The performance of the test models was evaluated using that of the control model as a baseline. Performance of the models was rated by evaluation of three variables: (1) the comparison of the maximum temperatures, (2) the temperature difference ratio (TDR), and (3) the percentage of overheated hours [10]. The temperature difference between a test model and the control model was indicative of the performance of the test model. According to Geros et al. [11], when the temperature reduction is high, the peak demand is delayed, and thus cooling load of a building is lower. The higher the drop in the peak indoor temperature of a test model, the better its performance.

The concept of TDR was proposed by Givoni and has been used with good results to compare passive cooling systems with different configurations [10]. TDR was calculated using the following expression:

$$\mathrm{TDR} = \frac{(T_{\mathrm{maxout}} - T_{\mathrm{maxin}})}{(T_{\mathrm{maxout}} - T_{\mathrm{minout}})}, \quad (1)$$

where TDR is temperature difference ratio; T_{maxin} is maximum temperature inside; T_{maxout} = maximum temperature outside; and T_{minout} is minimum temperature inside.

The numerator is the difference between the indoor maximum temperature and the outdoor maximum, and the denominator is the outdoor temperature swing. The higher the value of the TDR, the better the performance of a test model. A higher value indicates that there is a larger temperature difference between outdoors and indoors and there is more cooling. A negative TDR value indicates that the average maximum temperature inside the test cell is higher than that of outdoors. The relation between window-to-floor-ratio (WFR) and TDR was used to derive an expression to predict maximum indoor air temperature.

The number and percentage of overheated hours beyond the comfort band for each model were assessed. The comfort band proposed by Givoni [12] of between 18 and 29°C for tropical countries was used. Mean maximum temperature for a given day that was above 29°C were considered hot or uncomfortable. The number of hot and comfortable hours and their distribution permitted the examination of the overall pattern of temperature in the experimental models. A matrix made up of 24 rows that represent each hour of the day and a number of columns equal to the number of measured days in a series was prepared and used for the assessment of the performance of the models. Each of these cells indicated the average temperature for each hour of each day. The larger the difference in the number or percentage of overheated hours between a given test model and the control model the better the performance of the model.

2.4. Experimental Setup. An experimental setup was designed and constructed on the premises of the College of Architecture (CAP) at the Kwame Nkrumah University of Science and Technology (KNUST) in Kumasi, Ghana. The experimental system consisted of two cells (control and experimental cell); an active ventilation system consisting of two sets of 100 mm fan (in the south wall) and 100 mm outlet fans (in the north wall); HOBO H8 temperature data loggers to record air temperature; insulated external probe thermistor to monitor radiant temperature of internal surfaces. Figures 3 and 4, respectively, show the schematic drawing and the test cells with installations.

The test cells were carefully designed with the same specifications as the numerical models to achieve accurate experimental test results. The control and experimental cells had features of the base model and the most promising test model, respectively. The openings of both cells remained closed for 24 hours in a day.

The fans were operated to ventilate the experimental cell each night from 7 pm till 7 am (local time). The control cell was not ventilated at night. All the measured data were transferred through a USB to the laptop computer for the analysis. An area-weighted value for mean radiant temperature based on the five measured surfaces was calculated [13]:

$$T_{mr} = \frac{\sum_1^4 T_{sn} A_n}{\sum_1^4 A_n}, \quad (2)$$

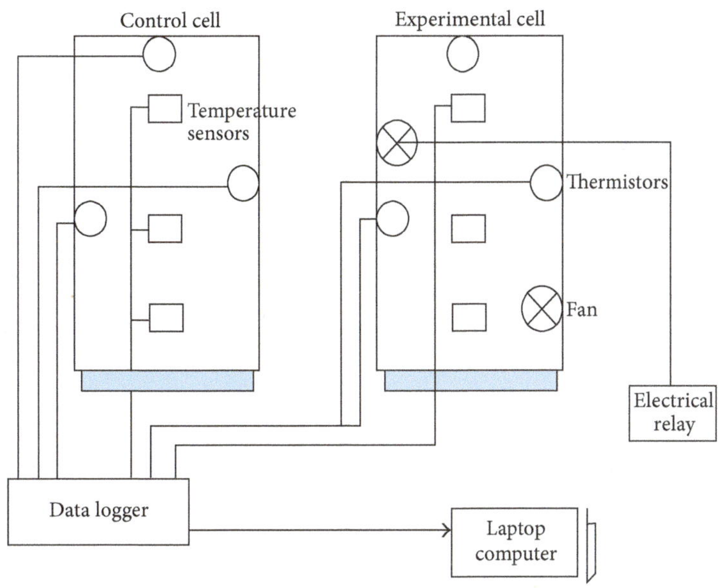

FIGURE 3: Schematic drawing of the experimental setup.

FIGURE 4: Test cells with installations.

where T_{mr} is the mean radiant temperature; T_{sn} is the temperature of a surface; and A_n is the area of a surface.

2.5. Validation Procedure. Measured indoor air and radiant temperature from the test cells were used to calibrate the predicted (simulated) indoor air temperature and mean surface temperature of a two-week period for a satisfactory agreement. Statistical analysis involving the root mean square difference (r^2) and the corresponding coefficient of variance of root-mean-squared error (CV (RMSE)) value between simulated and measured air temperature and mean radiant temperature was computed and checked to fall within allowable tolerances. The coefficient of variance of the root-mean-squared error, CV (RMSE) (%), is essentially the root-mean-squared error divided by the measured mean of all the data, which is a convenient way of reporting a nondimensional result [14]. CV (RMSE) allows one to determine how well a model fits the data; the lower the CV (RMSE), the better the calibration. According to Kreider and Habel [15], the acceptable range considered appropriate for hourly CV (RMSE) of empirical models is 10 to 20%

$$\text{CV (MSE)} = \frac{\left[\left(\sum y_{\text{pred},i} - \sum y_{\text{data},i}\right)^2 / (n-p)\right]^{1/2}}{\hat{y}_{\text{data}} \times 100}, \quad (3)$$

where $y_{\text{pred},i}$ is a predicted dependent variable value for the same set of independent variables; $y_{\text{data},i}$ is a data value of the dependent variable corresponding to a particular set of the independent variables; $\hat{y}_{\text{data}}$ is the mean value of the dependent variable of the data set; n is the number of data points in the data set; and p is the total number of regression parameters in the model (arbitrarily assigned as 1 for all models).

3. Discussion of Simulation Results

Analysis of the simulated results shows that all the models, both control and test, maintained indoor maximum air temperatures above the outdoor air temperature. However, all the test models had improved conditions than the control model. The effects of thermal mass and window size were assessed with maximum temperature difference between test models and the control model, and the effects of night-time ventilation were assessed with maximum temperature difference, temperature difference ratio, and percentage of overheated hours.

3.1. Effect of Thermal Mass. From the analysis, increasing the thermal mass by changing to materials of higher densities led to a reduction in peak indoor air temperature. Concrete had the highest effect in reducing PIAT, followed by baked bricks and then solid sandcrete blocks. Table 3 presents the peak indoor air temperature, maximum temperature differences with the control model, and peak delay hours of the various

TABLE 3: Outdoor and indoor air temperatures of model 9 and the control model.

	PIAT[a]	DBC[b]	PDH[c]
MAX_{out}	30.1	—	
SSB	35.6	—	2
BB	34.9	0.7	3
CONC	32.6	3.0	5

[a]Max. indoor air temperature.
[b]Difference below control Model.
[c]Peak delay hours.

thermal masses tested. The control model made of the solid sandcrete blocks had a peak indoor air temperature of 35.6°C. The test models of baked bricks and concrete envelopes had an average peak indoor air temperature of 34.9°C and 32.6°C. This means that BB reduced peak indoor air temperature (PIAT) by 0.7°C and concrete had the highest effect with a drop of 3°C below that of the control model.

Increased thermal mass also led to an increase in the number of hours of delay of PIAT occurrence after peak outdoor air temperature. SSB, BB, and CONC had number of hours of delay of 2, 3, and 5 hours, respectively. The above indicates that walls made of concrete and baked bricks can maintain more steady indoor temperature with lower fluctuations and postpone the time of occurrence when peak indoor air temperature appears with a longer time lag than solid sandcrete blocks. Heat storage properties of concrete and baked bricks can postpone the time of occurrence when peak indoor air temperature appears with a longer time lag than solid sandcrete blocks.

In models of solid sandcrete block walls, indoor air temperature rises dramatically during the morning (early) hours and reaches its maximum in late afternoons. It then cools down rapidly in the evening (late) hours, with a wider fluctuation in diurnal indoor temperature pattern. The temperatures in the solid sandcrete block wall models closely followed that of the outdoor conditions and did not offer any significant thermal storage. This is because they are of low mass material. This observation is similar to observations made by [16] in an architectural science inquiry on four environmental test chambers with different thermal mass level types in Nairobi, Kenya. The models with baked bricks walls have mean indoor patterns that follow closely that of the solid sandcrete blocks but with relatively smaller fluctuations.

The analysis revealed that higher thermal mass reduced indoor maxima and at the same time brought up indoor minima. However, the influence on maxima was larger than that on minima. In the test models with concrete walls, the moderating effect of thermal mass was noticeable. Because they are of a high mass material, the maximum indoor air temperatures are suppressed at daytime and rise slowly to achieve a relatively lower maximum in the late afternoons. The stored heat in concretes reradiates back into the space at night and causes much higher minimum temperature during night hours. The observed trend is characteristic of high mass materials, and similar observations were made in a study by [17] to investigate the effects of colour and building thermal mass in the hot humid climate of Hong Kong. Indoor temperature fluctuations associated with baked bricks and concrete are kept to a minimum.

3.2. Effect of Window Size. Reduced window-to-floor ratio for all the envelope materials tested leads to reduction in peak indoor air temperature for all the thermal masses (Table 4). Concrete wall envelope with no window had the best performance with the lowest PIAT. However, windows are an important source of natural lighting and views from nature that should not be eliminated in buildings.

The reductions were, however, not remarkable with none being up to 0.9°C drop in PAIT. This can be attributed to the effect of shading devices on the test models which eliminates solar and conductive heat gains through the windows. The effective shading led to marginal effect of the varying window sizes. A study by La Roche and Milne [10] to quantify the effects of modifying the amount of thermal mass and the window area on indoor comfort using smart controllers revealed that an increase rather in the size of unprotected window could have a more significant the performance.

3.3. Effect of Night-Time Ventilation. An analysis of maximum temperature difference between night-time ventilated models and the control model revealed that the largest difference of 4.3°C was obtained in the concrete model (model 9) that has no window and heavy weight mass with night ventilation (Table 5). The test models of series 3, 6, 7, and 8 also have appreciable differences between their average maximum and that of the control model that indicates that features in these test models have a significant effect on their performance. These features are the reduction of the window size, more thermal mass, or night ventilation. But it is the combination of these features that achieves the improved performance.

The impact of night ventilation was assessed for various air flow rates considering 10, 20, and 30 ACH. Increase in the air change rate did insignificantly reduce average peak indoor air temperature. For instance, for model 9, increase in air rate per hour from 10 to 20 ACH and from 20 to 30 ACH leads to reductions of 0.067°C and 0.044°C, respectively. Models 3, 6, 7, and 8 also had significant features with mean reductions of 0.046°C and 0.016°C with increased air rate per hour from 10 to 20 ACH and from 20 to 30 ACH. However, night ventilation of BB and CONC exposed them to cool outdoor air to offset, in part, the reradiated heat. This provides a relatively lower minimum indoor air temperature at night. The attained low indoor air temperature and cooled mass provided a heat sink that suppressed the attainable maximum indoor temperature during the daytime.

Unlike a study by [18] in the hot-humid climates at the Passive Cooling Laboratory (PCL) in the Florida Solar Energy Centre where increased air exchange rate could lower the average indoor air temperature, the difference in temperature as a result of night-time ventilation between the low mass and the high mass walls for this study was negligible. This could be explained by the relatively lower diurnal temperature range of Kumasi which does not effectively cool the building mass.

Table 4: Indoor air temperature models with varied window patterns.

Window-to-floor ratio	PIAT			DBC		
	SSB	BB	CONC	SSB	BB	CONC
54%	35.58	34.93	32.56	0.00	0.66	3.03
27%	34.77	34.16	32.06	0.81	1.42	3.52
No window	33.78	33.63	31.52	1.80	1.95	4.06

DAO: difference above outdoor maximum temperature.
DBC: difference below control model.

Table 5: Maximum temperature differences, temperature difference ratios, and percentage of overheated hours.

Models	Control	SSB			BB			CONC		
		1	2	3	4	5	6	7	8	9
MEAN MAX. (Nvent—10 ACH)	35.58	35.56	34.74	33.73	34.61	344.13	33.59	32.39	31.87	31.31
10 ACH	—	0.026	0.849	1.849	0.975	1.453	1.990	3.191	3.710	4.278
20 ACH	—	0.033	0.857	1.860	0.983	1.462	2.001	3.244	3.770	4.345
30 ACH	—	0.036	0.861	1.865	0.987	1.467	2.007	3.270	3.799	4.379
TDR at 10 ACH	—	−0.69	−0.59	−0.46	−0.57	−0.51	−0.44	−0.29	−0.23	−0.16
Overheated hours	48.2	41.1	38.7	36.9	39.3	38.3	37.5	38.7	37.5	35.7

Temperature difference ratio (TDR) was calculated and averaged for all the days for each model (Table 5). The best TDR amongst the test models is model 9, which performs far better, over four times, than the control model. Amongst all the thermal masses, as the north and south window-to-floor ratio increases, the TDR decreases. This implies that the large south- and-north facing windows reduce the performance of the test models. The relation between TDR and window-to-floor ratio (Figure 5) for the different thermal masses provided basis for the development of the predictive expression that can be used to predict TDR as a function of the floor to window ratio is derived (4). When TDR of a building is calculated for a given material, it is possible to predict the indoor maximum temperature using (1) and solving for T_{maxin}, where outdoor maximum and minimum temperatures, or daily temperature swing, must be known.

Since no trend line strikes the positive window-to-floor area axis, it means that for all masses, conditions inside the models would be higher (or worse) and above the average outdoor temperatures even when there are no windows. In La Roche and Milne [10] study, when the south window-to-floor ratio was of a minimum of 11% and above, the indoor air conditions in the control cell, with a fixed infiltration rate, were higher (worse) than that of the average outside air temperature. While the predictive expression developed for indoor maximum temperatures in this research is similar to that by La Roche [19] in Lyle Centre in the Cal Poly Pomona, differences exist in the gradient and the constants. The differences between the expressions could be attributed to differences in thermal properties of the materials and climatic conditions associated with this study and that of [10].

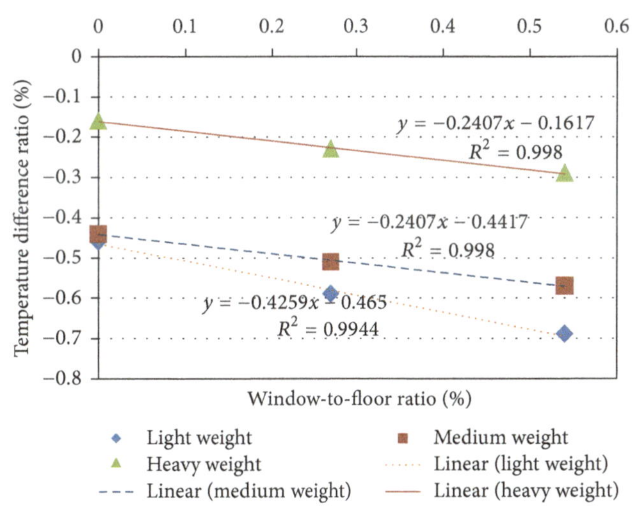

Figure 5: TDR as a function of the window-to-floor ratio.

Light-weight mass:

$$TDR = -0.4259 * WFR - 0.465 \quad [R^2 = 0.99] \quad (4)$$

Medium-weight mass:

$$TDR = -0.2407 * WFR - 0.4417 \quad [R^2 = 0.98] \quad (5)$$

Heavy-weight mass:

$$TDR = -0.2407 * WFR - 0.1617 \quad [R^2 = 0.99]. \quad (6)$$

In all the above, TDR is a temperature difference ratio; WFR is window-to-floor ratio (north and south window).

The variation in the expressions for this study suggests that it is applicable to buildings of envelopes of the respective thermal masses of 150 mm thick mass walls, with shaded north and south windows only, and 150 mm mass concrete floor construction, and night ventilated at 10 ACH in the warm-humid climates similar to that of the Ashanti Region of Ghana.

For all the expressions, as window-to-floor ratio increased, the temperature difference ratio decreased.

The number of overheated hours of each model for the simulation period is given in Table 5 and is compared with the number for the control model. With the base temperature of 29°C, the achieved reduction of the overheating hours for sandcrete blocks varies between 36% and 42% for air flow rate of 10 ACH. For baked bricks, the corresponding decrease varies between 37% and 39%, while for concrete, the reduction is between 35% and 39%.

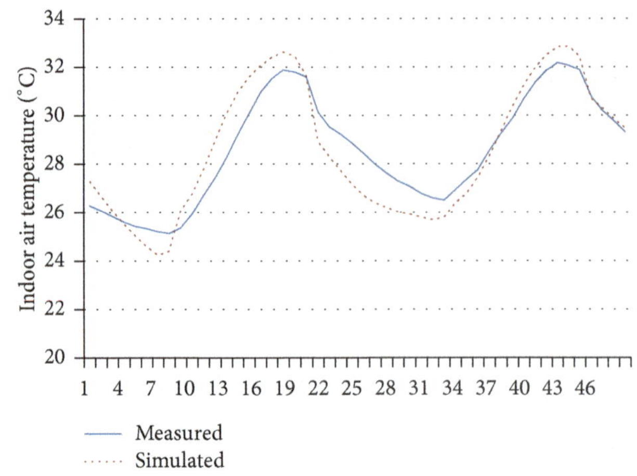

FIGURE 6: Measured versus simulated indoor air temperatures.

4. Validation (Calibration) of Simulation Results

Predicted values of the simulation models compared well with the measured values. Figures 6 and 7 illustrate segments of the match and relationship (in terms of regression lines) measured versus simulated indoor air temperature, respectively. The respective correlation coefficient values are summarized in Table 6. In terms of statistic analyses, root-mean-square difference (r^2) was 0.82 and 0.83 for indoor air temperature and mean radiant temperature, respectively, and coefficient of variation for the root-mean-squared error (CV (RMSE)) was 14.75% and 16.80% for indoor air temperature and mean radiant temperature, respectively, and is considered statistically inappropriate falling within the range suggested by [15]. The good match of predicted results with the measured data suggests that thermophysical properties of the materials used are consistent with that specified in the simulation program.

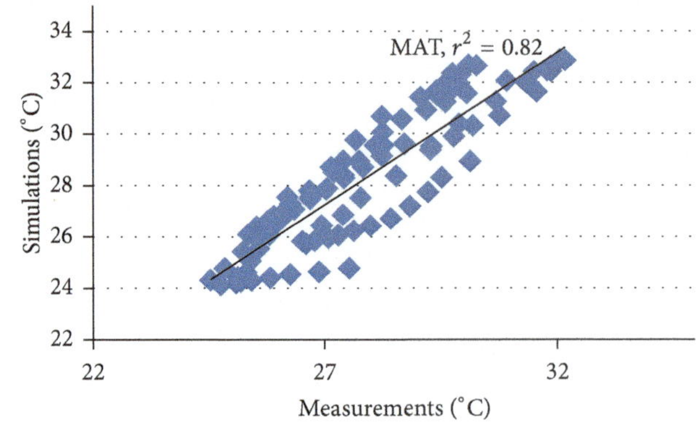

FIGURE 7: Measured versus simulated indoor air temperatures.

TABLE 6: Correlation coefficient values.

	r^2	CV (RMSE)
Indoor air temperature	0.82	14.75%
Mean radiant temperature	0.83	16.80%

5. Conclusion

Employing both sensitivity-based simulation and experimental approaches, the paper has presented the results of the effects of thermal mass, window size, and night-time ventilation on peak indoor air temperature in the warm-humid climate conditions with Ghana as a case.

From the analysis, increasing the thermal mass by changing to materials of higher densities led to a reduction in peak indoor air temperature. Concrete had the highest effect in reducing PIAT and maintained a drop of 3°C below that of the solid sandcrete blocks, with baked bricks having a drop of 0.7°C. Increased thermal mass also led to an increase in the number of hours of delay of PIAT occurrence after of peak outdoor air temperature. Solid sandcrete blocks, baked bricks, and concrete had number of hours of delay of 2, 3, and 5 hours, respectively. The above indicates that walls made of concrete and baked bricks have heat storage properties that can maintain more steady indoor temperature with lower fluctuations and postpone the time of occurrence when peak indoor air temperature appears with a longer time lag than solid sandcrete blocks.

A decrease in window size leads to reduction in peak indoor air temperature for all the envelope materials (thermal masses) tested. The reductions were, however, not remarkable with none being up to 0.9°C drop in PIAT. This can be attributed to the effect of shading devices on the test models which eliminates solar and conductive heat gains through the windows. The effective shading led to marginal effect of the varying window sizes. Concrete wall envelope with no window had the best performance with the lowest PIAT. However, windows are an important source of natural lighting and views from nature that should not be eliminated in buildings. Since no trend line strikes the positive window-to-floor area axis of the TDR-WFR correlation, it means that for all masses, conditions inside the models would be higher (or worse) and above the average outdoor temperatures even when there are no windows.

The analysis of indoor air temperature has shown that at an air change rate per hour of 10 ACH for various window-to-floor ratios, the expected reduction of the overheated hours was between 36% and 42%, 37% and 39%, and 37% and 39% for sandcrete blocks, baked bricks, and concrete, respectively. Increase in the air change rate per hour to 20 ACH and then to 30 ACH did not significantly reduce peak indoor air temperature.

The good match observed on validation of the of predicted results with the measured data confirms that the thermophysical properties of basic materials used for construction in Ghana are consistent with that used in international standard building simulation programs.

It can be concluded that the combined effects of thermal mass, window size, and night-time ventilation can synergistically reduce peak indoor air temperature in the warm-humid climates and improve thermal comfort. With the current leanings towards environmental sustainability of building design, the study is valuable towards energy-efficient building design in Ghana. These findings could guide building designers, prospective builders, and facility managers in the selection of building envelope material that will improve thermal performance and achieve thermal comfort. Future research is required to investigate the impact of other building envelop materials in Ghana on peak indoor temperature in buildings. Theoretically, the findings from this paper contribute further to the study of knowledge by providing sufficient evidence to support earlier studies that night-time ventilation is not effective in some warm-humid environments with narrow diurnal temperature range.

Acknowledgments

The project was supported with funding from the University Research and Conference Committee of the Kwame Nkrumah University of Science and Technology in Ghana. Mr. Sampson Deklu, an architecture four students of KNUST, is appreciated for his effort in regulating and monitoring the test cells.

References

[1] S. Amos-Abanyie, F. O. Akuffo, and V. Quagrain, "Unveiling energy saving techniques for cooling in residential buildings in Ghana," *International Journal of Ventilation*, vol. 8, no. 1, pp. 23–35, 2009.

[2] C. Koranteng, A. Mahdavi, K. Orehounig, and C. Pröglhöf, "Thermal performance of office buildings in Ghana," in *Proceedings of the 11th International IBPSA Conference*, Glasgow, Scotland, July 2009.

[3] A. K. Ofosu-Ahenkorah, "Potential for energy savings in Ghana," in *Energy Crises in Ghana: Drought, Technology or Policy?* A. Brew-Hammond and F. Kemausour, Eds., pp. 16–33, College of Engineering, Kwame Nkrumah University of Science and Technology, Kumasi, Ghana, 2007.

[4] V. Soebarto, *Calibration of hourly energy simulations using hourly monitored data and monthly utility records for two case study buildings [Ph.D. thesis]*, Department of Architecture, Texas A&M University, College Station, Tex, USA, 1997, http://www.ibpsa.org/proceedings/BS1997/BS97_P017.pdf.

[5] J. A. Clarke, "Assessing building performance by simulation," *Building and Environment*, vol. 28, no. 4, pp. 419–427, 1993.

[6] T. E. Bou-saada, *An improved procedure for developing a calibrated hourly simulation model of an electrically heated and cooled commercial building [M.S. thesis]*, Texas A&M University, College Station, Tex, USA, 1994.

[7] C. Reynolds, P. Komor, and M. Fels, "Using monthly billing data to find energy efficiency opportunities in small commercial buildings," in *Proceedings of the American Council for an Energy-Efficient Economy Summer Study on Energy Efficient Buildings*, vol. 10, pp. 221–231, Washington, DC, USA, 1990.

[8] US Departement of Energy (DOE), *Getting Started with Energy Plus. Energy Plus Program Documentation*, U.S. Departement of Energy, Washington, DC, USA, 2008.

[9] S. Amos-Abanyie, "Implications of climate change on human comfort in buildings: evidence from Nkontompo community of Sekondi-Takoradi, Ghana," *Journal of Science and Technology*, vol. 26, no. 1, 2006.

[10] P. La Roche and M. Milne, "Effects of window size and thermal mass on building comfort using an intelligent ventilation controller," *Solar Energy*, vol. 77, no. 4, pp. 421–434, 2004.

[11] V. Geros, M. Santamouris, A. Tsangrasoulis, and G. Guarracino, "Experimental evaluation of night ventilation phenomena," *Energy and Buildings*, vol. 29, no. 2, pp. 141–154, 1999.

[12] B. Givoni, *Passive and Low Energy Cooling of Buildings*, Van Nostrand Reinhold, New York, NY, USA, 1994, *Energy and Buildings*, vol. 29, no. 2, pp. 141–154, 1999.

[13] D. P. Finn, D. Connolly, and P. Kenny, "Sensitivity analysis of a maritime located night ventilated library building," *Solar Energy*, vol. 81, no. 6, pp. 697–710, 2007.

[14] N. Draper and H. Smith, *Applied Regression Analysis*, John Wiley & Sons, New York, NY, USA, 2nd edition, 1981.

[15] J. Kreider and J. Habel, "Predicting hourly biding energy usage: the great energy predictor and shootout: overview and discussion of results," *ASHRAE Transactions Technical Paper*, vol. 100, no. 2, pp. 56–70, 1994.

[16] D. M. Ogoli, "Predicting indoor temperatures in closed buildings with high thermal mass," *Energy and Buildings*, vol. 35, no. 9, pp. 851–862, 2003.

[17] V. Cheng, E. Ng, and B. Givoni, "Effect of envelope colour and thermal mass on indoor temperatures in hot humid climate," *Solar Energy*, vol. 78, no. 4, pp. 528–534, 2005.

[18] S. Chandra, "Ventilative cooling," in *Passive Cooling: Fundamentals and Applications*, J. Cook, Ed., MIT Press, Cambridge, Mass, USA, 1989.

[19] P. La Roche, "Passive cooling systems for sustainable architecture in developing countries-using the gap to bridge the gap," in *Proceedings of the World Sustainable Building Conference*, Tokyo, Japan, September 2005.

14

Material Flow Analysis of Fossil Fuels in China during 2000–2010

Sheng Wang,[1] Jing Dai,[2] and Meirong Su[2]

[1] *Chongqing Academy of Social Sciences, Chongqing 400020, China*
[2] *State Key Joint Laboratory of Environmental Simulation and Pollution Control, School of Environment, Beijing Normal University, Beijing 100875, China*

Correspondence should be addressed to Jing Dai, daijing0532@mail.bnu.edu.cn

Academic Editors: N. H. Afgan, H. Aras, and K. J. Chua

Since the relationship between the supply and demand of fossil fuels is on edge in the long run, the contradiction between the economic growth and limited resources will hinder the sustainable development of the Chinese society. This paper aims to analyze the input of fossil fuels in China during 2000–2010 via the material flow analysis (MFA) that takes hidden flows into account. With coal, oil, and natural gas quantified by MFA, three indexes, consumption and supply ratio (C/S ratio), resource consumption intensity (RCI), and fossil fuels productivity (FFP), are proposed to reflect the interactions between population, GDP, and fossil fuels. The results indicated that in the past 11 years, China's requirement for fossil fuels has been increasing continuously because of the growing mine productivity in domestic areas, which also leads to a single energy consumption structure as well as excessive dependence on the domestic exploitation. It is advisable to control the fossil fuels consumption by energy recycling and new energy facilities' popularization in order to lead a sustainable access to nonrenewable resources and decrease the soaring carbon emissions.

1. Introduction

Material flow analysis (MFA), quantified by material weight instead of currency, is built on the theories of industrial metabolism and social metabolism to pursue the translation path from nature to human ecosystem as well as the final regressive sinks [1]. The basic standpoint of MFA is that environmental effects which are brought about via social economic behaviors are mainly dependent on the quantity and quality of natural resources and materials devoted in the ecosystem, and the wastes translated from the consumption sectors back to the environment.

MFA is viewed as a significant measure to study the metabolism of material and resources. Ayres and Kneese firstly presented the first material flow accounts on the national level in 1969 [2]. During 1970s–1980s, the perfection of material balance and industrial metabolism theories laid a solid foundation for MFA application in the whole ecosystem. Austria, Japan, and Germany took the lead in calculating substance and natural resources in domestic economic scope in 1990s [3]. Subsequently, developed countries such as The Netherlands, America, and Australia finished their MFA in national boundary [1, 4]. In 2001, EU Statistics Department published a handbook about MFA research technique applied to ecosystem for the first time in the world [5], which enormously accelerated the promotion of MFA in economic field. Muñoz and Hubacek pointed out in 2008 that the economic growth was the major source of material changes [6] Chen proposed in his study in 2007 that the driving force of the social-economic-ecological complex system was the resource, which posed unparalleled challenges on each level of the society. The quantity and quality scarcities of the diverse resources require an efficient, effective, and interdependent utilization based on overall and unified accounting [7]. In recent years, China has also done necessary studies in MFA and gained achievements in scientific research of the relationship between total material input and consumption in the national base [8–10].

In the early 1990s, The concept of "ecological rucksacks" was firstly proposed which was commonly accepted as "hidden flows" afterward [11]. This concept refers to the wastes inevitably produced in the process of resource exploitation, though it is not devoted to the social production. Without creating commercial value, it will exert a huge influence on natural and social environment. In view of this, MFA is further modified and becoming an effective tool in

Table 1: Population and GDP information of China during 2000–2010.

Year	Total population (million)	Gross domestic product (¥ billion)
2000	1267.43	9921.5
2001	1276.27	10965.5
2002	1284.53	12033.3
2003	1292.27	13582.3
2004	1299.88	15987.8
2005	1307.56	18308.5
2006	1314.48	21087.1
2007	1321.29	24953.0
2008	1328.02	31404.5
2009	1334.50	34090.2
2010	1340.91	40120.2

measuring the balance between the resource depletion and the social development [12].

In this study, MFA's method is applied to the Chinese fossil fuels as a research case with certain modifications of hidden flows, which is considered as the more exact mode in accounting resource consumption. The results can make acceptable recommendations not only in transferring materialized into dematerialized consumption pattern, but also in building a low-carbon economy and coping with global warming.

2. Methodology and Data

The study period, lasting from 2000 to 2010, was an important stage for China's rapid economic progress in history. The rapid growth in the fields of both economy and resource exploitation is producing far-reaching impacts in social and environmental areas for today's decision making and stratagem implement. The basic information of China during the study period is listed as background information in Table 1.

Currently, a systemic MFA framework based on the national or regional ecosystem has been initially established all over the world, which has been comprehensively applied in occidental countries. The input stream of fossil fuels was divided into two parts, direct input flows, and hidden flows. The study boundary was confined into the domestic ecosystem. The input fossil fuels which come from the domestic production and abroad import contain raw coal, oil, and natural gas, excluding secondary energy input. Moreover, Wuppertal evaluated the average ratio of global hidden flow (GHF), and the results showed that crude oil, natural gas, and raw coal were 1 : 1.22, 1 : 1.66 [13], and 1 : 2.36 [14], respectively in view of the fact that most coal in China belongs to bastard coal. Usually the output stream of fossil fuels refers to direct output and contaminated discharge. Due to different burning efficiencies and regional variation in technologies, it is inaccurate to calculate direct output at the end of MFA. To express the end-result of the fossil fuels in the whole process of material flow, this paper chose three indexes, consumption and supply ratio (C/S Ratio), resource consumption intensity (RCI), and fossil fuels productivity (FFP) by considering population, GDP, and total energy input, to evaluate the influences of resource consumption in the fields of society and economy as well as environment. The three indexes are defined as follows:

$$\frac{S}{C} = \frac{\text{Consumption}}{\text{Production} - \text{Export} + \text{Import}} \times 100\%,$$
$$\text{RCI} = \frac{\sum \text{Consumption} \times \text{GHF}}{\text{Population}}, \tag{1}$$
$$\text{FFP} = \frac{\text{GDP}}{\sum \text{Consumption} \times \text{GHF}},$$

in which, the units are tsc/person ("tsc" refers to ton of standard coal (we use tsc for short in the rest part of this paper); standard coal, also known as coal equivalent, unified calorific value standard, different varieties, different energy contents of different calorific values converted to the calorific value of 7,000 kcal per kg of standard coal) for RCI, $/tsc for FFP (it is better for further comparison with other countries when we replace the currency unit from "¥" to "$").

All the data sources are acquired from publications, such as Energy Statistical Yearbooks of different years [15] and China Statistical Yearbooks [16] and websites [17]. In order to seek unity of economic value and avoid inflation or deflation in different times, we defined the year of 2000 as a base year, in which year the price was chosen as constant price; therefore, pure monetary value in all the other years should be converted into a standard value of 2000. The original yield and import/outport flows of three primary fossil fuels are listed in Table 2.

According to IPCC 2006 [18], different types of fuels have their own carbon emission coefficients (we use CEC in the following part of this paper), for coal, oil, and natural gas, the transition factors are listed in Table 4. Therefore, the CO_2 emission distributions will be obtained on the basis of fossil energy structure.

3. Results and Discussion

C/S ratio is of a paramount importance for the national or regional sustainability. A high value (>1) of C/S ratio means more resources will be depleted, and then more wastes or disturbuances will occur along with environmental deteriorations and nonrenewable energy shortage. If the C/S ratio is low, on the one part, it is good news that the present fossil energy supply is sufficient for demand. On the other part, it provides an important implication to readjust the energy consumption and supply relationship in the national level. The index called RCI can reflect the personal access to fossil resource in China. FFP denotes the transfer ability from raw fossil material occupation into economic value, which usually represents the efficiency of the resource consumption. The accounting of MFA and different indexes from 2000 to 2010 is shown in the following table (RMB exchange rate against the US dollar to compare this quota after conversion with different countries in a unified

TABLE 2: The original yield and import/outport flows of three primary fossil fuels during 2000–2010.

Year	Coal			Oil			Natural gas		
	Yield (million tsc)	Export (million ton)	Import (million ton)	Yield (million tsc)	Export (million ton)	Import (million ton)	Yield (million tsc)	Export (billion cu.m)	Import (billion cu.m)
2000	988.55	55.06	2.17	232.28	10.30	70.26	36.46	3.1	0
2001	1050.28	90.12	2.66	234.51	7.55	60.26	40.28	3.0	0
2002	1107.32	83.89	11.25	238.03	7.66	69.40	43.69	3.2	0
2003	1309.92	94.02	11.09	242.38	8.13	69.40	46.41	1.8	0
2004	1516.15	86.66	18.61	2517.01	11.46	122.72	55.06	2.4	0
2005	1677.85	71.68	26.17	259.46	8.07	126.82	64.86	2.9	0
2006	1806.25	63.23	38.25	262.34	6.34	145.18	78.93	2.9	1.0
2007	1921.35	53.17	51.01	267.06	3.98	163.17	91.49	2.6	4.0
2008	2001.03	45.43	40.40	273.57	4.16	178.88	106.56	3.2	4.6
2009	2122.80	22.40	125.83	271.87	5.07	203.79	112.59	3.2	7.6
2010	2271.40	19.03	164.78	290.97	3.03	239.31	127.67	4.0	16.5

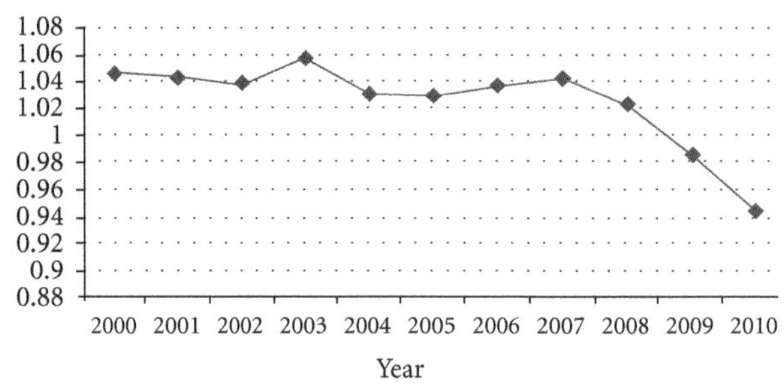

FIGURE 1: C/S ratio of fossil fuels in China during 2000–2010.

standard), and all the material flows are measured by the unit of "tsc," and the fossil fuels' consumption for three types of energy are collected in Tables 3(a) and 3(b).

3.1. Diversification of C/S Ratio See Figure 1. During the years from 2000 to 2007, it was nearly stable between 1 and 1.05 implying that a certain amount of previous fossil stock should be used to solve the supply and demand difference. During this period, there was a minor peak in 2003, reflecting that the obvious expansions of real estate construction, iron, and cement based heavy industry soaring development, which lead to the remarkable tension for energy supply and consumption relation. And after 2008, this ratio is fluctuate by most time lower than 1, which mainly attribute to the implementation of "energy saving and emission reduction" strategies in industries and domestic using. Generally speaking, from 2000 to 2010, C/S ratio of fossil fuels was decreasing roughly and the energy balance was trending into a capable supply state. Meanwhile, to better control the C/S ratio is an effective device to control the energy balance in the original production and final consumption.

Considering three kinds of fuels, coal yield was the main growth factor, which indicated that the energy structure in China was mainly based on raw coal as the dominant sector. Moreover, coal domestic production had occupied a share ranged 73–77% compared, and for oil was around 20% with total fossil energy during the past 11 years. The coal dominated energy structure has not dramatically reversed in that period. On average, the requirement of oil in occident countries account for 35% [11] of the total energy, that means we have a long way to amend the energy consumption structure, especially adding new energy into consideration, such as wind power, bioenergy, and tidal energy. Oil supply in China will not be sufficient forever, an urgent support policy is needed to encourage powerful companies and organizations to blend in the international arena and exploit a high quality resource. Furthermore, the domestic natural gas yield as well as the import proportion was enlarging, which indicated a good transition in enriching energy diversity and equitability, but with a too slight step. Therefore, we still have an adequate space to optimize the energy allocation so as to change the irrational existing state of fossil fuels consumption.

3.2. Diversification of RCI. RCI is an indicator that can express the level of per capita resource possess. Table 5 indicates that RCI was 2.1 times in 2010 than that in 2000, which implied that as individuals grow more affluent, their demands move beyond basic needs. The enormous requirements for private car, energy-consumption electronic products made life quality improve deeply on the one hand, and, on the other hand, extremely aggravated fossil resources depletion. The only solution to avoid energy crisis, especially

TABLE 3: (a) The fossil fuels' consumption from 2000 to 2010 in China (unit: million tsc). (b) The fossil fuels' supply[2] from 2000 to 2010 within China (unit: million tsc).

(a)

Year	Coal	Oil	Natural gas	Total consumption
2000	1007.07	323.08	32.02	1362.17
2001	1027.27	3278.89	36.10	1391.26
2002	1084.13	355.53	38.26	1477.93
2003	1282.87	389.64	45.95	1718.46
2004	1483.52	454.66	53.36	1991.54
2005	167.08	467.27	61.36	2199.49
2006	1839.19	499.24	75.02	2413.45
2007	1994.41	527.36	92.57	2614.33
2008	2048.88	533.35	107.84	2690.07
2009	2158.79	548.90	119.59	2827.29
2010	2209.59	617.38	142.97	2969.94

(b)

Year	Coal	Oil	Natural gas	Total
2000	950.77	317.94	32.65	1301.37
2001	987.81	309.82	36.60	1334.22
2002	1055.44	326.24	39.81	1421.48
2003	1250.69	329.92	44.14	1624.75
2004	1467.55	410.66	52.10	1930.30
2005	1645.35	429.11	61.26	2135.72
2006	1788.42	460.70	76.57	2325.68
2007	1919.81	494.48	93.22	2835.45
2008	1997.45	523.18	108.20	2628.83
2009	2196.68	555.76	117.96	2870.41
2010	2375.52	628.53	142.77	3146.82

[2] Supply = production − outport + inport.

TABLE 4: The carbon emission coefficients (CEC) for three types of fossil fuels.

Items	CEC
Coal	1.98 ton CO_2/ton
Oil	3.07 ton CO_2/ton
Natural gas	2.19 ton CO_2/ton

for nonrenewable energy, is to elevate utilization efficiency and attempt new renewable energy. In addition, China is the most populous developing country, whose contradiction between large energy-needed population and low per capita energy possession is always restricting the whole national development in social and economic fields. In addition, CEC from different types of the fossil fuel consumption from 2000 to 2010 in China is shown in Table 6 as a supplementary database.

3.3. Diversification of FFP. FFP, an economic value produced by the unit natural nonrenewable resource. Table 7 clearly illuminates that an obvious disparity is existing in fossil fuel productivity compared with the major energy-saving country. The level of fossil fuel productivity in China in 2010 was merely equivalent to that of The Netherlands in 1996, and Austria, Germany, and Great Britain had already owned this index nearly 1.5 times higher than that of China in 2010. The bog standard of FFP in China means not only a colossal waste of fossil fuels in production and consumption process, but also severe environmental pollution resulting in a great number of pollutants and wastes in all processes by using natural resources. The evident gap between China and other developed countries in creating economic value per unit fossil resource reflected low efficiency in energy input-output course. It is well accepted that economic development based on the extensive resource depletion is not the intrinsic pursuit of sustainable developing pattern, and it is needed to consummate a resource recovery system as soon as possible, promote the use of energy saving products in public, and participate in international cooperation that can develop bilateral and multilateral as well as regional collaboration in the field of new energy using and low-carbon technology [19]. Another step to solve the inefficiency in fossil fuel productivity is to change the approach of using fossil fuels into a closed material cycle mode [20] and enhance the GDP value created by per unit energy expending.

3.4. Comprehensive Analysis about Different Indexes. The variation trends of population, GDP, and RCI as well as FFP are illustrated in Figure 2. First, the population increased steadily which was fitting in with the national conditions during that time; second, GDP increased with an accelerating trend, which was not only due to the economic dynamic times in this period, but also affected by the appreciation of RMB since 2005; third, RCI was increased especially after the year 2002, meaning a society with the rapid consumption for fossil fuels; and finally, the variation of FFP was divided into two subperiods: the first half was nearly fair, however, the other half was soaring for a rapid fossil fuel economic productivity. Generally speaking, economy in China is operating well, but the extensive mode of growth could not change at once, with fossil fuel productivity impossibly boosted in a short time. However, infrastructures on fossil energy being or having been built will certainly carry weight in both economic development and the resource consumption.

3.5. CO_2 Emission Analysis. From Figure 3, it was clearly demonstrated the CO_2 emission distribution, and coal is the most carbon source. In China, the hidden flow ration for coal was the highest, and the energy structure could not be changed on the basis of coal consumption. That is why CEC from coal was nearly 2 times than that from oil and natural gas. From this result, using natural gas or the other low carbon emission energy is the most effective and urgent strategy for the government decision and public choice.

4. Conclusion

This paper analyzes the input of fossil fuels in China during 2000–2010 and proposed several indicators of MFA.

TABLE 5: C/S ratio, RCI, and FFP of China during 2000–2010.

Year	C/S ratio	Population million	RCI (tsc/person)	GDP (¥ billion)	Exchange rate (¥/100 $)	GDP ($ billion)	FFP ($/tsc)
2000	1.0467	1267.43	2.11	9921.4	827.84	1198.46	424.39
2001	1.0427	1276.27	2.22	10965.5	827.70	1324.82	459.32
2002	1.0397	1284.53	2.32	12033.2	827.70	1453.82	475.76
2003	1.0577	1292.27	2.68	13582.2	827.70	1640.97	458.47
2004	1.0317	1299.88	3.059	15987.8	827.68	1931.64	466.09
2005	1.0299	1307.56	3.35	18493.7	819.17	2257.62	489.17
2006	1.0377	1314.48	3.59	21631.4	797.18	2713.50	534.76
2007	1.0426	1321.29	3.79	26581.0	760.40	3495.67	635.13
2008	1.0233	1328.02	3.94	31404.5	694.51	4521.83	798.20
2009	0.9850	1334.50	4.14	34090.2	683.10	4990.53	836.92
2010	0.9438	1340.91	4.42	40120.2	696.95	5756.54	927.70

TABLE 6: CEC from different types of the fossil fuel consumption from 2000 to 2010 in China.

Year	Coal consumption (million ton)	Oil consumption (million ton)	Natural gas consumption (million ton)	CEC from coal (million ton CO_2)	CEC from Oil (million ton CO_2)	CEC from natural gas (million ton CO_2)
2000	1409.87	226.15	18.92	2791.55	694.28	41.45
2001	1438.15	229.51	21.34	2847.54	704.61	46.73
2002	1517.75	248.87	22.62	3005.15	764.02	49.54
2003	1795.97	272.74	27.16	3556.03	837.31	59.49
2004	2076.88	318.26	31.55	4112.23	977.04	69.09
2005	2339.15	327.09	36.27	4631.52	1004.15	79.44
2006	2574.80	349.46	44.35	5098.12	1072.85	97.13
2007	2792.12	369.14	54.72	5528.39	1133.26	119.85
2008	2868.37	373.33	63.75	5679.38	1146.14	139.62
2009	3022.25	384.22	70.70	5984.05	1179.55	154.84
2010	3093.35	432.16	84.53	6124.84	1326.73	185.12

In comparison with other countries, main conclusions are summed up as follows.

Generally speaking, the demand structure of coal, oil, and natural gas during the past 11 years has not changed significantly. The turning point of fossil fuel productivity appeared in 2002, and increased dramatically after 2005. Both phenomena are decided by international economic situations and domestic production levels. The former is due to good circulations of the world economy and China's entry into WTO, which lead to substantial materials input but a lack of efficiency advancements. The latter is owing to policy steering of the eleventh five-year plan unveiled in 2005, which drives the movement of technical improvement, renewable energy development, wastes recycling, and multilevel using. But we have to admit that China still has to go a long way to pursue fossil fuel productivity compared with countries like Japan and Germany.

MFA accounting system calculates environmental pressure produced by the resource depletion. Its standpoint is payments for import only cover the economic costs without environmental costs as well as hidden flows value in the process of energy production. By this token, one country can pass over the environmental cost by importing so as to reduce the domestic natural risk. From this point of view, it is necessary to shift energy strategy from export-oriented to import-oriented style and elevate the export price by considering environmental expense and enlarging import proportion, from which home resource and environment can be protected in a qualified sense. Nevertheless, it is myopia for a particular developing period in certain regions, but not beneficial for universal sustainable utility of resource. Natural wealth belongs to human beings, without boundary of different nations and territories.

5. Discussions and Prospects

For the existing energy supply and demand status quo, more efforts are needed to intensify energy structure reforms, though the expansion of gas domestic output and import quotation provided a new perspective. Furthermore, the unchanged conditions may be related to the existing exploitation patterns, the history of coal mining and technical problems in the development of pelagic natural gas, which are primarily reasons for the univocal energy current situation. Whereas, the future structure of energy utility will move forward to a rational and sustainable direction as long as technical progress, sufficient facilities, financing of safeguards, and national supporting policies are provided.

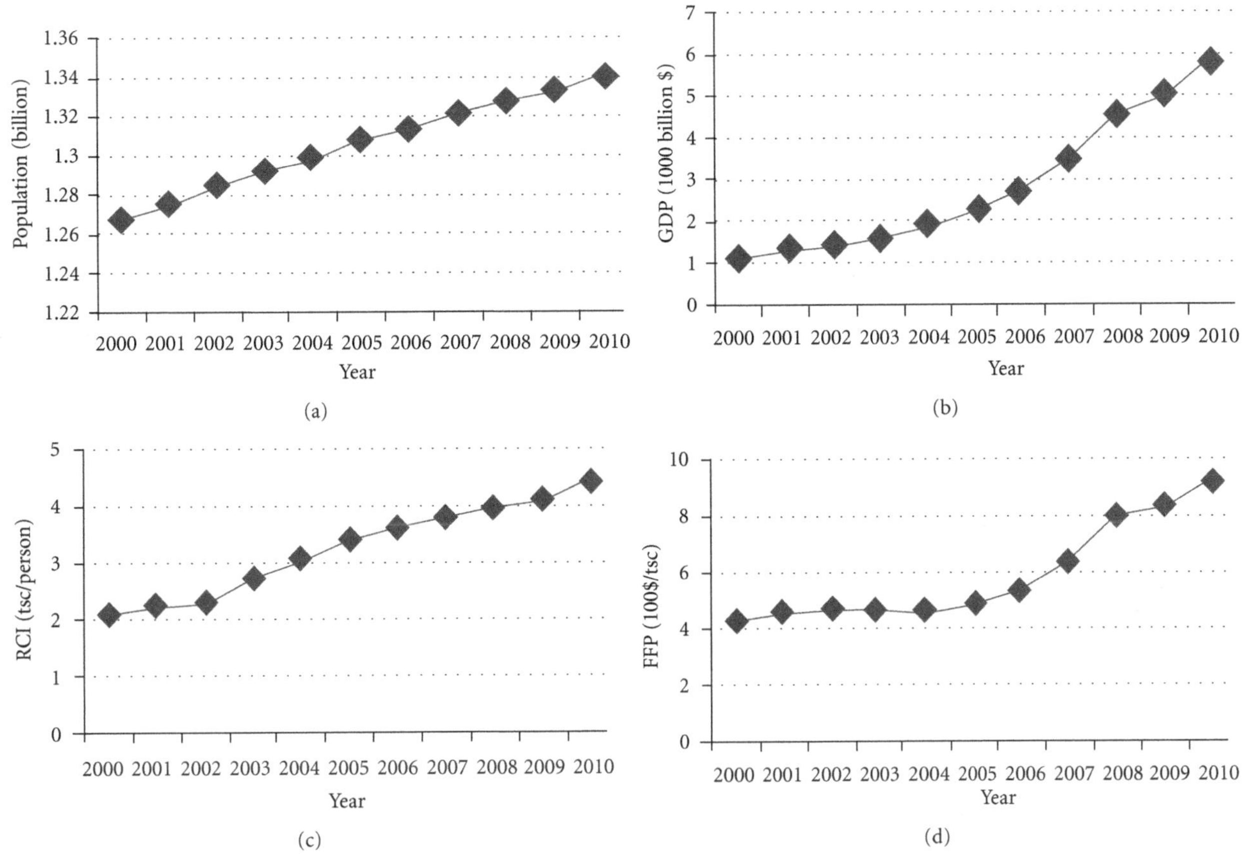

FIGURE 2: A comprehensive index comparison of fossil fuels in China during 2000–2010.

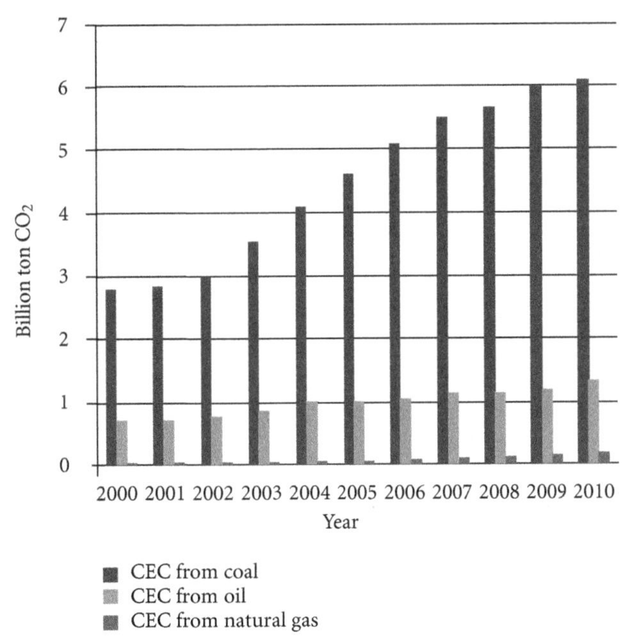

FIGURE 3: CEC amount from three types of fossil fuels from 2000 to 2010 in China.

TABLE 7: Statistics of FFP in other countries during recent years (Unit: \$/tsc).

Country	1975	1980	1985	1990	1995	1996
Austria	649.9	887.8	620.3	1197.9	1415.7	1362.2
Germany	649.8	984.6	751.6	1451.6	1296.6	1313.9
Japan	472.1	700.6	852.8	1410.2	2543.3	2255.7
Netherlands	460.5	670.9	437.1	884.2	1018	949.1
Great Britain	1487.01	1714.01	1012	1517.1	1482.2	1513.2

Meanwhile, the "closed material cycle mode" that with the process of "energy production → energy consumption → recycled energy → recycled energy reuse" should be extensively implemented in extensive energy based industries in order to decrease the C/S ratio value and create more economic value by consuming the previous amount of energy consumption. And the bundled infrastructures or strategies should be followed up, for example, the generalization of garbage classification with recycled waste energy appliance, popularization of new energy vehicle service facilities.

The transformation of GDP has necessary connections with the resource consumption intensity and the fossil fuel productivity. To accomplish the target of energy saving and emission reducing, and also to keep to the path of sustainable development, it is necessary to rely on domestic research and developments as well as global introductions of advanced technology in enhancing efficiency and reduced consumption for internal upswing.

Apparently, material and energy streams are dependent with each other in a different manner, so the simplex material flow analysis is not comprehensive in identifying the energy consumption in an environmental economic system. Next, we should combine the material and energy analysis to understand the social metabolism of fossil fuels from all angles. At the same time, the indexes and their variation trends provide reasonable explanations for future policy drafts in energy saving and emission reducing by evaluating the recent 11 years' resource consumption and production efficiency.

Acknowledgments

This study was supported by the National Science Foundation for Innovative Research Group (no. 51121003), Key Program of National Natural Science Foundation (no. 50939001 and 41271543), and Program for New Century Excellent Talents in University (NCET-09-0226). The authors are especially grateful for the financial support from Beijing Development Area Co., Ltd.

References

[1] E. Matthews, C. Amann, S. Bringezu et al., *Weight of Nations: Material Outflows From Industrial Economies*, World Resources Institute, Washington, DC, USA, 2000.

[2] U. R. Ayres and V. A. Kneese, "Production, consumption and externalities," *American Economic Review*, vol. 59, no. 3, pp. 282–297, 1969.

[3] M. Fischer-Kowalski, "Society's metabolism: the intellectual history of materials flow analysis, part I, 1860–1970," *Journal of Industrial Ecology*, vol. 2, no. 1, pp. 61–78, 1998.

[4] A. Adiranse, S. Bringezu, A. Hammond et al., *Resource Flows: the Material Basis of Industrial Economies*, World Resources Institute, Washington, DC, USA, 1997.

[5] European Communities, *Economy-Wide Material Flow Accounts and Derived Indicators: A methodological Guide*, Official Publications of the European Communities, Luxembourg, 2001.

[6] J. P. Muñzo and K. Hubacek, "Material implication of Chile's economic growth: combining material flow accounting (MFA) and structural decomposition analysis (SDA)," *Ecological Economics*, vol. 65, no. 1, pp. 136–144, 2008.

[7] G. Q. Chen and B. Chen, "Resource analysis of the Chinese society 1980–2002 based on exergy—part 1: fossil fuels and energy minerals," *Energy Policy*, vol. 35, no. 4, pp. 2038–2050, 2007.

[8] X. Q. Chen and L. J. Qiao, "Material flow analysis of Chinese economic-environmental system," *Journal of Natural Resource*, vol. 15, no. 1, pp. 17–23, 2000.

[9] M. Xu and T. Zhang, "Material flow analysis of fossil fuel usage in the Chinese economy," *Journal of Tsinghua University*, vol. 44, no. 9, pp. 1166–1170, 2004.

[10] C. Y. Xia, "Review on studies of economy, wide material flow analysis," *Journal of Natural Resources*, vol. 20, no. 3, pp. 415–421, 2005.

[11] B. F. Schmidt, *The MIPS-Concept: Bridging Ecological, Economic, and Social Dimensions With Sustainability Indicators*, United Nations University, Tokyo, Japan, 1999.

[12] Y. Chen and N. S. Deng, "Environmental management approach geared to the 21st century-matter and energy flow analysis," *Chongqing Environmental Science*, vol. 25, no. 3, pp. 1–5, 2003.

[13] X. Q. Chen, *Natural Geography*, Peking University Press, Beijing, China, 2001.

[14] W. Chen, W. Tang, and X. C. Zhang, "The resource and development study for China's Coal," *Coal Economic Research*, no. 7, pp. 6–11, 2003.

[15] CESY, *China Energy Statistic Yearbook*, China Environment Yearbook Press, Beijing, China, 2005–2011.

[16] CESY, *China Statistic Yearbook*, China Statistics Press, Beijing, China, 2001–2011.

[17] http://www.cngascn.com/html/news/show_news_w1_1_20258.html.

[18] IPCC, *IPCC Guidelines For National Greenhouse GasInventories*, Institute for Global Environmental Strategies, Hayama, Japan, 2007.

[19] H. P. Aubauer, "A just and efficient reduction of resource throughput to optimum," *Ecological Economics*, vol. 58, no. 3, pp. 637–649, 2006.

[20] N. Duan, "Material metabolism and circular economy," *China Environmental Science*, vol. 25, no. 3, pp. 320–323, 2005.

15

China and the Global Uranium Market: Prospects for Peaceful Coexistence

Pascale Massot[1] and Zhan-Ming Chen[2]

[1] *Department of Political Science, The University of British Columbia, Vancouver, BC, Canada V6T 1Z1*
[2] *School of Economics, Renmin University of China, Beijing 100872, China*

Correspondence should be addressed to Zhan-Ming Chen; chenzhanming@pku.edu.cn

Academic Editors: T. Kojima, J. Niu, and D. M. F. Santos

China's recent reemergence has resulted in a significant increase in the global demand of commodities and is already having major impacts on the dynamics of global commodity markets. In the case of the global uranium market, we stand at the very beginning of a period of change. However, interesting trends are already emerging. Whereas China has had many policy reversals, and some difficulties in taking control of its procurement strategy in other commodity markets, it is seemingly more successful in managing its uranium procurement strategy. Why? The argument presented here is that a mixture of domestic and international level variables has allowed China more room for maneuver in fulfilling its uranium procurement strategy. On the domestic level, a centralized industry, and, on the international level, a geographically dispersed and uncoordinated market have allowed China to forge ahead with an ambitious civilian nuclear power plan and triple its total uranium imports, all within the span of a few years. Many challenges remain, not the least that of negative public opinion, which has surged since the Fukushima disaster in 2011. Nevertheless, should uranium demand continue to grow, this paper will consider the potential for continued peaceful coexistence among uranium market participants worldwide.

1. Introduction

China's fast development has resulted in a significant increase in the global demand of almost all commodities. This increased demand has already had major impacts on the dynamics of certain global commodity markets [1–6].

In the case of the global uranium market, we stand at the very beginning of a period of change. China's uranium demand has only started to increase significantly in the past 10 years. Moreover, if many analysts are arguing that Chinese demand for certain commodities may plateau or even fall in the near future, the situation is clearly different in the case of uranium. China has 16 operating nuclear reactors as of mid 2012, but it has 27 reactors under construction and plans to build at least another 50 in the coming years [7].

Another reason why China's interaction with the global uranium market is interesting is the room for manoeuver that China was able to carve for itself in this global market. Whereas China has had many policy reversals, and some difficulties in taking control of its procurement strategy in the global market of iron ore, for instance, it is having a better time managing its uranium procurement strategy. Why is China seemingly more successful in its entry on the global uranium market? Is this emerging trend likely to persist? Considering the fact that uranium demand is projected to go up substantially in the coming years, what is the potential for continued peaceful coexistence and cooperation among uranium market participants worldwide? This paper seeks to answer these questions in turn.

In sum, the argument is made that specific components of the structure of the international uranium market allow China more room for maneuver than in many other commodity markets and are influencing the way in which China chooses to engage with it. The global structure of the uranium market is relatively dispersed geographically but, more importantly, exhibits a lack of coordination among the existing market stakeholders; it has also suffered from underinvestment and lack of interest for decades leading to the early 2000s and thus was ripe for welcoming increased investment and involvement by an emerging actor such as China. On the domestic side of things, the importance given to China's civilian nuclear program by the government and the concentration of decision-making among few actors have allowed the country to manage existing challenges and effectively advance its overseas procurement strategy.

In terms of prospects for future peaceful coexistence, they can be realized if the recent international cooperation initiatives, which encourage transparency and collaboration, can create a virtuous circle. In an unexpected way, nuclear safety and security concerns, which lead to international cooperation mechanisms in the first place, beyond enhancing nuclear safety in China, could provide the impetus for more cooperation internationally. These international efforts have also showcased China as a country ready to rise to the occasion and be a responsible player, as a member of the International Atomic Energy Agency for instance.

After a short review of the history of the uranium market since its emergence, the consequences of the Fukushima nuclear accident and the global structure of the uranium market will be discussed. Thereafter, China's current growing resource needs as well as the domestic and international challenges it faces will be reviewed. This will be followed by an assessment of the current Chinese uranium procurement strategy. In conclusion, an analysis of China's participation in recent cooperation initiatives bilaterally and internationally will be made.

2. A Short History of the Uranium Market

First of all, "There are at least four markets in the front-end of the nuclear fuel cycle that must be reviewed to determine assurance of supply: (1) uranium mining and milling, (2) uranium conversion, (3) uranium enrichment, and (4) nuclear fuel fabrication." [8]. See also Conde and Kallis [9]. This paper will concentrate mainly on the first step, uranium mining.

2.1. Emergence—Atoms for Peace—Cartel Period. The uranium industry emerged in the late 1940s early 1950s, mainly for military purposes [10]. In 1953, the race for dominance in the area of civilian nuclear power was set in motion by the US Atoms for Peace program.

In the 1960s, the American Energy Agency banned the use of foreign uranium in its domestic reactors and aggressively cut prices of its own uranium exports. This was a period of oversupply in the rest of the world. The Canadian government decided to support its domestic uranium industry while stockpiling its inevitable surplus of uranium production.

It was then that Canada and other major uranium producers of the world (Australia, France, South Africa, and Rio Tinto Zinc Ltd.), in the absence of the US, sought to mitigate the impacts of the American policy and resorted to the covert manipulation of the world market [11].

In June 1972, the secret international uranium cartel was formally established (arrangements included price fixing, bid rigging, and market sharing). The cartel was referred to as the Société d'Études de Recherches d'Uranium (SERU).

Westinghouse filed an antitrust action against the cartel members, including a Canadian company, in 1976, and the cartel subsequently dismantled. We have not been able to find evidence that uranium producers have continued to coordinate their operations since then.

2.2. Oversupply and Fall in Prices. During the 1980s and 1990s, for a combination of reasons, including the end of the Cold War (and thus the increased availability of secondary sources of uranium), a lull in the construction of new nuclear power plants worldwide (because of the consequences of the Three Mile Island and Chernobyl disasters and a reduction in expected growth of electricity demand) resulted in a fall of the uranium spot price, and thus of mining production.

Indeed, from the early 1980s until 2001, uranium prices trended downward and remained between USD $7 and USD $10 a pound.

2.3. Price Bubble. Beginning in 2001, the price of uranium began to rebound from historic lows and continued to rise through 2007. The real bubble occurred during the year 2007, triggered by shrinking weapons stockpiles (and thus the decreased availability of secondary sources), a flood at the Cigar Lake Mine in Canada, expected undersupply due to a slew of reactors coming online, compounded by the relatively recent news of an extensive nuclear program expansion in China, as well as speculative pressures.

As the uranium price shot to historical heights of USD $136 a pound (see Figure 1), the extent of the twenty previous years of underinvestment in uranium production became all the more obvious.

2.4. Impact of Fukushima Crisis. The 9.0 magnitude earthquake in Japan was a uniquely severe natural disaster. So severe that the safety standards in place at the Fukushima power plant did not allow for such a magnitude. This was compounded by the impact of the tsunami.

The international consequences of the Fukushima disaster were substantial, first and foremost in Japan. As of August 2012, only 2 out of the 54 Japanese reactors are operating again [12]. The government's basic stance on dealing with the power shortage has so far been to beef up its efforts to reduce power consumption and increase thermal power generation. But the combination of a fall in energy output and increased disruptions of supply is likely to have a lasting negative impact on growth, as argue Weinstein and Schnell [13]. Indeed, as Japanese communities continue to oppose the reopening of nuclear reactors, the effect of the earthquake has translated in a sustained drop in industrial production. The Japanese Minister of Industry (Yukio Edano) was quoted as saying that it is "necessary to restart nuclear reactors to avoid power shortages, provided that it can be done safely and with the agreement of local residents" [14].

The civilian nuclear industry seems poised to draw the lessons from this event, but it can be expected that the improvement in nuclear safety measures as well as the implementation of more efficient emergency procedures should have an inflationary impact on overall costs, just as it was

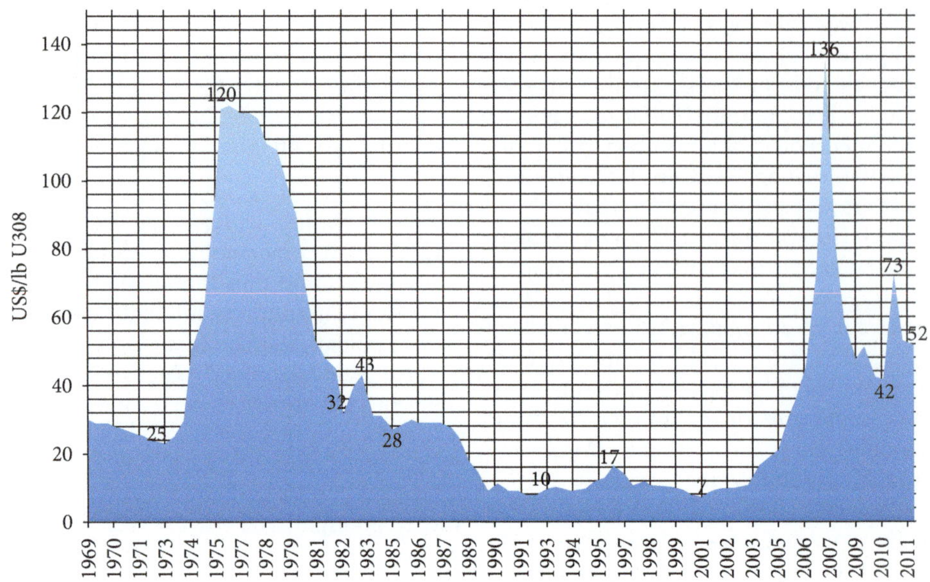

FIGURE 1: Uranium spot prices 1969–2012. Source: (1969–1989, NUEXCO Exchange Value, UX Consulting, constant 2007$; 1989–present, CAMECO, current$).

the case after previous major accidents [15]. Such expectations, as well as increased public opposition to nuclear power, resulted in a drop of uranium prices worldwide. Between February 2011 and August 2011, the spot indicator fell by around 30% from a high of USD $72.63 a pound to US $49.13 a pound.

In China, "five days after the earthquake and tsunami, the State Council suspended approval of new nuclear projects and started conducting comprehensive safety inspections of all nuclear projects—those in operation as well as those under construction. It also decided to halt four approved projects due to start construction in 2011" [16]. Chinese companies have responded to the changing global context by attempting to renegotiate outstanding bids and positions. In early May 2011, CGNPC withdrew its $1.24 billion bid to acquire a controlling interest in Kalahari Minerals PLC, which is developing a uranium mine in Namibia and then reopened the talks in the fall of 2011. A deal was finally struck again in February 2012 for close to $1 billion.

2.5. Current Prospects. In the end, however, many analysts argue that the impact of the Fukushima accident on uranium prices will be short-lived, since the projected drop in demand will most likely be more than compensated by growth in emerging countries. Indeed, it is expected that Asia will account for most of the growth in new nuclear reactors, of which 40 percent will come from China [17].

"Globally, (the CEO of CAMECO Tim) Gitzel said he expected uranium demand to grow about 3 percent a year in the "next few years." The "psychological" impact of the Fukushima nuclear accident in Japan will be in the "short-to-medium-term," he said [18].

Current analysts actually forecast a growing deficit in uranium supplies starting from now (shortfalls have been driven by either lower forecast prices compared to 2007, problems with existing operations or delays in new mine production). For instance, the Royal Bank of Canada Capital Markets argues that uranium price indicators are down from the 2007 bubble, but still up from the 2008 low, and foresees uranium demand growing by an average of 4% per year during the next 20 years. They project a price of $80 a pound for 2013 [19].

While the fact that Germany decided to phase out its civilian nuclear program by 2022 made the news following the Fukushima disaster, talks of a nuclear phase out of nuclear energy for commercial power generation purposes had been on the agenda in Germany since at least 2002. German uranium demand represents anywhere between 2300 [20] and 3800 tons annually [21], or less than 5% of global demand. All in all, the construction of reactors in China is expected to outweigh the decommissioning of plants in Germany, Japan, Belgium, Italy, and Switzerland.

Bullish price predictions are also supported by the fact that we are currently entering a key transition period for the uranium market. Current global uranium production meets around 75% of global demand, the rest being met by stockpiles (released from military sources), inventories of which are rapidly decreasing. Indeed, the US-Russian Highly Enriched Uranium (HEU) Purchase Agreement, which converts surplus HEU from Russian nuclear weapons into fuel for US commercial power reactors, is expiring in 2013 [19, 22].

This "Megatons to Megawatts Program is a unique, commercially financed government-industry partnership in which bomb-grade uranium from dismantled Russian nuclear warheads is being recycled into low enriched uranium (LEU) used to produce fuel for American nuclear power plants. USEC, as executive agent for the U.S. government, and Techsnabexport (TENEX), acting for the Russian government, implement[ed] this 20-year, $8 billion program. (. . .) In years past, up to 10 percent of the electricity produced in the United States has been generated by fuel fabricated using LEU from the Megatons to Megawatts program." [23].

The Director General of Rosatom says the contract will not be renewed in 2013. This will cause a gap of around 20,000 tons of uranium. Only around half of this gap will be filled through a new supply agreement that was signed between Russia's Techsnabexport (TENEX) and the United States Enrichment Corporation. Whereas the Megatons to Megawatts program was special in that it converted Russian nuclear weapon material, the new 2013–2022 supply agreement will provide the US with commercially enriched uranium from Russia [24]. Adding to the glut, many large uranium mines are currently close to depletion.

The uranium spot market reflected a persistent buyers' market over the 15-year period of 1980 to 1994, and again between 1998 and 2003 [25]. The situation is reversed to a significant degree now.

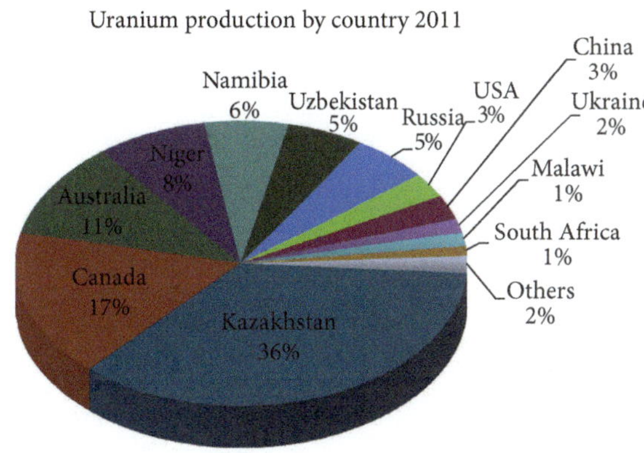

FIGURE 2: Uranium production by country in 2011. Source: Euratom Supply Agency, World Nuclear Association.

2.6. Uranium Spot Market. Due to its special history, uranium is not widely traded on an organized commodity exchange, such as the London Metal Exchange. Spot prices are responsible for no more than 15% of global trades. For instance, during 2008, about 48 million pounds of U_3O_8 changed hands in the spot auctions [26]. This is a relatively small quantity compared to the 250 million pounds of uranium contracted by world utilities in 2005.

It is true that the long-term uranium market and spot prices have a tendency to move together, because of "market-related" price mechanisms built in long-term contracts, and the fact that long-term contracts have quantity flexibilities built in them [25]. However, as Gene Clark, formerly with the US Department of Energy, said in an interview for UraniumSeek, a lack of liquidity and sophisticated market mechanisms remain features of the global uranium market [25].

It remains to be seen whether the emergence of trading mechanisms that we have seen in other markets (coal and iron ore come to mind) will extend to the uranium market. In 2007, the New York Mercantile Exchange signed a 10-year agreement with UX Consulting Company, to introduce U_3O_8 swap futures on CME and NYMEX platforms [27].

As of 2006, the 6 biggest uranium mining companies occupied 77% of the global market (Areva-17%, Cameco-16%, Rio Tinto-16%, Kazatoprom-13%, ARMZ/Rosatom-9%, and BHP Billiton-6%) [28]. In 2011, 4 countries contributed 72% of global uranium production (see Figure 2). Therefore, on the one hand, this indicates that the global uranium market is relatively concentrated.

However, two characteristics of this global market deserve further attention. First, the uranium market has neither a liquid spot market nor a producer's organization (its 1970s cartel having been disbanded), it is relatively fragmented geographically, and the largest producer of uranium really only emerged within the last decade. So on the other hand, the global uranium market is a thinly institutionalized market. Second, while the purpose of this paper is not to delve in the technical details of nuclear fuel production, suffice it to say that uranium is an unusual commodity in that it necessitates highly capital-intensive transformations (enrichment) prior to being used as fuel, a step that can be (and often is) conducted in a different country than the country of uranium ore extraction (see Table 1), and a step that is closely related to security issues. China does possess enrichment capabilities, so it mainly only needs uranium ore, but this still complicates the picture further.

Perhaps because of historical security concerns, and the absence of a widely used spot market, the global uranium market also remains a market where state-to-state relations continue to play an important role. Some recent major developments in Chinese procurement contracts were the result of state-to-state negotiations, including, recently, with the Canadian Government (more details below). The fact that China has ongoing long-term relationships with its close neighbours in Central Asia, such as with Kazakhstan and Uzbekistan, has also played a part in its ability to sign procurement agreements with them.

3. China's Projected Uranium Demand and Procurement Strategy

3.1. Current Needs and Supply-Demand Gap. Whereas a key dimension of China's energy security aims [29] has been to rely as much as possible on domestic production [30], more recently, an emphasis on rapid expansion of electricity production, diversification of the energy mix, as well as environmental protection has contributed to the emergence of China as a civilian nuclear power [31–33]. In 1991, China connected its first nuclear reactor to the electricity grid [34]. In 2002, only 2 nuclear reactors had been built in China, but the country was already firmly looking ahead towards a future where nuclear energy would produce between 40 and 80 GW [31]. "During the 10th 5 Year Plan (2001–2005) period, the key part of China's energy policy [was] to 'guarantee energy security, optimize energy mix, improve energy efficiency, protect ecological environment, continue to open up wider, and speed up the development of the west regions' [sic]." [31].

Five days after the earthquake and tsunami, the State Council suspended approval of new nuclear projects and started conducting comprehensive safety inspections of all nuclear projects. However, if there were any doubts as to whether China would continue to go ahead with its ambitious

TABLE 1: Top 5 exporters of uranium ore, natural uranium, and enriched uranium between 2008 and 2011.

(a)

Top 5 exporters of uranium and thorium ores (2008–2011)	Trade value in USD	Top 5 exporters of natural uranium and its compounds (2008–2011)	Trade value in USD
Namibia	$2,803,841,482	Canada	$7,152,512,173
Niger	$1,549,622,598	USA	$2,578,682,476
Australia	$1,152,033,863	France	$2,568,788,240
Malawi	$243,107,642	Kazakhstan	$1,593,772,641
United Kingdom	$17,024,333	Russian Federation	$322,983,583
Total world	$5,777,745,980	Total world	$15,131,504,216

(b)

Top 5 exporters of enriched uranium (2008–2011)	Trade value in USD
France	$10,755,771,990
USA	$5,493,937,151
Germany	$4,016,817,724
Netherlands	$3,631,753,994
China	$305,168,764
Total world	$24,623,102,401

Source: UN Comtrade.

TABLE 2: The 8 biggest civilian nuclear powers: planned reactors and uranium requirements in 2012.

Country	Reactors operable	Reactors under construction	Reactors planned and proposed within 15 years	Uranium requirements in 2012 (Tonnes U)
China	15	26	171	6 550
France	58	1	2	9 254
Japan	51	2	15	4 636
Russia	33	10	41	5 488
South Korea	23	3	6	3 967
Ukraine	15	0	13	2 348
United Kingdom	17	0	13	2 096
United States	104	1	30	19 724
World	435	62	489	67 990

Source: World Nuclear Association, April 2012.

civilian nuclear program following the Fukushima disaster and the year-long safety review, these doubts were dispelled early this year. In his speech for the Nuclear Security Summit in Seoul in March 2012, President Hu Jintao underlined the "irreplaceable role of nuclear energy in ensuring energy security and climate change" [35]. This was a signal that echoed Wen Jiabao's comments made a couple of months earlier in Abu Dhabi, where he said that "Nuclear power is a safe, reliable, mature technology providing clean energy. The safe and efficient development of nuclear power is the solution to future energy supply strategy." [36].

Then, "the former head of the NEA [National Energy administration] said that full-scale construction of nuclear plants would resume in March 2012" [37]. This confirmed that China is going ahead with its extensive expansion of civilian nuclear power plant program, albeit potentially at a slower rate: the target set by the National Development and Reform Commission (NDRC) in 2007 to have 40 GWe (Gigawatt-electric) online by 2020 [38] was upgraded to 70–80 GWe in 2010 and revised to 60–70 GWe in the aftermath of the Fukushima accident [37]. Currently at least 27 reactors are under construction [36] (or 26, according to the World Nuclear Association, see Table 2) and 50 more are planned according to the China Nuclear Energy Association [7].

As a consequence, whereas China's share of the market is still relatively low, planned construction of nuclear power plants is ambitious and China may be the first ranked importer of uranium globally by 2020. Qian Zhimin (China National Energy Administration) argued that by 2020, nuclear power could be contributing 7%-8% of China's energy needs, a higher rate than the official government target of 5% [39].

Table 3: Natural uranium production in 2010, compared with 2009 (tons of uranium).

Country	Production 2011 (tU)	Production 2010 (tU)	Production 2009 (tU)	Share in 2011 (%)	Share in 2010 (%)	Share in 2009 (%)	Change 2011/2009 (%)
Kazakhstan	19 451	17 803	14 020	36%	33%	28%	39%
Canada	9 145	9 783	10 173	17%	18%	20%	−10%
Australia	5 983	5 900	7 982	15%	11%	16%	−25%
Niger	4 351	4 198	3 243	8%	8%	6%	34%
Namibia	3 258	4 496	4 626	6%	8%	9%	−3%
Uzbekistan	3 000	2 400	2 429	6%	4%	5%	24%
Russia	2 993	3 562	3 564	6%	7%	7%	−16%
USA	1 537	1 660	1 453	3%	3%	3%	6%
China	1 500	827	750	3%	2%	1%	100%
Ukraine	890	850	840	2%	2%	2%	6%
Malawi	846	670	104	2%	1%	0%	713%
South Africa	582	583	563	1%	1%	1%	3%
Others	1 074	931	1 025	2%	2%	2%	5%
Total	54 610	53 663	50 772				8%

Source: Euratom Supply Agency, World Nuclear Association.

But such an ambitious civilian nuclear program coupled with very limited Chinese uranium reserves will only exacerbate China's import dependency ratio in this area, which is already high. As emphasized by Xiao Xinjian in China Energy, China has no choice but to develop a strong foreign procurement strategy in light of the country's poor uranium resources [40].

China's known uranium resources are insufficient: China total possesses at most 1 percent of the world's known recoverable uranium resources or about 68,000 tons [41]. The country's uranium output in 2011 was only 1500 tons (about 3% of global production, see Table 3), while its annual consumption had been at around 4,500 tons up to 2012 (or about 2% of global consumption) [42]. Its output is expected to eventually rise to 2,500 tons a year according to UX Consulting [7], but the quality of China's uranium resources is poor [43]. Clearly though, additional supplies are required and increasing dependence on imports is unavoidable in light of current development plans. China's imports may rise to about 17% of global consumption, or about 40,000 tons every year by 2020 [44].

Presumably in anticipation of a rapid increase in its uranium demand, China is already importing more uranium than it needs in a given year. For instance, China Daily indicated that China imported 17,135 tons of uranium in 2010, and 16,126 tons in 2011, according to the Chinese Customs [7], more than triple the amount in 2009.

3.2. Domestic Challenges. At the Chinese domestic level, many institutions with widely varying responsibilities are charged with ensuring nuclear safety, but these institutions all have different mandates and pursue different objectives. Similarly, at the regulatory level, a set of independent agencies exists, and this can create overlap issues. In addition, there remains work to be done on the level of the human capital needed to ensure the safety of the civilian nuclear program.

Hu Jintao emphasized this in his March 26th speech in Seoul: "we should establish and improve nuclear safety legal and regulatory system, strengthen the building of a nuclear emergency team, increase investment in research, staff training, and provide an institutional guarantee for strengthening nuclear safety, in response to the emergency mechanism to protect, and provide technical support to raise the level of nuclear safety human resources and support to enhance nuclear safety" [35].

These efforts, while enhanced since the Fukushima disaster, were built on earlier commitments to nuclear safety. In 2006 Kadak argued that

> *"China has developed top-level nuclear safety regulations on site location, safety in design, operations, and quality assurance. They annually set up inspection plans for each power station which focus on key areas of the regulations to assure compliance. They also have special inspections and reviews based on events that may occur at the plant. While the organizational framework is quite similar to the U.S. system, the intrusiveness of the regulator in day-to-day operations is not. The onsite inspectors follow the overall plans for inspections but are not as involved in day-to-day oversight of normal operations and outages. (...) Chinese regulatory personnel do perform inspections and oversee major activities during outages, however.*
>
> *(...) The Chinese government has made it quite clear that they will not tolerate injuries or radiation release. Given the power of the Chinese government, this clarity in expectations of the regulator and the government makes operational decisions of the plant very safety-focused. The challenge is for management to reinforce these*

expectations to all plant employees and contractors to be sure that no unsafe conditions exist [32].

In recent years—and this trend has been compounded by the Fukushima disaster—we have seen the rise of another domestic challenge for the government in China, that of increased public awareness of the risks involved with nuclear power. This increased awareness, coupled with better access to information and means to express opposition via the Internet, has produced much activity online. A case in point is the opposition to the Gaozuang power plant in Nanyang, Henan province [45].

Despite domestic challenges, China has been able up to now to forge ahead with its national nuclear energy plan (as we have seen, albeit at a somewhat slower pace following the Fukushima disaster).

This has much to do with the fact that the Chinese nuclear industry is centralized domestically. Indeed, the civilian nuclear industry and the uranium mining industry are overwhelmingly controlled by two state-owned enterprises—China National Nuclear Corporation and China Guangdong Nuclear Corporation—that report directly to the State Council [41]. The structure of the domestic uranium market might therefore have afforded the Chinese government a stronger hand in fulfilling its policy goals (contrary to other more fragmented markets, such as iron ore, where the government has struggled). This may have allowed China more room for maneuver to concentrate on fulfilling its procurement strategy abroad.

3.3. International Challenges. Chinese companies also face challenges in fulfilling their uranium procurement needs internationally. This point was emphasized by Xie Qingxia, Hua-ming, and Wu Ping in their 2011 China Mining Magazine article [46]. The authors say that challenges lie in the form of a dearth of experience in the management of global uranium extraction companies, lack of knowledge in the legislation and domestic policies of foreign countries, the presence of political risks in host countries and the fact that China is a late player in the game.

In terms of political risks, besides possible diplomatic tensions, China must also manage potential issues to do with corruption and political instability with the source countries, including in Central Asia and Africa.

Further, because of the predicted excess of global demand over global supply in coming years, the Central Asian uranium market will remain very competitive. China is competing for access with Russia and India in Kazakhstan and Mongolia, while South Korea and Japan also buy significant amounts of uranium there and Iran is looking to raise its import as well. India is competing with China in Namibia and Niger as well. Thus, China is not operating alone in markets it is arguably the most comfortable in. Therefore, although China has been trying to diversify its supply sources and shown caution not to overbid for resource acquisition, in the end, strong competition has compelled it to make use of economic and diplomatic tools to gain an edge. Indeed, "the Chinese have shown they will often pay above market prices for those mines, companies and other assets that are genuinely rich in natural resources" [47]. In many ways, uranium market pressures continue to be resolved within a state-to-state framework.

Another type of political risk is the risk of falling afoul of domestic public opinion in source countries. This risk exists in developed source countries. For instance, the Australian population reticence towards nuclear power domestically and abroad has had an impact on the country's export capacity. Since the 1970s, the country has seen ongoing debates between the uranium mining and nuclear industry and environmentalists and indigenous land rights activists. Australia also suffers from other usual obstacles to uranium mining, including in the form of shortage of labour and infrastructure [48]. But it is Australia's 1984 "Three Mines Policy" [9] and subsequent "No New Mines" policy that really limited the scope of uranium mining in the country, until a recent, and timid, loosening.

Social opposition and resulting stricter environmental regulations are thus partly responsible for the transfer of uranium mining away from developed countries in recent years.

However, public opinion risks do not come exclusively from developed economies. Indeed, when it comes to planning investments in developing countries, Chinese companies need to successfully manage the public perception of their actions as well. Following the Fukushima incident, public opposition has risen in Kazakhstan, already exacerbated by years of environmental and safety mismanagement during the Soviet period. "The Chinese have apparently sought to decrease this risk by partnering with the state-owned Kazatomprom in exchange for equity in domestic facilities." [47].

This situation is not unlike that found in other commodity markets. Indeed, China was not a major stakeholder when global commodity market institutions emerged during the second half of the twentieth century and has found itself a newcomer in many markets as it rose to prominence in the past two decades. But in the case of uranium, the fact that the market does not have a producer's cartel, and was in need of investment and expansion as China entered the market, has given China more room to maneuver.

3.4. Procurement Strategy. China has been basing its uranium procurement strategy on the national "Two markets, Two resources" policy, coined in the 1990s, which works to develop both domestic supply sources (this includes developing advanced nuclear power systems and/or alternative nuclear power methods to save fuel) and international supply sources (through foreign acquisition, investment as well as long-term contracts). "By 2020, 1/3 of China's supply of natural uranium will come from domestic uranium production, 1/3 from direct procurement from foreign suppliers, and 1/3 from the overseas holdings of uranium production." [46]. Such a diversified strategy can spread the risk associated with a high import dependency ratio.

Therefore, on the one hand, China is increasing its uranium production domestically (see Table 3). On the other hand, as part of the "Going Out" strategy (which fits under the "Two Markets, Two Resources" policy), China is taking steps to acquire uranium resources internationally. It has equity in Niger and Kazakhstan mines, is investigating Uzbekistan, Mongolia, Namibia, Algeria, and Zimbabwe, and other sources are progressively being added. A bilateral safeguards agreement will also allow imports from Australia, and more recently, Canada.

Beijing is also using creative ways to engineer procurement contracts. As mentioned above, in November 2006, two Chinese firms have established a joint venture with Kazakhstan's state-owned Kazatomprom in a uranium mining project (49 percent stake), in exchange for stakes in either Chinese nuclear power plants or fuel reprocessing facilities for Kazatomprom [49]. China has also "provided interest-free soft loans to the governments of Uzbekistan, Niger and other uranium-rich countries (World Nuclear News, June 10, 2010; November 4, 2008; Reuters, April 24, 2010)" [47].

We stand at the beginning of China's engagement with the global uranium market. China has thus an opportunity to learn from past experiences acquired while dealing with high import-dependency ratios in other types of commodities. Up to now, China has developed a multi-pronged strategy of engagement in the uranium market, that has allowed it to triple its total uranium imports, and forge ahead with the world's most ambitious civilian nuclear power development plan, all within the span of a few years.

4. Potential for International Cooperation

4.1. A Case in Point: Bilateral Cooperation between China and Canada

Uranium Ore. Uranium trade relations between Canada and China show that there remains a large state-to-state angle to this trade. Among the largest uranium producers, Canada has the highest-quality uranium. Indeed, "only Canada has a significant amount of ore above 1 percent—up to about 20 percent of the country's total reserves. In Australia, on the other hand, some 90 percent of uranium has a grade of less than 0.06 percent. Much of Kazakhstan's ore is less than 0.1 percent" [48]. While the Fukushima accident reduced demand for new nuclear power plants in the near-term, CAMECO has said it is sticking to its target to double uranium production to 40 million pounds by 2018. Canada's uranium production is projected to increase at an average rate of 9 per cent a year to 15 300 tonnes in 2016.

During the Canadian Prime Minister's visit to China in February 2012, a protocol amending Canada's nuclear cooperation agreement with China to allow the export of uranium concentrate was announced. This was big news both for China's energy security and potential to diversify supply sources, and Canada, which can now export uranium directly to China. These sales were banned up until now. In 1976, Canada barred exports of uranium and nuclear reactors to countries that had not signed the Non-Proliferation Treaty (China signed in 1992). A 1994 agreement allowed the sale of reactors, but until last year's amendment to that pact, Canada hadn't yet relaxed its restriction on selling nuclear fuel to China. So until last year, Cameco Corp. had to ship the uranium that it was selling to China from other countries, such as Namibia and Kazakhstan.

In November 2011, Cameco signed an agreement with China Guangdong Nuclear Power Holding to import 29,000 tons of the mineral through 2025. Canada's Cameco has also contracted to sell 23,000 tons of uranium concentrate trough 2020 to China Nuclear Energy Industry Corporation (directly owned by China National Nuclear Corporation) [44].

Following this, Saskatchewan Minister of Energy Bill Boyd signed a Memorandum of Understanding (MOU) in February 2012, on scientific and technical research co-operation on uranium geology with the Beijing Research Institute of Uranium Geology, a research establishment of the China National Nuclear Corporation.

Cameco Corp. Chief Executive Officer Tim Gitzel said "China is becoming the leader of the world" for nuclear energy [50]. Gitzel said Cameco continues to discuss partnerships with Chinese companies on the possibility of jointly developing uranium projects in China, Canada, and other countries.

4.2. Technology. This relationship extends to technical levels of cooperation as well. Indeed, technological cooperation in civilian nuclear technology has occurred between Canada and China in recent years. Indeed, in 2009, the capacity of CANDU's heavy water reactors to reuse spent fuel recycled from other light water reactors was explicitly recognized by a panel appointed by the China National Nuclear Corporation. It "cited the design's 'enhanced safety and good economics' as reasons it could be deployed in China in the near term" [51] and recommended that two units be built.

China is also pursuing reprocessing capabilities in partnership with Canadian CANDU designer Atomic Energy of Canada. China has built two CANDU reactors (and is considering building additional units), which are utilizing reprocessed fuel from its nine light water reactors. CANDU reactors can also run on thorium fuel, and China has been working on developing a thorium fuel cycle with its Canadian partners. Thorium is more abundant in China, cheaper to mine, produces less waste and, if successful, will enhance Chinese energy security.

As the case of Canada-China uranium and nuclear relationship shows, the uranium trade is nested within a thicker web of state-to-state relations that include safety and technological exchange issues. As will be developed further below, this embeddedness may have a positive impact on the nature of China's participation in the global uranium market [52, 53].

4.3. Cooperation on the Enrichment Level. Uranium being only the first step towards the development of nuclear fuel, cooperation initiatives regarding enriched fuel also play a role in the broader civilian nuclear fuel market. The World Nuclear Association lists operating uranium enrichment facilities in 11 countries, 6 with large capacity (France, the US, the UK, Germany, Russia and the Netherlands) and 5 of smaller capacity (China, Japan, Brazil, Pakistan, and Iran) [54].

There are multiple initiatives underway which seek to coordinate enrichment activities worldwide, and the International Atomic Energy Agency (IAEA) actively encourages these "Multilateral Nuclear Approaches" [54]. These programs contribute to enhancing the global tracking of enriched fuel, and thus fulfill nonproliferation goals, while facilitating the coordination of the global market for enriched fuel. A case in point is the Eurodif enrichment center in France, jointly owned by five countries (France, Belgium, Italy, Spain, and Iran), operating under the IAEA oversight, and giving the participants some controlled access to the final product without sharing any technology [54].

Similar initiatives are underway in Russia, where under the IAEA's supervision, an International Uranium Enrichment Centre is being created in Siberia, with Russian, Kazakh, Ukrainian, and Armenian equity. Other such projects are underway in the US, the UK, the Netherlands, and Germany.

The IAEA is also calling for the establishment of international banks of enriched fuel. A deal was signed for Russia to make available 120 tons of nuclear fuel, and a similar arrangement was being discussed with Kazakhstan in 2010. Such initiatives, say former IAEA Director General Mohamed ElBaradei, "make sure that every country that is a bona fide user of nuclear energy, and that is fulfilling its nonproliferation obligations, is getting fuel" [48].

This only goes to show that uranium procurement is a part of broader civilian nuclear issues. China has displayed willingness to collaborate on security issues that pertain to civilian nuclear programs and nonproliferation, and additional participation in a global cooperation mechanism on uranium enrichment would further cement China's role as a constructive participant in such multilateral initiatives.

4.4. Cooperation among Developing Countries. At the level of safety standards, China has indicated its willingness to play a leading role in fostering cooperation initiatives among developing countries. Indeed, the country has indicated that it is interested in providing nuclear safety assistance, and focuses on helping developing countries establish and improve their nuclear safety infrastructure, as well as improve nuclear safety and technical standards [35].

China has already been involved in multiple regional nuclear security training courses providing training to nearly 100 people from more than 10 countries in the Asia-Pacific region [35]. It also looks to help developing countries improve technical levels of nuclear safety.

This is yet another way in which the multidimensionality of the civilian nuclear industry manifests itself. China's engagement internationally in the realm of safety issues, and with developing countries (a setting where China has long played a role), may provide it with the experience and a platform to eventually broaden its engagement to other issues, for example, to issues of natural/enriched uranium procurement.

4.5. Impact of the Nuclear Security Summits. Enhanced cooperation on issues of nuclear security at an international level had already started with the Washington Nuclear Security Summit held in 2010, but the Fukushima accident brought back to the fore the need to enhance state capacity to cope with the unexpected and the need to address issues of nuclear safety.

Indeed, the second Nuclear Security Summit held in Seoul in March 2012 broadened its agenda to include nuclear safety and proved to be an important step following the Fukushima accident and the need to discuss this issue in a multilateral setting. There is much to learn from the Japanese accident, and China has shown its willingness to learn from this experience.

The summit is useful despite the fact that sovereignty concerns and economic and technological differences hinder the establishment of binding safety standards across the board. There are difficulties in harmonization of safety standards, but China has demonstrated a will to work in cooperation with other developed nations on this regard. It shows great confidence on China's behalf to fully engage developed countries in this sensitive issue.

"In July 2010 a 22-strong IAEA team from 15 countries carried out a two-week Integrated Regulatory Review Service mission to review of China's regulatory framework for nuclear safety. The IAEA made a number of recommendations but said that the review had provided 'confidence in the effectiveness of the Chinese safety regulatory system and the future safety of the vast expanding nuclear industry.'" [37]. China has also "requested and hosted 12 Operational Safety Review Team (OSART) missions from IAEA teams to October 2011" [37].

In the past years, China has made substantial efforts in this regard. China has also made relatively noticeable efforts in reaching out regarding nuclear safety, as it has done with the US (implemented in 1998 and reinforced in 2005 by a Memorandum of Understanding that granted Westinghouse the contract to build four commercial nuclear reactors in China). China is part and parcel of the international community and its efforts to enhance nuclear security and safety globally [16].

A case in point, China is a fully fledged member of the International Atomic Energy Agency, committed to international nonproliferation efforts and cooperating on issues of civilian nuclear technology with France (Areva), Canada (Atomic Energy of Canada—CANDU), and the US (Westinghouse) among others, as well as participating in related international frameworks.

China's energy needs are growing at such speeds that the parallel growth of its civilian nuclear program appears inevitable. But to do it, as President Hu Jintao emphasized in his Seoul speech, China needs to "face the risk of nuclear safety, to learn the lessons of the nuclear accident, and take effective measures to enhance security and reliability of nuclear energy, to promote the safety of nuclear energy, sustainable development" [35].

All in all, at the domestic level, a centralized industry, and at the international level, a geographically dispersed and uncoordinated market allowing some space for expansion have provided China with relatively comfortable conditions to roll out its procurement strategy abroad. But the fact that the uranium market remains one step away from nuclear security and safety issues, which have spawned a web of international cooperation initiatives, may foster a cooperative environment for uranium as well.

It remains that whereas the structure of the global uranium market have allowed China to carve itself a place and roll out its procurement policy, the safety and security dimensions of the nuclear industry have provided China with the opportunity for a cooperative and confident engagement at a multilateral level.

Many uncertainties remain, namely, the potential for a supply squeeze, delayed production due to mining accidents, and public opinion protests among others, so peaceful coexistence, or even cooperation, is not a foregone conclusion. Indeed, it remains to be seen whether a conceptual frontier will remain between natural uranium ore mining and enriched uranium production and other civilian nuclear issues. Further research could look at the likelihood of spilling effects. However, the particular conditions under which this market is evolving currently allow for such a possible development.

5. Conclusion

The first signs of China's likely future impact on the global uranium industry have already been felt. Despite China's uranium requirements being less than half than that of the US, China still imported almost as much uranium as the US did in 2010, three times more than the amount it imported in 2009. This is despite the fact that the US has over 100 reactors in operation, against China's 15.

Early signs also point to a reorganization of the global uranium market, with the emergence of new players such as Kazakhstan. This sudden increase and investment in Kazakhstan's uranium mining industry has contributed to the rise of this country to number one uranium producer in the world, whereas it was all but absent from global uranium trade in 2003 [19].

The answer as to why China has succeeded in establishing itself as a confident player pushing forward a multi-pronged international uranium procurement strategy has to do with both domestic and international variables. On the one hand, at the domestic level, the centralized structure of China's uranium procurement industry has allowed industry and government stakeholders to develop and implement a coherent strategy (contrary to the situation in the fragmented iron ore market for instance). On the other hand, at the international level, the absence of an exporters' cartel or oligopoly established before China's emergence as a large purchaser have enabled it to carve itself a place on the global market.

On the top of that, recent international cooperation initiatives, which encourage transparency and collaboration, even if they are concentrating on nuclear safety and security issues, have provided China with the opportunity for a cooperative and confident engagement. These international efforts have showcased China as a country ready to rise to the occasion and be a responsible player, including as a fully-fledged member of the International Atomic Energy Agency.

As the country builds confidence in dealing with these issues in an international setting, could it lead it to spearhead further initiatives, such as an international uranium demand management initiative, that could be managed through the Shanghai Cooperation Organization? China seeks to have a say in global market institutions, which for the most part have been created prior to its recent reemergence, that is commensurate to the share of the global demand that it now occupies. Demand management initiatives like this could be an area where there is room for China to innovate, perhaps in its own regional setting at first [55].

All in all, there is potential for peaceful coexistence—and even international cooperation—in the global uranium market, something towards which China can contribute meaningfully.

Acknowledgments

This paper constitutes a preliminary work of a larger research project by P. Massot on The Impact of China on Global Commodity Markets, funded by the Social Sciences and Humanities Research Council of Canada. Sincere thanks are given to Matthew Gravelle for his helpful comments on an earlier draft of this paper. The research of Z.-M. Chen has been supported by the Ministry of Education Foundation of China (no. 12YJCZH021). Some sections of this paper are built on an op-ed published in Chinese by P. Massot: "How can China deal with the dependence on imported uranium (Zhongguo ruhe yingdui haiwai you yilai)," China Business News Daily, April 23, 2012, A6, http://www.yicai.com/news/2012/04/1654156.html.

References

[1] H. Wang and J. N. Rosenau, "China and global governance," *Asian Perspective*, vol. 33, no. 3, pp. 5–39, 2009.

[2] R. Garnaut and L. Song, "Rapid industrialization and market for energy and minerals: China in the East Asian context," *Frontiers of Economics in China*, vol. 1, no. 3, pp. 373–394, 2006.

[3] Y. Song and F. Li, "How can China build purchase alliances in teh world market of resource commodities?" *Journal of Chinese Economic and Foreign Trade Studies*, vol. 2, no. 2, pp. 121–130, 2009.

[4] D. Zweig and B. Jianhai, "China's global hunt for energy," *Foreign Affairs*, vol. 84, no. 5, pp. 25–38, 2005.

[5] Y. Tiberghien, *L'Asie Et Le Futur Du Monde. Nouveaux Débats*, Presses de Sciences Po, Paris, France, 2012.

[6] K. Matthies, "Commodity prices remain high," *Intereconomics*, vol. 42, no. 2, pp. 109–112, 2007.

[7] Q. Ding and L. Yiyu, *Nation Plans to Import More Uranium*, China Daily, 2012.

[8] G. Rothwell, "Market power in uranium enrichment," *Science and Global Security*, vol. 17, no. 2-3, pp. 132–154, 2009.

[9] M. Conde and G. Kallis, "The global uranium rush and its Africa frontier. Effects, reactions and social movements in Namibia," *Global Environmental Change*, vol. 22, no. 3, pp. 596–610, 2012.

[10] L. M. Lidsky and M. M. Miller, "Nuclear power and energy security: a revised strategy for Japan," *Science & Global Security*, vol. 10, no. 2, pp. 127–150, 2002.

[11] L. R. Stewart, "Canada's role in the International Uranium Cartel," *International Organization*, vol. 35, no. 4, pp. 657–689, 1981.

[12] *Edano: Not Up to Government to Decide reactor Restarts*, The Asahi Shimbun, 2012.

[13] D. E. Weinstein and M. K. Schnell, *Evaluating the Economic Response To Japan's Earthquake*, vol. 301 of *Columbia Business School Working Paper Series*, Columbia University, New York, NY, USA, 2012.

[14] A. Sharp and T. Fujioka, *Nuclear-Free Summer Poses Growth Risk For Japan's West*, Bloomberg, 2012.

[15] I. N. Kessides, "The future of the nuclear industry reconsidered: risks, uncertainties, and continued promise," *Energy Policy*, vol. 48, pp. 185–208, 2012.

[16] Y.-C. Xu, "Nuclear power in China: how it really works," *Global Asia*, vol. 7, no. 1, pp. 32–43, 2012.

[17] Y. Liu, *New Nuclear Projects Ready to Power Ahead*, China Daily, 2012.

[18] *Cameco Seeing "Strong" Growth in China, Chief Executive Says*, Bloomberg, 2012.

[19] *Metal Prospects: Uranium Market Outlook—Second Quarter 2011*, RBC Capital Markets, 2011.

[20] *Uranium 2009: Resources, Production and Demand*, International Atomic Energy Agency (IAEA), 2010.

[21] *Nuclear Power in Germany*, World Nuclear Association, 2012.

[22] O. Bukharin, "Understanding Russia's Uranium Enrichment Complex," *Science & Global Security*, vol. 12, no. 3, pp. 193–214, 2004.

[23] "Megatons to Megawatts," 2012, http://www.usec.com/russian-contracts/megatons-megawatts.

[24] "US-Russian uranium agreement takes effect," 2012, http://www.neimagazine.com/story.asp?sc=2061460.

[25] J. Finch, *Spot Uranium Price Climb Stalls*, UraniumSeek, 2006.

[26] "Uranium 101," 2012, http://www.cameco.com/uranium_101/markets/#five.

[27] "CME/NYMEX Partners with Ux Consulting to Offer Uranium Futures Conracts," 2012, http://www.uxc.com/data/nymex/NymexOverview.aspx.

[28] J. J. Licata, "Going nuclear with uranium stocks. Futures: News, Analysis and Strategies for Futures," *Options and Derivatives Traders*, vol. 40, no. 2, pp. 30–32, 2011.

[29] C. Winzer, *Conceptualizing Energy Security*, University of Cambridge. Electricity Policy Research Group, 2011.

[30] Z. Daojiong, "China's energy security: domestic and international issues," *Survival*, vol. 48, no. 1, pp. 179–190, 2006.

[31] W. Dazhong and L. Yingyun, "Roles and prospect of nuclear power in China's energy supply strategy," *Nuclear Engineering and Design*, vol. 218, no. 1-3, pp. 3–12, 2002.

[32] A. C. Kadak, "Nuclear power, 'Made in China,'" *Brown Journal of World Affairs*, vol. 13, no. 1, 2006.

[33] A. Boey, "Nuclear power and China's energy future: limited options," *The Asia-Pacifc Journal*, vol. 10, no. 9, 2012.

[34] Y. Zhou, "Why is China going nuclear?" *Energy Policy*, vol. 38, no. 7, pp. 3755–3762, 2010.

[35] J. T. Hu, "President Hu Jintao Delivers a Speech at the Nuclear Security Summit in Seoul (Hu Jintao zai Shouer heanquan fenghui shang de jianghua)," *Xinhua News*, 2012 (Chinese).

[36] J. Wen, "China firmly set on a path towards green and sustainable development (Zhongguo jianding zou lüse he kechixu fazhan daolu)," *Economic Daily*, 2012 (Chinese).

[37] "Nuclear Power in China," World Nuclear Association, 2012, http://www.world-nuclear.org/info/inf63.html.

[38] Q. Wang, "China needing a cautious approach to nuclear power strategy," *Energy Policy*, vol. 37, no. 7, pp. 2487–2491, 2009.

[39] S. Hall, R. M. Stewart, and H. Michael, *China Hunts For Uranium*, Wall Street Journal, 2011.

[40] X. Xiao, "Chinese nuclear energy development in 2011, future trends and policy suggestions (Nian Zhongguo hedian fazhan zhuangkuang, weilai qushi ji zhengce jianyi)," *China Energy*, vol. 34, no. 2, pp. 18–23, 2012 (Chinese).

[41] R. Weitz, "China's uranium quest part 1: domestic shortages fuel global ambitions," *China Brief*, vol. 11, no. 15, pp. 6–10, 2011.

[42] Y. Liu, *Uranium Imports in Marked Decline*, China Daily, 2011.

[43] Y. Guang and H. Wenjie, "The status quo of China's nuclear power and the uranium gap solution," *Energy Policy*, vol. 38, no. 2, pp. 966–975, 2010.

[44] T. Gitzel, "Cameco," in *Proceedings of the Whistler Institutional Investor Conference (CIBC '11)*, Whistler, Canada, 2011.

[45] "The Reason Why Nanyang Is Not Suitable for the Construction of a Nuclear Power Plant (Nanyang bu shihe jian hedianzhan liyou)," *Tianya*, 2011 (Chinese).

[46] Q. Xie, M. Hua, and P. Wu, "China's uranium industry, "going out" policy: opportunities and challenges (Woguo youkuangye "zouchuqu" suo mianlin de jiyu yu tiaozhan)," *China Mining Magazine*, vol. 20, no. 12, pp. 16–19, 2011 (Chinese).

[47] R. Weitz, "China's uranium quest part 2: the turn to foreign markets," *China Brief*, vol. 11, no. 16, pp. 12–15, 2011.

[48] T. Johnson, *Global Uranium Supply and Demand*, Council on Foreign Relations, 2010.

[49] M. Komesaroff, "Energy: China goes nuclear," *China Economic Quarterly*, vol. Q4, pp. 10–13, 2007.

[50] W. Bi, *Cameco Seeing 'Strong' Growth in China, Chief Executive Says* , Bloomberg, 2012.

[51] "Expert panel recommends CANDU as best choice for China's alternative nuclear fuel program," AECL Atomic Energy of Canada Limited, 2009.

[52] G. J. Ikenberry, "The rise of China and the future of the West," *Foreign Affairs*, vol. 87, no. 1, pp. 23–37, 2008.

[53] E. Economy and O. Michel, *China Joins the World: Progress and Prospects*, Council on Foreign Relations Press, New York, NY, USA, 1999.

[54] "Uranium Enrichment," World Nuclear Association, 2013, http://www.world-nuclear.org/info/inf28.html.

[55] X. Xiaojie, *Energy BRICS: Towards a New Cooperation Framework*, Chinese Academy of Social Sciences/CASS NEWS, 2012.

Exergetic Assessment for Resources Input and Environmental Emissions by Chinese Industry during 1997–2006

Bo Zhang,[1,2] Beihua Peng,[3] and Mingchu Liu[4]

[1] *School of Management, China University of Mining and Technology (Beijing), Beijing 100083, China*
[2] *State Key Laboratory of Coal Resources and Safe Mining, China University of Mining and Technology (Beijing), Beijing 100083, China*
[3] *Shenhua International Trading Co. Ltd., Beijing 100011, China*
[4] *Pingxiang College, Jiangxi Province, Pingxiang 337000, China*

Correspondence should be addressed to Bo Zhang, zhb@pku.edu.cn

Academic Editors: C. Bin and C. Zhan-Ming

This paper presents an overview of the resources use and environmental impact of the Chinese industry during 1997–2006. For the purpose of this analysis the thermodynamic concept of exergy has been employed both to quantify and aggregate the resources input and the environmental emissions arising from the sector. The resources input and environmental emissions show an increasing trend in this period. Compared with 47568.7 PJ in 1997, resources input in 2006 increased by 75.4% and reached 83437.9 PJ, of which 82.5% came from nonrenewable resources, mainly from coal and other energy minerals. Furthermore, the total exergy of environmental emissions was estimated to be 3499.3 PJ in 2006, 1.7 times of that in 1997, of which 93.4% was from GHG emissions and only 6.6% from "three wastes" emissions. A rapid increment of the nonrenewable resources input and GHG emissions over 2002–2006 can be found, owing to the excessive expansion of resource- and energy-intensive subsectors. Exergy intensities in terms of resource input intensity and environmental emission intensity time-series are also calculated, and the trends are influenced by the macroeconomic situation evidently, particularly by the investment-derived economic development in recent years. Corresponding policy implications to guide a more sustainable industry system are addressed.

1. Introduction

The natural resources depletion has been considered as one of the main constraints for sustainable development [1, 2]. Resources, especially nonrenewable resources, are required to supply the basic human needs and to improve the quality of life [3]. At the same time, a majority of nonrenewable resources are consumed in the industry sector, which provides most energy and matter used in modern society. Resources production and consumption by industrial activities are therefore reckoned as a strong positive determinant ingredient of air pollution and climate change [4].

In China, the industry sector accounts for approximately 70% of the total energy resources input and consumes the largest amount of mineral resources such as iron ores in the world [5]. Industrial activities along with huge resources input and low resource use efficiency have engendered striking environmental emissions such as greenhouse gases (GHG). Averagely, 81.4% of SO_2, 80.9% of soot, and 47.8% of waste water in China were emitted by the industry sector during 1997–2006 [6]. In the year of 2006, 24.0 billion ton industrial waste water, 22.3 million ton SO_2, and 13.0 million ton solid waste were discharged into the environment. The share of CO_2 emissions from the industry sector accounted for more than two-thirds of China's total energy-related CO_2 emissions [7]. During a period of rapid economic growth in China, the challenge confronted with the industry sector is the ever-increasing pressure on natural environment due to large amounts of nonrenewable resources consumption with urgent regard for environmental consequences.

Without explicit throughput measures, the scale question of the physical resource base and human conditions cannot be analyzed and reflected adequately [8, 9]. An efficient understanding of the resources use and environmental impact of the Chinese industry against drastic socioeconomic transitions demands systematic biophysical assessment with a unified measure. Exergy is defined using thermodynamics principles as the maximum amount of work which can be produced by a system as it comes to equilibrium with a reference environment [10–12]. The potential usefulness or ability to perform work for a natural resource is its exergy content [13, 14], and then exergy quantifies the quantity and quality scarcities of diverse resources effectively [15]. Distinguishing from the traditional economic analysis, exergy accounting provides a unified way to measure different materials and energy with solid scientific basis [15, 16] and provides a wide and clear vision of the use and degradation of energy and subsequently of natural resources [17, 18]. As an overall and unifying assessing tool, exergy analysis has been widely employed to evaluate the resources use at different scales [19–23], and particularly to perform the resource exergy analysis of different countries [16, 24–40].

Meanwhile, uses of exergy are increasing in fields related to environmental impact. All utilization of resources and disposal of waste products affect nature and the effect is strongly related to the amount of exergy in the utilized resource or the disposed waste [11, 19, 41]. The exergy amount of an emission is the physicochemical work absorbed by the environment in order to equilibrate the substances of the emission with the standard environment [42]. All emissions have definable, calculable, and additive exergy contents with respect to the defined reference environment, and then exergy can be regard as a suitable unifying measure of environmental emissions. Rosen and Dincer [43–45] further stressed that the exergy embodied in waste emissions represents a potential for environmental change. The concept of exergy has been gradually accepted as a "direct" measure or at least as a proxy stated by Ayres [46] for the potential environmental impact of waste emissions [41, 42, 47–50].

Closely relevant to exergy-based insight into resources use and environmental impact of the Chinese industry, Chen and his fellows have carried out a series of studies in their social exergy analysis of resources use and environmental emissions at the national scale covering the industry sector [38–40, 50, 51]: Chen and Qi [38] presented systems account for the resources exergy utilization of China society 2003; G. Q. Chen and B. Chen [39] provided an extend-exergy analysis of the resources conversion and waste emissions of the China society in 2005; Zhang and Chen [40] provided an exergy-based systems account for the resources use and environmental emissions (including GHGs and "three wastes") of China society 2006; Zhang et al. [50] provided a chemical exergy-based unifying assessment of the "three waste" emissions by Chinese industry during 1997–2006. However, the overall status and trend of the resources use and environmental impact by Chinese industry remain to be revealed systematically with an objective measure to quantify and evaluate various resources and wastes in more aggregated levels.

The aim of this paper is to present an exergetic assessment for the resources input and environmental emissions of the Chinese industry during 1997–2006. By accounting the fundamental utility of resource inflows into Chinese industry including fossil fuels, mineral resources, agricultural and forest products, and other industrial raw materials based on a unified measure, resources use of the Chinese industry is elucidated. Meanwhile, environmental impact of the Chinese industry and in particular, main environmental emissions covering GHGs and "three wastes" are evaluated. Exergy intensities in terms of resource input intensity and environmental emission intensity time-series are also calculated. Corresponding discussion and policy implications coupled with China's macroeconomic situation are presented. In sum, insights provided by exergy analysis in this study can be added to the poor knowledge between industrial economic profitability and ecological sustainability and contribute to resources management and environmental regulation for the policymakers in China.

2. Methodology and Data Sources

2.1. System Boundary and Data Sources. Chinese industry refers to the material production sector which is engaged in the extraction of natural resources and processing and reprocessing of minerals and agricultural products [6], including (1) extraction of natural resources, such as mining and salt production (excluding hunting and fishing); (2) processing and reprocessing of agricultural products, such as rice husking, flour milling, wine making, oil pressing, silk reeling, spinning and weaving, and leather making; (3) manufacture of industrial products, like steel making, iron smelting, chemicals manufacturing, petroleum processing, machine building, timber processing; water and gas production, and electricity generation and supply; (4) repairing of industrial products such as the repairing of machinery and means of transport (including cars).

For the national-scale system, the resources input into the Chinese industry contains the imported, gathered, constrained, and extracted commodities as exergy carriers [15, 16]. For avoidance of repetitive and cross calculations, the entrance boundary points are set at the same level of the exergy inflow. Most of relevant environmental resources and economic data for the mainland China are adopted or derived from the official databases and public issued official statistical yearbooks, such as Almanac of China Paper Industry [52], China Environment Yearbook [53], China Food Industry Yearbook [54], China Industrial Economic Statistical Yearbook [55], China Steel Yearbook [56], China Yearbook of Nonferrous Metal Industry [57], and Statistical Yearbook of China [6].

2.2. Exergy Methodology. In this study, all the thermal exergy of the materials are neglected, for the difference between the temperatures of the materials and the environment is small and therefore the thermal exergy is much less than the chemical exergy of the materials according to the basic definition of exergy [38]. Extensive illustrations for estimating exergy coefficients for different resources in China have been provided by B. Chen and G. Q. Chen [30] and Chen and Qi [38]. Concrete exergy coefficients of the accounted resources are listed in Table 1.

Table 1: Exergy coefficient of various resources.

Item	Exergy coefficient	Unit	Source
Fossil fuels			
Coal	22.16	PJ/Mton	[38]
Oil and oil product	44.32	PJ/Mton	[38]
Natural gas	4.13	PJ/10^8 cu·m	[38]
Minerals			
Iron ore (55% Fe)	0.46	PJ/Mton	[38]
Iron ore fine (70% Fe)	0.84	PJ/Mton	[38]
Sulphur iron ore (35% S)	9	PJ/Mton	[38]
Copper ore (0.65% Cu)	0.03	PJ/Mton	[30]
Lead ore (3.5% Pb)	0.02	PJ/Mton	[30]
Zinc ore (5.9% Zn)	0.05	PJ/Mton	[30]
Copper ore fine (23.8% Cu)	1.1	PJ/Mton	[38]
Alumina (63.7% Al)	2	PJ/Mton	[38]
Phosphorite (25% P_2O_5)	0.1	PJ/Mton	[38]
Raw salt (NaCl)	0.2	PJ/Mton	[38]
Limestone	0.01	PJ/Mton	[30]
Metal scraps			
Steel (Fe)	6.8	PJ/Mton	[38]
Copper (Cu)	2.1	PJ/Mton	[38]
Aluminum (Al)	32.9	PJ/Mton	[38]
Forest products			
Wood	10	PJ/Mton	[30]
Bamboo	18.67	PJ/Mton	[58]
Turpentine	37.4	PJ/Mton	Calculated by authors
Oil-tea camellia seed	35.3	PJ/Mton	Calculated by authors
Tung oil	38.9	PJ/Mton	Calculated by authors
Agricultural products			
Sugarcane	5	PJ/Mton	[38]
Cotton	16.4	PJ/Mton	[30]
Hemp	16.35	PJ/Mton	[30]
Rapeseed	37	PJ/Mton	[30]
Beet	5	PJ/Mton	[30]
Soybean	3.9	PJ/Mton	[30]
Cocoon	4.5	PJ/Mton	[30]
Wool	3.7	PJ/Mton	[30]
Peanut	24.6	PJ/Mton	[30]
Sesame	23.4	PJ/Mton	[30]
Tubers	3.7	PJ/Mton	[30]
Bean	3.9	PJ/Mton	[30]
Rice	15.56	PJ/Mton	[30]
Wheat	15.4	PJ/Mton	[30]
Corn	8.6	PJ/Mton	[30]
Tobacco leaf	10.7	PJ/Mton	[30]
Pork	25	PJ/Mton	[30]
Beef	11.5	PJ/Mton	[30]
Mutton	16	PJ/Mton	[30]
Poultry	4.5	PJ/Mton	[30]
Milk	4.9	PJ/Mton	[30]
Egg	6.2	PJ/Mton	[30]
Fruit	1.9	PJ/Mton	[30]

Table 1: Continued.

Item	Exergy coefficient	Unit	Source
Aquatic product	5.77	PJ/Mton	[30]
Straw	14.3	PJ/Mton	[30]
Other raw materials			
Pulp	17	PJ/Mton	[30]
Rubber	32.48	PJ/Mton	[30]
Synthetic rubber	45.53	PJ/Mton	[59]
Ethylenc glycol	19.34	PJ/Mton	[60]
Terephthalic acid	24.8	PJ/Mton	[60]
Polyethylene in primary forms	48.26	PJ/Mton	[60]
Polypropylene in primary forms	47.7	PJ/Mton	[59]
Polystyrene in primary forms	50.2	PJ/Mton	[60]
Polyvinyl chloride in primary forms	20.35	PJ/Mton	[59]

Note: The exergy coefficients of water potential energy and nuclear energy were deduced from their product of electricity (0.36 PJ/10^8 kWh) with the transformation factor of 1.17 and 3.51, respectively [38]. Some chemical materials, nonmetallic mineral, and other raw material are not included due to their negligible exergy input or scarcity data.

Table 2: Exergy coefficient of various emissions.

Item	Exergy coefficient	Unit	Source
CO_2	0.45	PJ/Mton	[60]
CH_4	51.98	PJ/Mton	[60]
COD	13.6	PJ/Mton	[40]
SO_2	4.9	PJ/Mton	[60]
Soot	3.5	PJ/Mton	[50]
Dust	1.5	PJ/Mton	[50]
Solid waste	0.5	PJ/Mton	[50]

As to the emission account for the industry system as a macroeconomy, it is reasonable to adopt a global standard environment model to resemble the atmosphere, ocean and earth's upper crust with average geophysical chemical characteristics as the reference environment [60, 61]. The chemical exergy of an emission, as the dominant exergy component is considered in this paper. In China, industrial environmental emissions were not covered in statistics until 1997, and the environmental statistic items only cover the main emissions of the conventional "three wastes." We extract all the available data for the period from 1997 to 2006 and chose the most remarkable environmental emissions to do a trend analysis. Owing to the data availability, seven major emissions (i.e., CO_2, CH_4, COD, SO_2, soot, dust, and solid waste) are included in our calculations. The emission data of CO_2 and CH_4 are taken from Zhang [62] and other emission data from the official published statistical yearbooks [53]. Detailed exergy coefficients of the accounted emissions are presented in Table 2.

3. Results

3.1. Resources Input. As the sum of all input fluxes outside the system boundary, a detailed exergy accounting for the resources input of the Chinese industry is performed. Compared with 47568.7 PJ in 1997, resources input in 2006 increased by 75.4% and reached 83437.9 PJ. Concretely, the input of resources exergy kept steady around 46196.5–48187.9 PJ during 1997–2001; while afterwards, it increased from 51777.7 PJ in 2002 to 83437.9 PJ in 2006, with an average annual growth rate of 12.7%. Two categories of resources input are divided, that is, nonrenewable and renewable resources, with corresponding results of exergy accounting shown in Figure 1. The greater part of resource inflows into the industry sector were seen to come from nonrenewable resources, which accounted for 75.6%–82.5% of the total. A rapid increment of the nonrenewable resources input in the recent 5 years can be found, from 40183.3 PJ in 2002 to 68878.6 PJ in 2006, owing to the increasing input of raw coal, crude oil, natural gas, metal and nonmetal minerals into the industrial subsectors. Details are shown in Table 4.

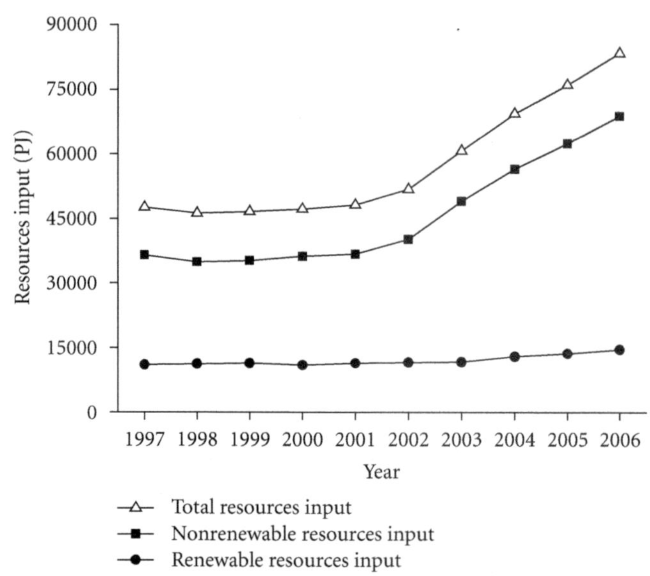

Figure 1: Resources input by Chinese industry.

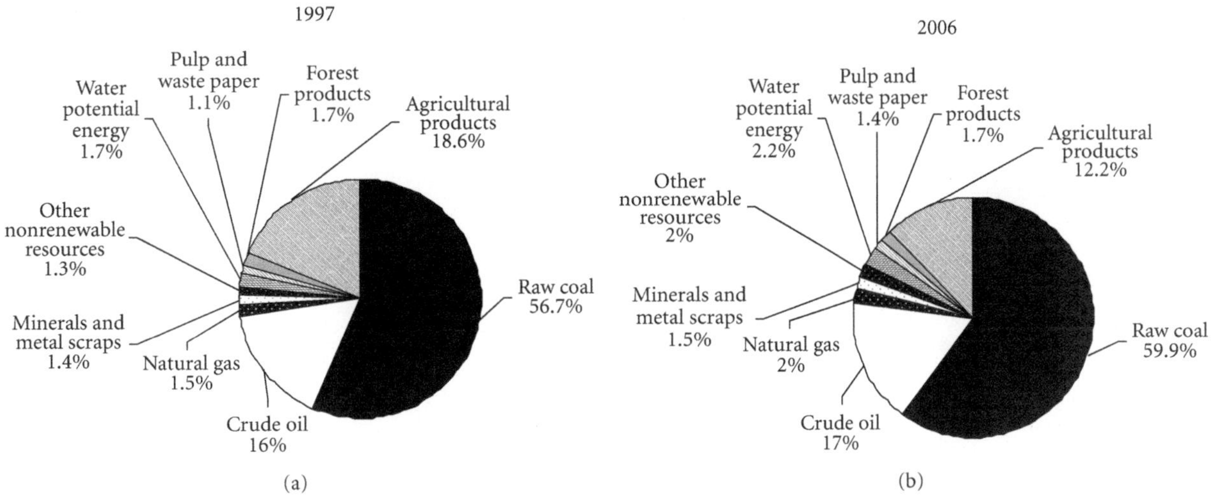

Figure 2: Components of resources input by Chinese industry in 1997 and 2006.

Of all the nonrenewable resources, coal inflow was the largest, contributing to 52.4%–59.9% of the total resources input. In particular, the coal input decreased from 26962.3 PJ in 1997 to 25175.5 PJ in 2001, which can be contributed to rectification and readjustment of coal production performed to balance the wide gap between the supply and demand [16]. During 1998–2001, 58000 small village coal mines were shut down and their production capacity with 2.7×10^8 ton was stopped [63]. Since 2002, the coal production rebounded, restored and continued to increase due to the rapid rise of coal consumption and electricity demand. The input of crude oil also increased 1.9 times in 2006 of that in 1997 and accounted for averagely 18.0% of the total resources input over this period. Natural gas input amounted to 1709.7 PJ, 2.5 times of that in 1997. The iron ore and scrap steel resources input in iron and steel industry increased by 178.7% in the past decade, from 359.5 PJ in 1997 to 1001.9 PJ in 2006. Particularly, the imported iron ore fine and steel product rose rapidly and amounted to 400.0 PJ in 2006, compared with 136.2 PJ in 1997. Nonferrous ores and scrap resources input had increased by more than 3.9 times from 35.4 PJ in 1997 to 136.0 PJ in 2006. As the primary raw material for the cement industry, limestone also expanded 2.4 times during the past ten years.

Only a small part of resource inflows from renewable resources, for example, within agriculture and forestry, were used in the industry sector. Renewable resources inflows increased by 31.9% from 11041.4 PJ in 1997 to 14559.2 PJ in 2006, owing to the increasing input of water potential energy, the grains and meats into the food processing industry, and other industrial materials (e.g., wood, pulp, and waste paper). For instance, water potential energy input increased rapidly from 830.0 PJ in 1997 to 1845.7 PJ in 2006, and the imported wood and bean rose by 585.6% and 816.7% in 2006, respectively. Totally, the share of renewable resources in the total resources input decreased from 23.2% (11041.4 PJ) in 1997 to 17.4% (14559.3 PJ) in 2006. Detailed components of resources input by Chinese industry in 1997 and 2006 are shown in Figure 2.

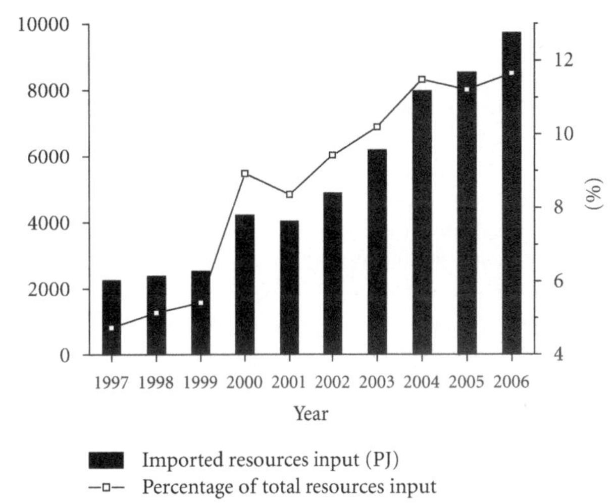

Figure 3: Imported resources input by Chinese industry (Note: Right y-axis refers to percentage of total resources input).

Furthermore, the domestic supply of energy and mineral resources always cannot meet the huge and ever-increasing demands in China, and then a large amount of industrial raw materials need to be imported. The total amount of imported resources input into the Chinese industry increased rapidly from 2249.9 PJ (4.7% of the total resources input) in 1997 to 9720.9 PJ in 2006 (11.7% of the total), as shown in Figure 3. As the largest imported resource, crude oil accounted for 67.2% of the total imported resources input for the period between 1997 and 2006 on average.

Resource input intensity (RII), as the ratio of the total exergy input of resources to the total industrial value added (IVA), is a critical parameter for resource policies that aims to reduce resource consumption while maintaining or even boosting economic growth. The lower the ratio, the fewer the resources input to yield per unit IVA and the higher macroeconomic efficiency of resources use in the industrial economy. Macroeconomic output of the Chinese industry

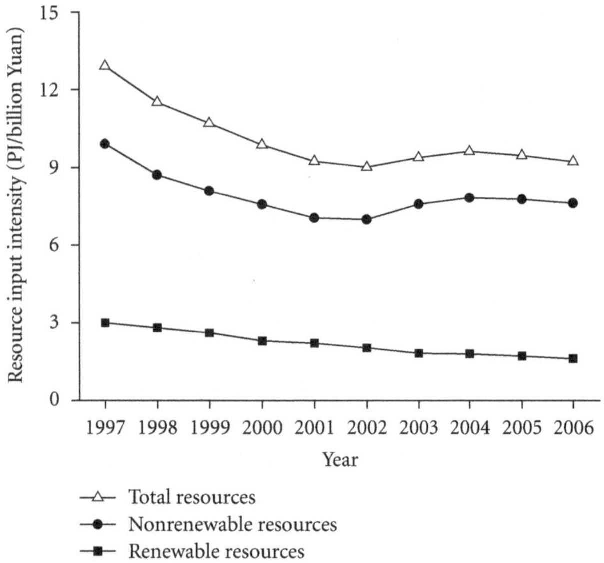

Figure 4: Resource input intensity by Chinese industry.

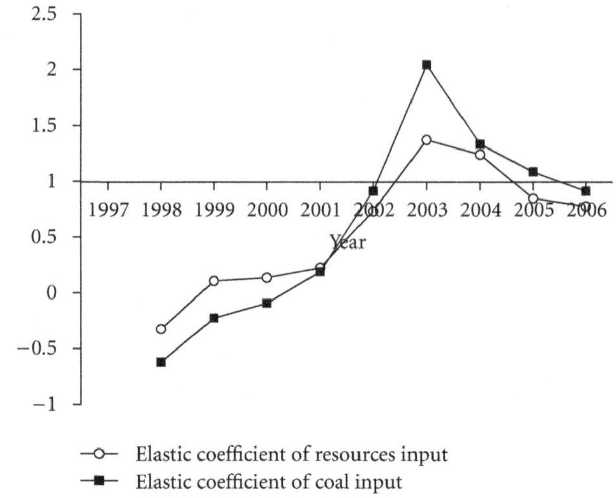

Figure 5: Elastic coefficients of resources input and coal input by Chinese industry.

along with a large amount of resources input has experienced spectacular uprising with 10.6% average annual growth rate in the total industrial value added (at 2006 constant price, similarly hereafter) over 1997–2006. Figure 4 presents the resource input intensity of the Chinese industry in this period. The total RII decreased from 12.9 PJ/billion Yuan in 1997 to 9.0 PJ/billion Yuan (1 US$ = 7.7087 RMB Yuan in 2006) in 2002. However, it started to increase by 6.8% over 2003-2004, and declined by 1.6% in 2005 and 4.1% in 2006. As noted previously, the nonrenewable resource input had the dominated share (75.6%–82.5%) in the total resources input. Then the trends of the nonrenewable resource input intensity and the RII show little difference, while the renewable resource input intensity decreased gradually during 1997–2006. Since fossil fuels are the largest resources input, energy intensity measured by the fossil fuels input (including coal, oil, and natural gas) per unit IVA is also calculated. During this period, the energy intensity decreased by 30.2% in 1997–2002, however it rose by 8.1% in 2003 or 11.7% in 2004 and then slightly declined by 2.3% over 2005-2006.

Displayed in Figure 5 is the resource input elastic coefficient (RIEC) measured by the ratio of the growth rate of resources input to the growth rate of industrial value added [64]. In the detail years, the growth of IVA was faster than the growth of total resources input during 1997–2002 with the average value of the RIEC 0.17; however, the growth of total resources input exceeded the growth of IVA in 2003 and 2004, and the RIEC reached 1.37 in 2003 and 1.24 in 2004; but the RIEC declined to 0.85 in 2005 and 0.77 in 2006. The drastic change of RIEC values is largely due to the change of fossil fuels input. Prominently, the elastic coefficient of resources input changed simultaneously with that of coal input during 1997–2006, as shown in Figure 5.

3.2. Environmental Emissions. Environmental emissions can be categorized into GHG emissions and "three wastes" emissions (i.e., waste water, waste gas, and solid waste) in conventional sense. Industrial environmental emissions in terms of GHG emissions and "three wastes" emissions in the past decade (1997–2006) are shown in Figure 6. The total exergy of environmental emissions by Chinese industry amounted to 2107.4 PJ in 1997; however, this figure rose by 66.0% and jumped to 3499.3 PJ in 2006. From the exergetic perspectives, higher exergetic value of the emission reflects the larger deviation in chemical composition from the reference environment and indicates its essential effect on environmental change. In exergy, the GHG emission dwarfs the "three wastes" emission by an order of magnitude and determined the trend of industrial environmental emissions in the whole period to a remarkable extent. In 2006, the total exergy of all the seven primary emissions in 2006 was estimated to be 3499.3 PJ, of which 93.4% was from GHG emissions and only 6.6% from "three wastes" emissions. A rapid growth of the GHG emissions took place for the period between 2002 and 2006, increasing from 1915.7 PJ in 2002 to 3267.2 PJ in 2006 with an average annual growth rate of 14.3%. Meanwhile, the total exergy of "three wastes" emissions did not change remarkably over 1997–2006.

Displayed in Figure 7 is a further comparison of the emission shares in 1997 and 2006. As the largest emission category, the share of the CO_2 emissions in the total emissions increased from 54.6% (1150.0 PJ) in 1997 to 65.8% (2302.5 PJ) in 2006, followed by the CH_4 emissions, contributing to 26.2% and 27.6% of the total in 1997 and 2006, respectively. As to the GHG emissions concretely, CO_2 emissions accounted for 67.6%–74.9% of the total GHG emissions and CH_4 emissions 30% on average in the past decade. It is worth noting that SO_2 and COD were the two main pollutants in "three wastes" emissions. The exergy of COD emissions of the Chinese industry decreased from 145.9 PJ in 1997 to 69.3 PJ in 2004, afterward it increased by 8.8% in 2005 and declined by 2.4% in 2006. Meanwhile, SO_2 emissions decreased by 15.7% in 1997–2002 and then increased rapidly from 76.5 PJ in 2002 to 109.5 PJ in 2006.

TABLE 3: The three components of GDP by expenditure approach.

Year	Final consumption expenditure Share (%)	Final consumption expenditure Contribution to the GDP growth (percentage points)	Gross capital formation Share (%)	Gross capital formation Contribution to the GDP growth (percentage points)	Net exports of goods and services Share (%)	Net exports of goods and services Contribution to the GDP growth (percentage points)
1997	37.0	3.4	18.6	1.7	44.4	4.2
1998	57.1	4.4	26.4	2.1	16.5	1.3
1999	74.7	5.7	23.7	1.8	1.6	0.1
2000	65.1	5.5	22.4	1.9	12.5	1.0
2001	50.0	4.1	50.1	4.2	−0.1	0
2002	43.6	4.0	48.8	4.4	7.6	0.7
2003	35.3	3.5	63.7	6.4	1.0	0.1
2004	38.7	3.9	55.3	5.6	6.0	0.6
2005	38.2	4.0	37.7	3.9	24.1	2.5
2006	39.2	4.3	41.3	4.6	19.5	2.2

Data source: [6].

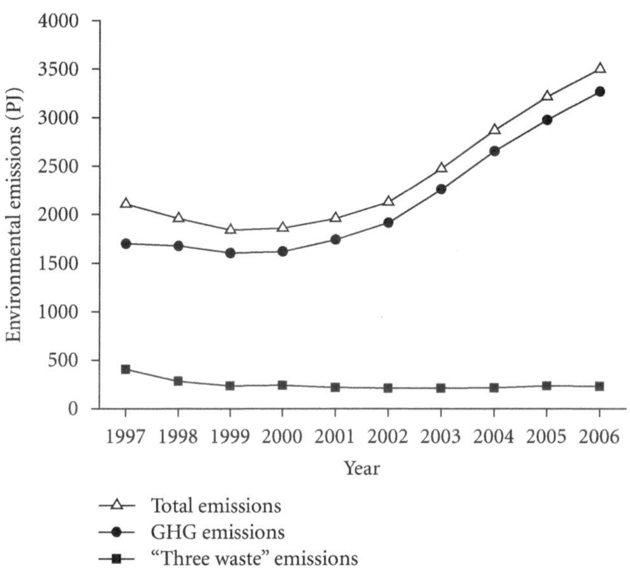

FIGURE 6: Environmental emissions by Chinese industry.

The emissions of soot, dust, and solid waste experienced a significant drop during 1997–2006. Detailed results of environmental emissions of the Chinese industry during 1997–2006 are shown in Table 5.

Environmental emission intensity (EEI) defined as the environmental emission exergy per unit of the total industrial value added indicates the environmental effect along with industrial economic output. The lower the EEI, the better environmental performance of industrial activities can be conducted. Figure 8 displays that the total EEI decreased from 0.57 PJ/billion Yuan in 1997 to 0.37 PJ/billion Yuan in 2002, and then fluctuated slightly during 2002–2006. The GHG emission intensity determined the trend of environmental emission intensity over this period to some extent, increasing its share from 80.7% (0.46 PJ/billion Yuan) in 1997 to 93.4% (0.36 PJ/billion Yuan) in 2006. It is worthy of noting that the time-series trend of the environmental emission intensity is in line with that of the resource input intensity, largely owing to the coal-dominated energy structure in China.

4. Discussion

It is worth noting that a majority of the industrial subsectors with high resources input level are the energy-intensive sectors. According to the China Energy Statistical Yearbook [65], the primary end-use energy consumption sectors in industrial system in 2006 were the manufacturing sectors, which accounted for 85.4% of the total industrial energy consumption. Among the manufacturing sectors, the sector of *Smelting and Pressing of Ferrous Metals* made up 25.8% of the total end-use energy consumption, followed by *Manufacture of Raw Chemical Materials and Chemical Products* with 14.9%, and *Manufacture of Nonmetallic Mineral Products* with 12.1% [65]. Correlation analysis shows that the correlation coefficients between mineral resource inflows into the iron and steel industry and energy resource inflows (i.e., coal, petroleum, natural gas) over 1997–2006 were higher than 0.9. Similar results can be found in the nonferrous industry.

China is adopting energy-intensive technology and investing the excessive expansion of high-energy consuming sectors, such as iron and steel, cement, and electrolytic aluminum. The outputs of main industrial products, especially most energy-intensive products, increased rapidly during 1997–2006. For instance, the outputs of crude steel, ten major nonferrous metals, motor vehicles, ethylene, cement, plate glass, electricity, chemical fiber, and primary plastic in 2006 were 3.9, 3.3, 4.6, 2.6, 2.4, 2.8, 2.5, 4.4, and 3.8 times of those in 1997, respectively [6, 55]. Some studies in energy intensity (measured by energy consumption with mass units per unit of GDP) reported that the primary driving force for the decline in China's energy intensity during 1997–2002 was efficiency effect rather than sectoral structural shifting [66–68]. It implies, therefore, that technical progress made a notable contribution in the industry during 1997–2002. However, since 2003, the industry sector has raised

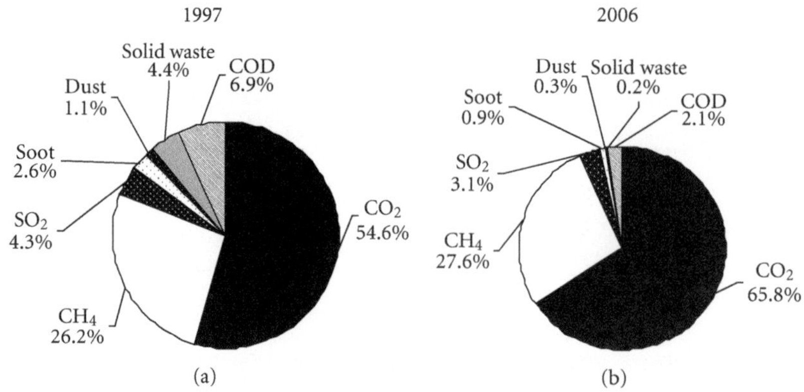

FIGURE 7: Components of environmental emissions by Chinese industry in 1997 and 2006.

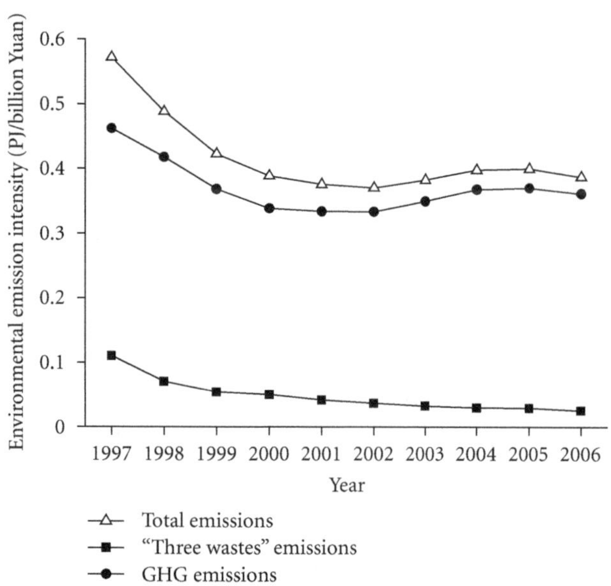

FIGURE 8: Environmental emission intensity by Chinese industry.

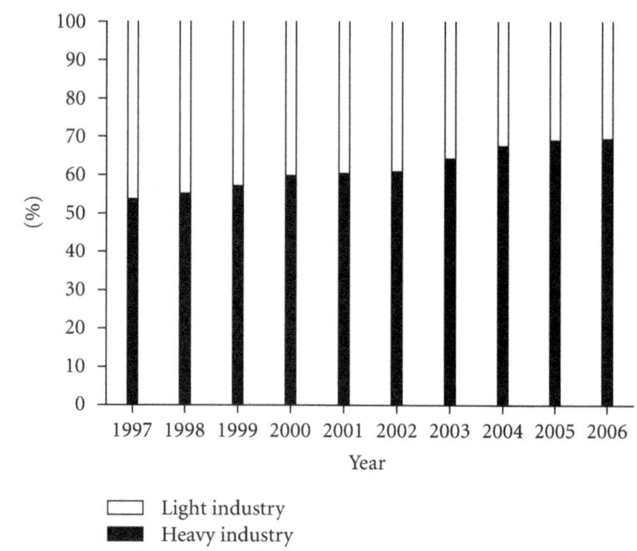

FIGURE 9: Component of the industrial value added by year.

its production levels and expanded energy-intensive subsectors rapidly. Liao et al. [69] also found that the excessive expansion of high-energy consuming sectors and the high investment ratio were foremost sources of the increasing energy intensity during 2003–2005. Figure 9 further shows that the heavy industry contributed the increasing share to the total industrial value added over 1997–2006 [6]. In 2006, the ratio of the industrial value added of heavy industry to that of light industry reached 70 : 30.

In fact, the resource utilization level in China still has large gaps in production process, technology, and management, compared with the international advanced level. The average resource extraction efficiency in China is lower than 20%–30% of the global advanced average [70]. As to the production process, the average energy consumption level of equipment and technology in China's manufacturing sectors is more than 10% of that in the OECD countries in general [71]. For instance, the overall energy consumption for per ton of steel, cement, oil refining, ethylene, and calcium carbide output in 2004 were higher than 15.6%, 23.3%, 53.4%, 59.6%, and 19.4% of those in the OCED countries, respectively [72]. The GDP energy intensity in China's industry is also distinctly higher than international levels. According to Yuan et al. [73], the average energy intensity for main products in eight industry sectors of electric power, oil, nonferrous metals, construction material, textile, and others is 40% higher than the world average. Therefore, the potential for promoting resource utilization level is substantial and urgent, especially in some resource-intensive or energy-intensive sectors. At the same time, the industry faces the tremendous challenges of limit resources supply in domestic reserves. It is well known that a large amount of industrial raw materials consumed in China comes from imported goods from the rest of the world. For instance, 50% of the domestic iron ore demand, 33% of alumina, 40% of crude oil, and 44% of wood in 2004 were met through international trade [63]. The pressure for seeking sustained resource supply and improving resource use efficiency is unprecedented.

Table 4: Resources input by Chinese industry, 1997–2006 (Unit: PJ).

Resource category	1997	1998	1999	2000	2001	2002	2003	2004	2005	2006
Nonrenewable resources	36527.3	34953.4	35259.0	36264.4	36956.1	40183.3	49028.7	56466.1	62427.3	68878.6
Raw coal	26962.3	25473.4	24987.0	24759.4	25175.5	27521.6	34606.9	39917.9	44861.6	49979.4
Crude oil	7622.1	7633.0	8321.2	9330.3	9381.8	9908.8	10977.4	12686.8	13278.0	14218.5
Natural gas	697.6	708.3	744.2	834.3	899.5	939.6	1106.0	1212.6	1461.2	1709.7
Iron ores and scraps	359.5	344.1	337.1	359.7	411.3	466.7	572.4	703.4	855.1	1001.9
Nuclear energy	182.2	178.1	188.8	211.5	220.7	317.3	547.5	637.5	670.6	692.7
Nonferrous metal ores and scraps	35.4	32.3	34.9	57.0	47.1	66.8	79.2	106.3	128.3	136.2
Nonmetal minerals	247.8	128.7	121.9	101.6	95.8	100.9	97.3	136.9	127.4	136.0
Other nonrenewable resources	420.5	455.6	523.8	610.8	724.4	861.6	1042.0	1064.8	1045.1	1004.1
Renewable resources	11041.4	11243.2	11353.2	10984.2	11231.9	11594.5	11695.2	12928.6	13659.0	14559.3
Water potential energy	830.0	880.9	863.2	941.9	1174.9	1219.8	1201.6	1497.2	1681.5	1845.7
Pulp and waste paper	535.2	536.9	589.7	596.6	619.1	718.6	808.7	907.4	1022.8	1168.0
Forest products	831.5	835.1	798.9	795.0	818.0	924.8	1037.5	1124.3	1207.2	1382.2
Agricultural products	8844.6	8990.2	9101.3	8650.7	8620.0	8731.4	8647.5	9399.8	9747.5	10163.4
Total	47568.7	46196.6	46612.2	47248.6	48187.9	51777.7	60723.9	69394.7	76086.3	83437.9
Imported resources	2249.9	2374.5	2521.9	4216.6	4027.8	4882.7	6192.2	7967.1	8525.7	9720.9

Furthermore, the rapid growth of materials production and the energy demand for electricity and coal in some major industrial sub-sectors (e.g., steel, electrolytic aluminum, cements, and paper industry) with high-energy consumption and heavy environmental emissions determine the emission profile of the Chinese industry [50, 74–76]. The energy or raw materials utility subsectors are the major sources of industrial environmental emissions. For the period between 1997 and 2006, the sectors of electric power production and coal mining were the leading emitters of CO_2 and CH_4 among all the industrial subsectors, respectively [62]. The electric power production, iron and steel production, manufacture of nonmetallic mineral products, nonferrous smelting accounted for about 90% of industrial SO_2, soot and solid waste emissions in 2004 [77]. It is well known that the quantities of industrial GHG emissions and air pollutants in China are closely related with energy consumption, especially coal consumption [50, 74, 75]. Inefficient and coal-dominated energy production and consumption are at the core of China's environmental emissions. Along with the rapid growth of industrial value added and resources use, the total exergy of industrial "three wastes" emissions has seen a steady decline, though a slight increase of SO_2 and COD emissions in some years. This effect can be attributed to the effective emission control policies made by the central and local governments. However, GHG emissions of the Chinese industry increased rapidly along with a new rising period of Chinese economy since 2002. It is important to note that China's emissions control programs focus specifically on "three wastes" emissions rather than targeting at greenhouse gases such as carbon dioxide [78]. To tackle the problems of industrial environmental emissions, a more international way of thinking instead of a regional approach should be taken, with specially emphasis on the greenhouse gases rather than the regional pollutants merely [50].

Prominently, the resources use and environmental impact of the Chinese industry have been notably influenced by the macroeconomic situation in the last decade. Table 3 presents the three components of GDP by expenditure approach during 1997–2006. Totally, final consumption expenditure and gross capital formation shared the majority proportion of the GDP over this period. During 1998–2001, the economic growth was largely derived by the domestic demand. After 2002, the situation started to overturn: the contribution of gross capital formation in China's total GDP exceeded that of the final consumption expenditure. Investment has become an important motor for China's economic growth in recent years [76, 79]. Most of the investment flows into manufacturing, infrastructure, and real estate related sectors [80], which enormously pushes up the demand for certain resource- and energy-intensive products, such as steel, nonferrous metals, cement, glass, and machine. Since the second half year of 2003, the government had implemented a series measures to strengthen macro-control, with specially emphasis on the control of the investment in fixed assets, land supply management and environmental regulation [63]. There was significant decline in the growth rate of investment and total investment for new planned projects, especially heavy industry investment after 2005, while slight decline of resource input intensity and environmental emission intensity by Chinese industry over 2005-2006 can be found.

5. Concluding Remarks

Natural resources from the ecological system are required for producing and supplying goods and service in the industry system. Environmental emission assimilation as an additional ecological input into the industry sector can also be regarded as the use of an "ecological service." For sustainable

Table 5: Environmental emissions by Chinese industry, 1997–2006 (PJ).

Emission category	1997	1998	1999	2000	2001	2002	2003	2004	2005	2006
GHGs	1701.3	1676.0	1604.6	1620.8	1740.7	1915.7	2261.7	2653.5	2976.0	3267.2
CO_2	1150.0	1169.6	1178.5	1213.4	1259.5	1349.3	1582.3	1837.2	2077.9	2302.5
CH_4	551.4	506.4	426.1	407.3	481.2	566.4	679.4	816.4	898.2	964.7
Three wastes	406.1	283.1	236.0	240.5	218.2	211.4	212.0	215.4	236.9	232.0
SO_2	90.7	78.1	71.5	79.0	76.8	76.5	87.8	92.7	106.3	109.5
Soot	54.8	41.1	33.4	33.4	29.4	28.1	29.6	31.0	33.2	30.3
Dust	22.6	19.8	17.6	16.4	14.9	14.1	15.3	13.6	13.7	12.1
Solid waste	92.1	35.2	19.4	15.9	14.5	13.2	9.7	8.8	8.3	6.5
COD	145.9	108.9	94.1	95.8	82.6	79.4	69.6	69.3	75.5	73.6
Total	2107.4	1959.2	1840.6	1861.2	1958.8	2127.1	2473.8	2868.9	3212.9	3499.3

development, natural resources, especially nonrenewable resources should not run out and environmental emissions should not endanger the ecological system [40]. Given China's rapid industrial expansion, policy-makers require a more detailed understanding of the complex linkages between industrial activities and natural environment if the resultant resource use and environmental impact are to be minimized. In this paper, an exergy-based physical assessment is performed to measure the resources use and environmental impact of the Chinese industry for the period between 1997 and 2006.

The resources input into the Chinese industry reached 83437.9 PJ in 2006, and increased by 75.4% compared with that in 1997. For the time-series trend, resources input showed little variation during 1997–2001 and the initial trend had changed since 2002 with the input levels showing a great rebound. Nonrenewable resources accounted for 75.6%–82.5% of the total and determined the trend of resources input to a certain extent. A rapid increment of the nonrenewable resources input in the recent 5 years can be found, from 40183.3 PJ in 2002 to 68878.6 PJ in 2006 with an average annual growth rate of 14.5%. Coal input was the largest contributor, accounting for 52.4%–59.9% of the total resources input during the period, followed by crude oil and natural gas. The imported resources input increased its share from 4.7% of the total resources input (2249.9 PJ) in 1997 to 11.7% in 2006 (9720.9 PJ), mainly coming from crude oil import.

The environmental emissions by Chinese industry increased from 2107.4 PJ in 1997 to 3499.3 PJ in 2006. In exergy, the GHG emission dwarfs the "three wastes" emission by an order of magnitude and determined the trend of industrial environmental emissions in the whole period to a remarkable extent. The total exergy of all the seven primary emissions in 2006 amounted to 3499.3 PJ, of which 93.4% was from GHG emissions and only 6.6% from "three wastes" emissions. A rapid growth of total GHG emissions took place for the period between 2002 and 2006, increasing from 1915.7 PJ in 2002 to 3267.2 PJ in 2006 with an average annual growth rate of 14.3%. As the largest emission category, the CO_2 emissions increased its share from 54.6% of the total emissions in 1997 to 65.8% in 2006, followed by CH_4 emissions contributing averagely 26% to the total. The exergy of "three wastes" emissions did not change remarkably over 1997–2006, and SO_2 and COD were the two main pollutants.

Exergy intensities in terms of resource input intensity and environmental emission intensity time-series are calculated. The resource input intensity declined for the period between 1997 and 2002, but it started to increase over 2003-2004 and then declined slightly in 2005 and 2006. The environmental emission intensity in the whole period shows a similar trend. Moreover, the development of macroeconomic efficiencies of resources input and environmental emissions can be split into two main periods with different characteristics: the first period from 1997 to 2001 corresponding to a more notable improvement in resource and environmental efficiency; the second period from 2002 onwards with faster increased nonrenewable resources input into resource—or energy-intensive subsectors under slower yield of industrial value added. The excessive expansion of high-energy consuming industrial subsectors and the high investment ratio in the macroeconomic structure were foremost sources of the increasing exergy intensities. To obtain the industrial value added of one billion Yuan (129 million US$) in 2006, the resources input and environmental emissions by Chinese industry were estimated to be 9.2 and 0.4 PJ, respectively.

Industry plays an important role in Chinese economy. The contributions of industrial value added to the increase of the GDP in China reached 47.0%–58.3% and the shares in the GDP were around 40% during 1997–2006 [6], which means that China relied on manufacturing industry to an unusually great extent. Nevertheless, skyrocketing resources input and environmental emissions of the Chinese industry mean a surging and huge pressure into the ecosystem. The depletion of the resources brings on the economic increase, and the resulting wastes are returned to the environment where they induce environmental pollution and climate change. Also the development of the Chinese industry can hardly become more resource and energy intensive that it is now, along with the limited resource reserves and adverse environmental quality. Therefore, increasing GDP based on traditional industrialization mode on the expense of natural environment is unsustainable. Continued strong emphasis on sustainability requires that future industrial economic growth must rely much more on environmental friendly and be less dependent on material products and natural resources than in the past. A large effort has to be made to promote industrial structure adjustment, strength resources management and resource efficiency improvement, rationalize resource prices, and implement more stringent energy saving and emission control policies. More importantly, the policies of restructuring and transformation of the resource-intensive economic growth pattern in China will affect and improve the whole situation of resources use and environmental impact of the industry sector.

Appendix

See Tables 4 and 5.

Acknowledgments

This study has been supported by the Humanities and Social Sciences Fund, Ministry of Education of China (Grant no. 12YJC790255), the Foundation of State Key Laboratory of Coal Resources and Safe Mining, China University of Mining & Technology (Grant no. SKLCRSM10KFA06), and the Fundamental Research Funds for the Central Universities (Grant no. 2011QG03).

References

[1] J. Dewulf, H. van Langenhove, J. Mulder, M. M. D. van den Berg, H. J. van der Kooi, and J. de Swaan Arons, "Illustrations towards quantifying the sustainability of technology," *Green Chemistry*, vol. 2, no. 3, pp. 108–114, 2000.

[2] H. X. Nguyen and R. Yamamoto, "Modification of ecological footprint evaluation method to include non-renewable resource consumption using thermodynamic approach," *Resources, Conservation and Recycling*, vol. 51, no. 4, pp. 870–884, 2007.

[3] M. A. Rosen and I. Dincer, "Exergy as the confluence of energy, environment and sustainable development," *Exergy*, vol. 1, no. 1, pp. 3–13, 2001.

[4] M. A. Cole, R. J. R. Elliott, and S. S. Wu, "Industrial activity and the environment in China: an industry-level analysis," *China Economic Review*, vol. 19, no. 3, pp. 393–408, 2008.

[5] B. Zhang, G. Q. Chen, Q. Yang, Z. M. Chen, B. Chen, and Z. Li, "How to guide a sustainable industrial economy: emergy account for resources input of Chinese industry," *Procedia Environmental Sciences*, vol. 5, pp. 51–59, 2011.

[6] National Bureau of Statistics of China, *Statistical Yearbook of China*, China Statistical Publishing House, Beijing, China, 1998–2007.

[7] M. Zhang, H. Mu, and Y. Ning, "Accounting for energy-related CO_2 emission in China, 1991–2006," *Energy Policy*, vol. 37, no. 3, pp. 767–773, 2009.

[8] L. Lewan, "Why human societies need sustainability analyses based on biophysical assessments," *Ecological Economics*, vol. 29, no. 1, pp. 57–60, 1999.

[9] R. U. Ayres, "Eco-thermodynamics: economics and the second law," *Ecological Economics*, vol. 26, no. 2, pp. 189–209, 1998.

[10] G. Wall and M. Gong, "On exergy and sustainable development—part 1: conditions and concepts," *Exergy*, vol. 1, no. 3, pp. 128–145, 2001.

[11] G. Q. Chen, "Scarcity of exergy and ecological evaluation based on embodied exergy," *Communications in Nonlinear Science and Numerical Simulation*, vol. 11, no. 4, pp. 531–552, 2006.

[12] E. Sciubba and G. Wall, "A brief commented history of exergy from the beginnings to 2004," *International Journal of Thermodynamics*, vol. 10, no. 1, pp. 1–26, 2007.

[13] R. U. Ayres, "The minimum complexity of endogenous growth models: the role of physical resource flows," *Energy*, vol. 26, no. 9, pp. 817–838, 2001.

[14] R. U. Ayres and B. Warr, "Accounting for growth: the role of physical work," *Structural Change and Economic Dynamics*, vol. 16, no. 2, pp. 181–209, 2005.

[15] G. Wall, "Exergy—a useful concept within resource accounting," Report no. 77-42, Institute of Theoretical Physics, Goteborg, Sweden, 1977.

[16] G. Q. Chen and B. Chen, "Resource analysis of the Chinese society 1980–2002 based on exergy—part 1: fossil fuels and energy minerals," *Energy Policy*, vol. 35, no. 4, pp. 2038–2050, 2007.

[17] A. Valero, "Exergy accounting: capabilities and drawbacks," *Energy*, vol. 31, no. 1, pp. 164–180, 2006.

[18] A. Valero and A. Valero, "Physical geonomics: combining the exergy and Hubbert peak analysis for predicting mineral resources depletion," *Resources, Conservation and Recycling*, vol. 54, no. 12, pp. 1074–1083, 2010.

[19] G. Wall, *Exergy—a useful concept [thesis]*, Chalmers University of Technology, Göteborg, Sweden, 1986.

[20] I. Dincer, "The role of exergy in energy policy making," *Energy Policy*, vol. 30, no. 2, pp. 137–149, 2002.

[21] J. Dewulf and H. van Langenhove, "Integrating industrial ecology principles into a set of environmental sustainability indicators for technology assessment," *Resources, Conservation and Recycling*, vol. 43, no. 4, pp. 419–432, 2005.

[22] B. de Meester, J. Dewulf, S. Verbeke, A. Janssens, and H. van Langenhove, "Exergetic life-cycle assessment (ELCA) for resource consumption evaluation in the built environment," *Building and Environment*, vol. 44, no. 1, pp. 11–17, 2009.

[23] J. Dewulf, G. van der Vorst, N. Versele, A. Janssens, and H. van Langenhove, "Quantification of the impact of the end-of-life scenario on the overall resource consumption for a dwelling house," *Resources, Conservation and Recycling*, vol. 53, no. 4, pp. 231–236, 2009.

[24] G. Wall, "Exergy conversion in the Japanese society," *Energy*, vol. 15, no. 5, pp. 435–444, 1990.

[25] G. Wall, E. Sciubba, and V. Naso, "Exergy use in the Italian society," *Energy*, vol. 19, no. 12, pp. 1267–1274, 1994.

[26] I. S. Ertesvåg, "Society exergy analysis: a comparison of different societies," *Energy*, vol. 26, no. 3, pp. 253–270, 2001.

[27] R. U. Ayres, L. W. Ayres, and B. Warr, "Exergy, power and work in the US economy, 1900–1998," *Energy*, vol. 28, no. 3, pp. 219–273, 2003.

[28] I. S. Ertesvåg, "Energy, exergy, and extended-exergy analysis of the Norwegian society 2000," *Energy*, vol. 30, no. 5, pp. 649–675, 2005.

[29] B. Chen, G. Q. Chen, and Z. F. Yang, "Exergy-based resource accounting for China," *Ecological Modelling*, vol. 196, no. 3-4, pp. 313–328, 2006.

[30] B. Chen and G. Q. Chen, "Exergy analysis for resource conversion of the Chinese Society 1993 under the material product system," *Energy*, vol. 31, no. 8-9, pp. 1115–1150, 2006.

[31] B. Chen and G. Q. Chen, "Resource analysis of the Chinese society 1980–2002 based on exergy—part 2: renewable energy sources and forest," *Energy Policy*, vol. 35, no. 4, pp. 2051–2064, 2007.

[32] B. Chen and G. Q. Chen, "Resource analysis of the Chinese society 1980–2002 based on exergy—part 3: agricultural products," *Energy Policy*, vol. 35, no. 4, pp. 2065–2078, 2007.

[33] B. Chen and G. Q. Chen, "Resource analysis of the Chinese society 1980–2002 based on exergy—part 4: fishery and rangeland," *Energy Policy*, vol. 35, no. 4, pp. 2079–2086, 2007.

[34] G. Q. Chen and B. Chen, "Resource analysis of the Chinese society 1980–2002 based on energy—part 5: resource structure and intensity," *Energy Policy*, vol. 35, no. 4, pp. 2087–2095, 2007.

[35] B. Warr, H. Schandl, and R. U. Ayres, "Long term trends in resource exergy consumption and useful work supplies in the UK, 1900 to 2000," *Ecological Economics*, vol. 68, no. 1-2, pp. 126–140, 2008.

[36] B. S. Warr, E. Williams, and R. U. Ayres, "Efficiency dilution: long-term exergy conversion trends in Japan," *Environmental Science and Technology*, vol. 42, no. 13, pp. 4964–4970, 2008.

[37] A. Gasparatos, M. El-Haram, and M. Horner, "Assessing the sustainability of the UK society using thermodynamic concepts: part 2," *Renewable and Sustainable Energy Reviews*, vol. 13, no. 5, pp. 956–970, 2009.

[38] G. Q. Chen and Z. H. Qi, "Systems account of societal exergy utilization: China 2003," *Ecological Modelling*, vol. 208, no. 2–4, pp. 102–118, 2007.

[39] G. Q. Chen and B. Chen, "Extended-exergy analysis of the Chinese society," *Energy*, vol. 34, no. 9, pp. 1127–1144, 2009.

[40] B. Zhang and G. Q. Chen, "Physical sustainability assessment for the China society: exergy-based systems account for resources use and environmental emissions," *Renewable and Sustainable Energy Reviews*, vol. 14, no. 6, pp. 1527–1545, 2010.

[41] X. Ji, G. Q. Chen, B. Chen, and M. M. Jiang, "Exergy-based assessment for waste gas emissions from Chinese transportation," *Energy Policy*, vol. 37, no. 6, pp. 2231–2240, 2009.

[42] A. R. Ometto and W. N. L. Roma, "Atmospheric impacts of the life cycle emissions of fuel ethanol in Brazil: based on chemical exergy," *Journal of Cleaner Production*, vol. 18, no. 1, pp. 71–76, 2010.

[43] M. A. Rosen and I. Dincer, "On exergy and environmental impact," *International Journal of Energy Research*, vol. 21, no. 7, pp. 643–654, 1997.

[44] M. A. Rosen and I. Dincer, "Exergy analysis of waste emissions," *International Journal of Energy Research*, vol. 23, no. 13, pp. 1153–1163, 1999.

[45] M. A. Rosen, I. Dincer, and M. Kanoglu, "Role of exergy in increasing efficiency and sustainability and reducing environmental impact," *Energy Policy*, vol. 36, no. 1, pp. 128–137, 2008.

[46] R. U. Ayres, L. W. Ayres, and K. Martínas, "Exergy, waste accounting, and life-cycle analysis," *Energy*, vol. 23, no. 5, pp. 355–363, 1998.

[47] L. Q. Huang, G. Q. Chen, Y. Zhang, B. Chen, and S. J. Luan, "Exergy as a unified measure of water quality," *Communications in Nonlinear Science and Numerical Simulation*, vol. 12, no. 5, pp. 663–672, 2007.

[48] A. Gasparatos, M. El-Haram, and M. Horner, "A longitudinal analysis of the UK transport sector, 1970–2010," *Energy Policy*, vol. 37, no. 2, pp. 623–632, 2009.

[49] X. Ji and G. Q. Chen, "Unified account of gas pollutants and greenhouse gas emissions: Chinese transportation 1978–2004," *Communications in Nonlinear Science and Numerical Simulation*, vol. 15, no. 9, pp. 2710–2722, 2010.

[50] B. Zhang, G. Q. Chen, X. H. Xia, S. C. Li, Z. M. Chen, and X. Ji, "Environmental emissions by Chinese industry: exergy-based unifying assessment," *Energy Policy*, vol. 45, pp. 490–501, 2012.

[51] Z. F. Cai, *Water resources and greenhouse gas emissions covered ecological exergy account [Ph.D. thesis]*, Peking University, Beijing, China, 2009.

[52] ACPI, *Almanac of China Paper Industry*, Light Industry Publishing House, Beijing, China, 2006.

[53] CEY, *China Environment Yearbook*, China Environment Yearbook Press, Beijing, China, 1998–2000, 2007.

[54] CFIY, *China Food Industry Yearbook*, China Food Industry Yearbook Press, Beijing, China, 1998–2007.

[55] CIESY, *China Industrial Economic Statistical Yearbook*, China Statistical Publishing House, Beijing, China, 2007.

[56] CSY, *China Steel Yearbook*, Metallurgical Industry Press, Beijing, China, 1998–2007.

[57] CYNMI, *China Yearbook of Nonferrous Metal Industry*, Metallurgical Industry Press, Beijing, China, 1998–2007.

[58] F. C. Zhou, "Chemical characteristic and heat value of bamboo," *Bamboo Research*, vol. 1, pp. 58–71, 1991 (Chinese).

[59] O. Ignatenko, A. van Schaik, and M. A. Reuter, "Exergy as a tool for evaluation of the resource efficiency of recycling systems," *Minerals Engineering*, vol. 20, no. 9, pp. 862–874, 2007.

[60] J. Szargut, *Exergy Method, Technical and Ecological Applications*, WIT Press, Southampton, UK, 2005.

[61] D. R. Morris and J. Szargut, "Standard chemical exergy of some elements and compounds on the planet earth," *Energy*, vol. 11, no. 8, pp. 733–755, 1986.

[62] B. Zhang, *Methane emissions in China: time series inventories and systems embodiment analysis [Ph.D. thesis]*, Peking University, Beijing, China, 2011.

[63] J. G. Chen, *China economic research report (2005-2006) [Ph.D. thesis]*, Economy & Management Publishing House, Beijing, China, 2006.

[64] J. W. Li, H. S. Li, H. Zhang, and C. R. Yu, "Pilot study on appraisal index for energy usage situation in China," *Energy Environmental Protection*, vol. 20, pp. 10–13, 2006 (Chinese).

[65] CESY, *China Energy Statistical Yearbook*, China Statistical Publishing House, Beijing, China, 2007.

[66] Z. X. Zhang, "Why did the energy intensity fall in China's industrial sector in the 1990s? The relative importance of structural change and intensity change," *Energy Economics*, vol. 25, no. 6, pp. 625–638, 2003.

[67] C. B. Ma and D. I. Stern, "China's changing energy intensity trend: a decomposition analysis," *Energy Economics*, vol. 30, no. 3, pp. 1037–1053, 2008.

[68] D. L. Zha, D. Q. Zhou, and N. Ding, "The contribution degree of sub-sectors to structure effect and intensity effects on industry energy intensity in China from 1993 to 2003," *Renewable and Sustainable Energy Reviews*, vol. 13, no. 4, pp. 895–902, 2009.

[69] H. Liao, Y. Fan, and Y. M. Wei, "What induced China's energy intensity to fluctuate: 1997–2006?" *Energy Policy*, vol. 35, no. 9, pp. 4640–4649, 2007.

[70] C. Tong, G. B. Song, B. R. Chen, and W. H. Ye, "Macroeconomic efficiency of use of non-renewable resources in the industrial economy during a period of rapid economic growth in China," *Resources, Conservation and Recycling*, vol. 52, no. 5, pp. 737–746, 2008.

[71] Y. H. Lang, P. Zhou, and L. Shen, "Analysis on the saving potentials of mineral resources in China," *Resource Science*, vol. 27, pp. 23–27, 2005 (Chinese).

[72] M. Q. Wang, *China: Forwards New Models of Economic Growth*, Social Science Academic Press, Beijing, China, 2007.

[73] J. H. Yuan, J. G. Kang, C. H. Zhao, and Z. G. Hu, "Energy consumption and economic growth: evidence from China at both aggregated and disaggregated levels," *Energy Economics*, vol. 30, no. 6, pp. 3077–3094, 2008.

[74] K. M. Zhang and Z. G. Wen, "Review and challenges of policies of environmental protection and sustainable development in China," *Journal of Environmental Management*, vol. 88, no. 4, pp. 1249–1261, 2008.

[75] G. Q. Chen and B. Zhang, "Greenhouse gas emissions in China 2007: inventory and input-output analysis," *Energy Policy*, vol. 38, no. 10, pp. 6180–6193, 2010.

[76] G. P. Peters, C. L. Weber, D. Guan, and K. Hubacek, "China's growing CO_2 emissions—a race between increasing consumption and efficiency gains," *Environmental Science and Technology*, vol. 41, no. 17, pp. 5939–5944, 2007.

[77] State Environmental Protection Administration of China, *Annual Statistic Report on Environment in China 2004*, China Environmental Science Press, Beijing, China, 2005.

[78] Y. C. Chang and N. Wang, "Environmental regulations and emissions trading in China," *Energy Policy*, vol. 38, no. 7, pp. 3356–3364, 2010.

[79] B. Zhang, S. Peng, X. Xu, and L. Wang, "Embodiment analysis for greenhouse gas emissions by Chinese economy based on Global Thermodynamic Potentials," *Energies*, vol. 4, no. 11, pp. 1897–1915, 2011.

[80] B. Zhang and G. Q. Chen, "Methane emissions by Chinese economy: inventory and embodiment analysis," *Energy Policy*, vol. 38, no. 8, pp. 4304–4316, 2010.

17

Experimental Research of Dynamic Instabilities in the Presence of Coiled Wire Inserts on Two-Phase Flow

Gokhan Omeroglu,[1] Omer Comakli,[2] Sendogan Karagoz,[2] and Bayram Sahin[3]

[1] *Department of Mechanical Engineering, Faculty of Engineering, Bayburt University, 69000 Bayburt, Turkey*
[2] *Department of Mechanical Engineering, Faculty of Engineering, Atatürk University, 25240 Erzurum, Turkey*
[3] *Department of Mechanical Engineering, Faculty of Engineering and Architecture, Technical University of Erzurum, Erzurum, Turkey*

Correspondence should be addressed to Gokhan Omeroglu; gomeroglu@bayburt.edu.tr

Academic Editors: T.-C. Jen and R.-H. Yeh

The aim of this study is to experimentally investigate the effect of the coiled wire insertions on dynamic instabilities and to compare the results with the smooth tube for forced convection boiling. The experiments were conducted in a circular tube, and water was used as the working fluid. Two different pitch ratios (H/D = 2.77 and 5.55) of coiled wire with circular cross-sections were utilised. The constant heat flux boundary condition was applied to the outer side of the test tube, and the constant exit restriction was used at the tube outlet. The mass flow rate changed from 110 to 20 g/s in order to obtain a detailed idea about the density wave and pressure drop oscillations, and the range of the inlet temperature was 15–35°C. The changes in pressure drop, inlet temperature, amplitude, and the period with mass flow rate are presented. For each configuration, it is seen that density wave and pressure drop oscillations occur at all inlet temperatures. Analyses show that the decrease in the mass flow rate and inlet temperature causes the amplitude and the period of the density wave and the pressure drop oscillations to decrease separately.

1. Introduction

In many industrial systems in which boiling heat transfer exists, the flow instabilities occurring based on the boiling heat transfer cause certain parts where heat is transferred to breakdown. The fact that the system pressure, flow rate, and similar parameters create oscillations shortens the life of the process systems. It is possible to order the events decreasing the existence of the thermal systems operated by two-phase flows as thermal fatigues, mechanical vibrations, the difficulty of control caused by high transient temperatures, and the burn-out occurring on the test pipe surface [1, 2].

The main property categorising the type of two-phase flow is the shapes that interfaces occurring between two-phases take. The effect of the flow direction on these occurring shapes is highly distinctive. Two-phase flows are classified as horizontal, vertical, and inclined. The flow regimes and the characteristics occurring in horizontal, vertical, and inclined tubes are different from each other. The direction that the gravitational force affects according to the flow direction brings together the main classification [1, 3, 4]. Menteş et al. [5] and Kakaç et al. [6, 7] studied two-phase flows in different flow directions.

Considerable efforts have been made to investigate two-phase flow instabilities by researchers for many years because the instabilities and thus the oscillations shorten the life of the systems. In different types of test tubes and tubes with different cross-sections, Coleman and Garimella [8] and Leung et al. [9] investigated the instabilities of two-phase flow systems. The forced convection boiling in a horizontal tube was studied by Çomaklı et al. [10]. All types of dynamic instabilities were observed at all conducted temperatures, and the appearance boundaries of the oscillations were determined. It was observed that the instability of the system increased with the increase of the inlet temperature. Besides this trend, the periods, and the amplitudes of the pressure drop and density wave type oscillations decreased with the decreasing mass flow rate and increased with the decreasing inlet temperature. Furthermore, it was noticed that the channel length had an important effect on two-phase dynamic flow

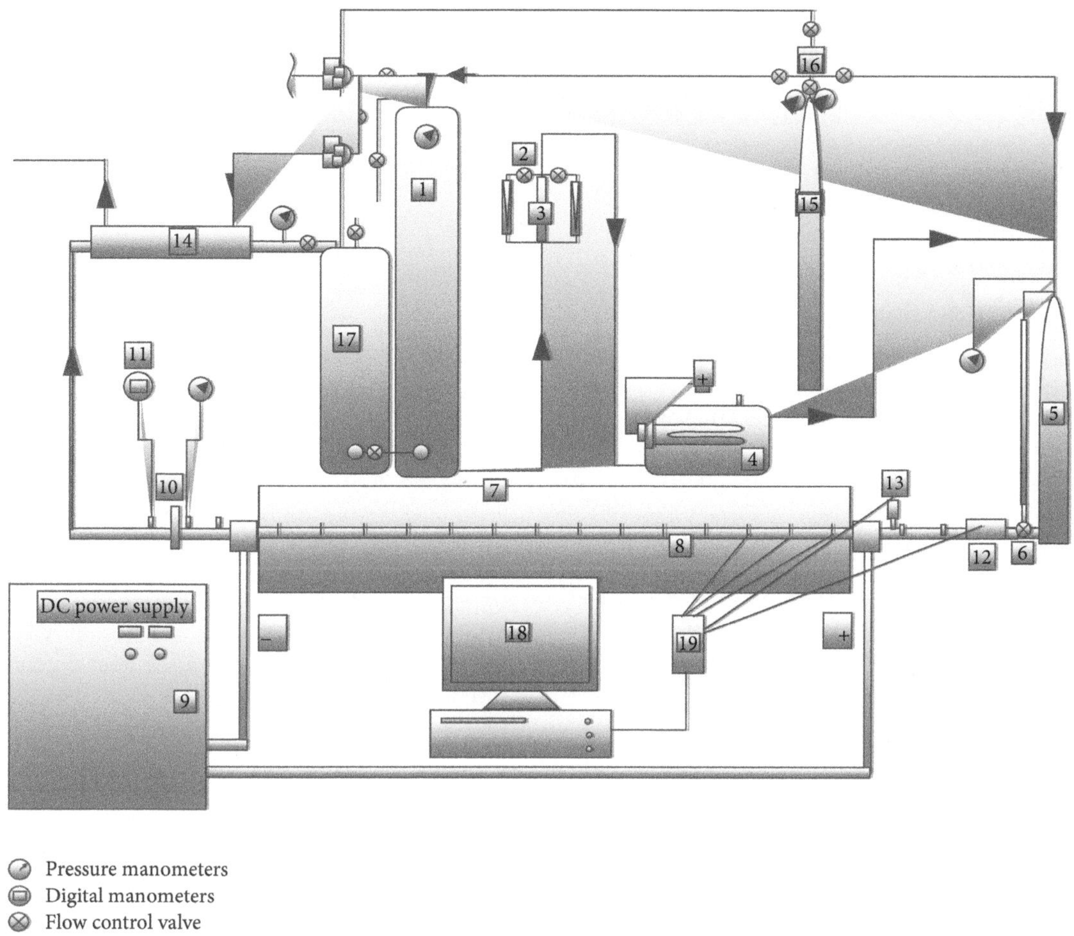

Fluid supply section	Test section	Fluid recovery section
1-Main supply tank	5-Surge tank	14-Condenser
2-Flow control valve	6-Inlet fluid control valve	15-Nitrogen tank
3-Rotameter	7-Test plenum	16-Regulator
4-Subcooling heat exchanger	8-Test tube	17-Storage tank
	9-DC power supply	18-Computer
	10-Exit restriction	19-Data reading card
	11-Digital manometer	
	12-Mass transducer	
	13-Pressure transducer	

FIGURE 1: Schematic of the experimental set-up.

TABLE 1: The tube characteristics.

Tube name	Tube characteristics	D_e (mm)	Pitch ratio
Tube-1	Smooth tube	8.95	—
Tube-2	Ring diameter: 1.8 mm, the pitch: 5 mm	7.96	3.33
Tube-3	Ring diameter: 1.8 mm, the pitch: 10 mm	8.44	2.77

instabilities. In narrow channels, Tadrist [11] conducted an experimental study to examine two-phase flow instabilities. Parallel channel and single channel oscillations were clearly determined. Bao et al. [12] investigated experimentally the gas-liquid two-phase flow in a narrow channel. Heat transfer characteristics and pressure drop were studied for nonboiling single phase flows.

In single phase and two-phase flow systems, water is the most commonly used working fluid. Except for water, there are some other fluids and fluid mixtures. The R-11 working

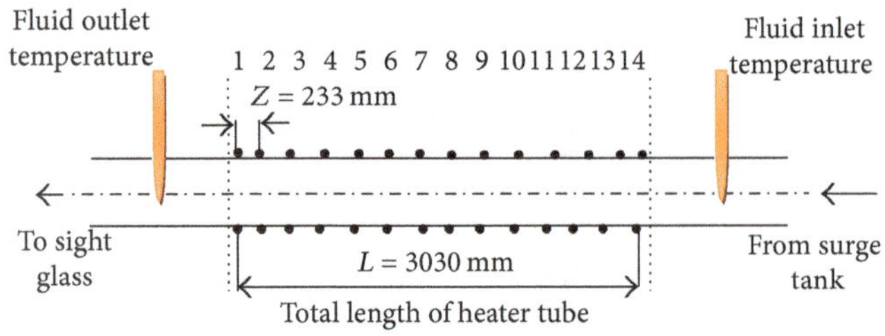

Figure 2: Thermocouples location.

fluid in a two-phase horizontal boiling system was used by Çomaklı et al. [13] and Kakaç and Cao [14]. The amplitudes and the periods of the pressure drop type and the density wave type oscillations of R-11 were higher than those of water [13]. Thermal oscillations and pressure drop type oscillations occurred for all thermal power levels. The periods and the amplitudes of the oscillations increased with the increase of thermal power and inlet subcooling [14]. Oil-water mixtures were used by Poesio [15] and Sotgia et al. [16] in two-phase flow systems to analyse the combined effect of both fluids on dynamic instabilities.

In order to investigate the two-phase flow dynamic instabilities, different approaches have been employed besides experimental methods. Homogenous and drift flow models and the Wojtan-Ursenbacher-Thome flow model were used by Moreno Quibén and Thome [17] and Kakac and Bon [18] in a horizontal test tube, and the model results were confirmed with experimental models. Yu et al. [19] investigated two-phase pressure drop, forced convection boiling heat transfer and the critical heat flux of water in a horizontal test tube. The results were successfully related to the correlations improved for different types of working fluids and boiling water in both narrow and wide channels.

In order to enhance the heat transfer in two-phase flows, there are numerous techniques as in the single phase flows. At the head of these techniques, inserting inner elements with different geometry and configuration as turbulence promoter and surface area enhancer is the most preferred among the passive heat transfer enhancement techniques. Yılmaz et al. [20, 21] examined the heat transfer enhancement methods used in two-phase flow in detail and presented basic findings related to heat transfer enhancement in two-phase flows. The effects of enhanced surfaces on two-phase flow instabilities in a horizontal boiling system were investigated by Widmann et al. [22]. Dynamic instability types occurred in all different tube types. Çomakli et al. [4] and Karagoz et al. [23] studied the dynamic flow instabilities in a forced convection boiling system using a smooth tube and a tube with different insertions. By the addition of the insertions into the flow field, it was seen that the periods and the amplitudes of the pressure drop type and density wave type oscillations of the test tube with inner elements were higher than those of the smooth tube, and the smooth tube had a very stable structure in terms of flow instability. It was also noticed that the stability of the system increased with the decrease of the equivalent diameter for the same type enhanced surface element. In this experimental work, the coiled wire insert with two pitch ratios was used as turbulator in addition to the smooth tube in order to investigate the effects on two-phase flow instabilities. Besides the inner element and thus the pitch ratio, the effect of the inlet subcooling and the mass flow rate were also examined.

2. Experimental Set-Up

The experimental set-up is schematically shown in Figure 1, which has been designed to create three main dynamic instabilities (pressure drop type, density wave type and thermal oscillations). In this study, the effects of inlet subcooling, the insertion type, pitch ratio, and the mass flow rate on two-phase flow instabilities have been investigated. As seen in Figure 1, the experimental system consists of three main parts: fluid supply section, test section, and fluid recovery section. The working fluid (water) in liquid form supplied from the fluid supply section enters the circular tube and turns into a mixture of liquid and vapour by the effect of the heat input from the tube walls. The fluid that is nearly in the vapour phase at the end of the test section is sent to the fluid recovery section. This flow process goes on consistently as a close loop.

2.1. Fluid Supply Section. The fluid supply section is composed of a main tank (1), flow control valve (2), flow meter (3), and heater (4). The main tank made of stainless steel stores the water used during the experiments. The cylindrical tank has 3 metres of height and 0.7 m^3 volume rates. The flow control valve is used to set the rate of the flow to the desired level. The flow rate is controlled by two flow meters. The ranges of the flow meters is 0–400 and 0–1000 l/hour, respectively. The first flow meter is only used in small flow rates to obtain more sensitive measurements. The inlet subcooling level is remarkably important for two-phase flow experiments. Therefore, a shell-tube heat exchanger is used to send the water at the desired temperature to the test section. The working fluid is heated by electrical heaters while passing through the tube. The temperature of the fluid heated by two heaters (each heater is 4 kW) is controlled with a digital thermometer.

2.2. Test Section. The test section where the dynamic flow instabilities are generated included a surge tank (5), inlet fluid control valve (6), test plenum (7), test tube (8), DC power supply (9), orifice (10), digital manometer (11), flow meter (12), and pressure transducer (13). In order to create the compressible volume for two-phase flow of water, a surge tank with 0.05 m^3 was used. A level viewing glass, which is durable against high pressure up to 30 bar, was added to the surge tank in order to see the variations in both water and the compressible volume. The surge tank also contained a level range to measure the level and a manometer to measure the pressure inside the tank. A turbine type flow meter to measure the oscillations in water flow rate, a Bourdon type manometer to measure the pressure of the water at the tube inlet, and a pressure transducer to measure the fluid pressure oscillations at the tube inlet were installed between the surge tank and

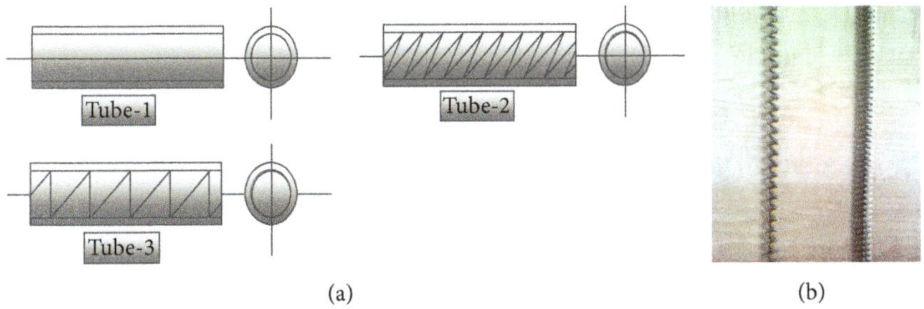

FIGURE 3: (a) Heat transfer surface augmentation elements, (b) the pictures of the elements.

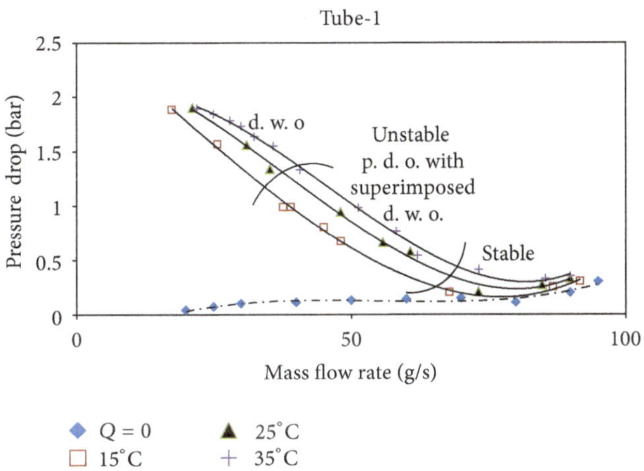

FIGURE 4: The steady state characteristic curves for Tube-1.

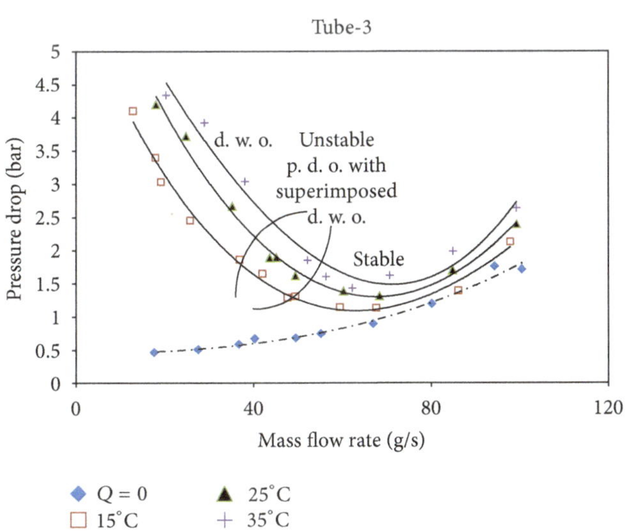

FIGURE 5: The steady state characteristic curves for Tube-3.

the test tube. The inlet temperature of the working fluid at the tune inlet was measured with a T-type thermocouple. An orifice plate was installed on the exit side of the tube in order to define the effects of exit restriction on flow oscillations. A Bourdon type manometer was used to measure the pressure difference caused by the orifice plate.

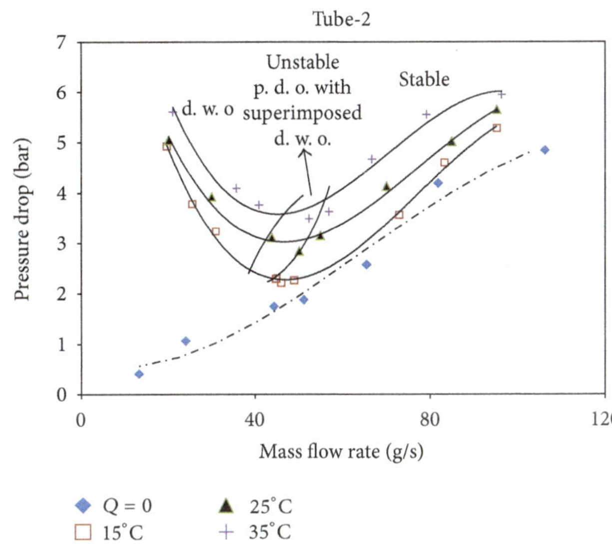

FIGURE 6: The steady state characteristic curves for Tube-2.

A stainless-steel pipe of 0.017 m outer diameter, 3.4 mm wall thickness, and 3 m length was heated with uniform electrical heat input from a DC generator (24 kW power supply). To measure the oscillations of the wall temperature, 28 copper-constantan thermocouples were fixed on the outer surface of the test tube (Figure 2). Half of these thermocouples, that is, 14 thermocouples, were fixed along the top of the test tube; the other half were fixed along the bottom of the test tube. The fluid outlet bulk temperature was measured by placing another thermocouple midstream inside the tube immediately after the test section. The exit restriction (10) created the necessary pressure drop. In this system, an orifice plate was used as exit restriction and the diameter ratio of the exit restriction equal to 0.448 was used. The diameter ratio is defined as the ratio of the inner diameter of the orifice plate to the inner diameter of the tube ($\beta = d/di$). Downstream of the exit restriction, the exit pressure was measured with a pressure gauge. The test tube was electrically heated uniformly and insulated with glass wool that can withstand a temperature of 1000°C. The heat input was determined by measuring the current and the voltage drop across the heated section. In order to avoid floating voltage effects, the thermocouple bead was insulated from the electrically heated tube wall surface with a dab of

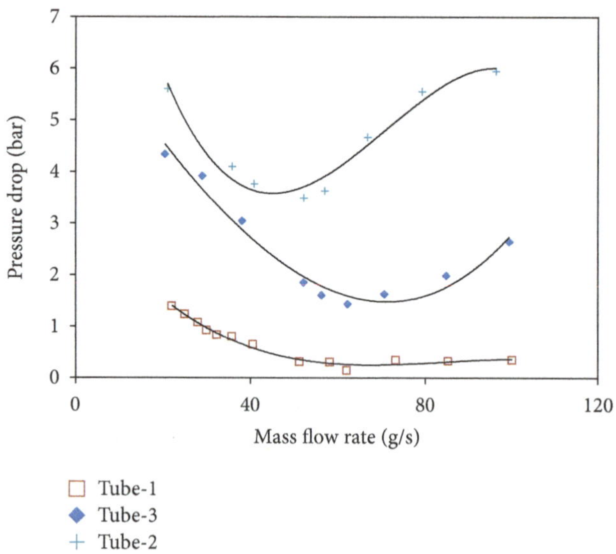

FIGURE 7: The comparison of the steady state characteristics for all tubes. p_{in} = 7.5 bar, Q = 24 kW, T_{in} = 25°C, β = 0.45.

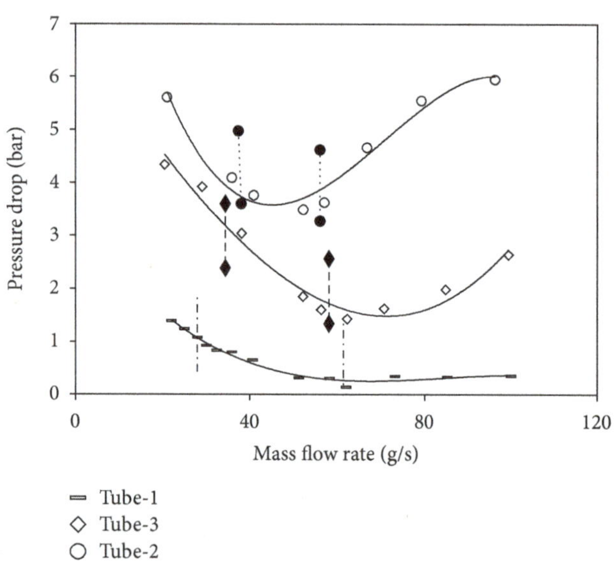

FIGURE 8: The comparison of the oscillation boundaries for all tubes T_{in} = 25°C.

electrically nonconductive paste. The test tube was followed by a sight glass for visual inspection of the flow.

2.3. Fluid Recovery Section. After passing through the test tube, the working fluid comes into the fluid recovery section in the vapour phase. The fluid recovery section mainly consists of four components: condenser (14), nitrogen tank (15), regulator (16), and fluid storage tank (17). The working fluid, nearly all in the vapour phase at the end of the test section, is condensed with a water cooled condenser. The condensed water is sent to the storage tank. The storage water is again sent to the main tank by pressuring with the help of nitrogen gas.

2.4. Experimental Procedure. The experiments were conducted in two different categories: steady and unsteady experiments. In steady experiments, the steady state characteristics were defined; likewise, the two-phase flow dynamic instabilities were investigated in unsteady experiments. In order to investigate the effects of inlet subcooling on steady and unsteady state characteristics, the experiments were conducted for three different inlet temperatures of 15°C, 25°C, and 35°C and three different test tubes under constant heat input (24 kW), system pressure (7.5 bar), and exit restriction (diameter ratio of 0.448). The first mass flow rate for each group is 110 g/s, because the flow is single phase flow at high flow rates, and the mass flow rate was decreased by 10–12 g/s slight steps to define the characteristic curve. The lowest flow rate was taken as 20 g/s in the experiments. Because of the burn-out possibility, flow rates lower than 20 g/s cannot be achieved. In Figure 3(a), the heat transfer enhancement surfaces and characteristics are presented. Figure 3(b) shows the heat transfer enhancement elements: rings. The insertions are generally characterised with the effective diameter. Table 1 presents the effective diameter and pitch ratios for all types of tube. The experiments were first conducted with a smooth tube (Tube-1) and then repeated for Tube-2 and Tube-3.

At the beginning of the experiments, the main tank was pressured with high pressure nitrogen gas, and the system pressure was set by the pressure regulator on the nitrogen tube. In steady state experiments, the nitrogen gas was not used, so the gas inside the surge tank was drained. As for the unsteady experiments, the surge tank was pressured with the nitrogen gas in order to create a constant compressible volume. The water level inside the tank was controlled and observed by the aid of a transparent tube level gauge. For both steady and unsteady experiments, the flow rate was set to the highest value of 110 g/s with a control valve, and to achieve the desired temperature at the tube inlet the temperature of water leaving the main tank was controlled with a digital thermostat. Later, the cooling water was sent to the condenser, and then the constant heat was applied to the tube walls by a DC power supply. Finally, the system was operated and brought to a steady state. It was decided that the steady state was achieved when no higher change than 0.5°C was observed on the test tube surface temperatures. After the steady state was reached, all measurements were taken and the same procedures are repeated for each other mass flow rates up to 25 g/s.

The fact that quick variations in the pressure value of the surge tank and water level were observed means that the oscillations began. At the unsteady experiments, the mass flow rate was decreased until the oscillation boundary was reached. In order to define the boundary where the pressure drop type oscillations break up, the periods of the oscillations were observed. The slight periods in the oscillations show the boundary where the pressure drop type oscillations finished and the independent density wave type oscillations started. By the decreasing mass flow rate, the wall temperatures were followed carefully and the creation of the thermal oscillations was provided. The experiments were suddenly halted when the burn-out began.

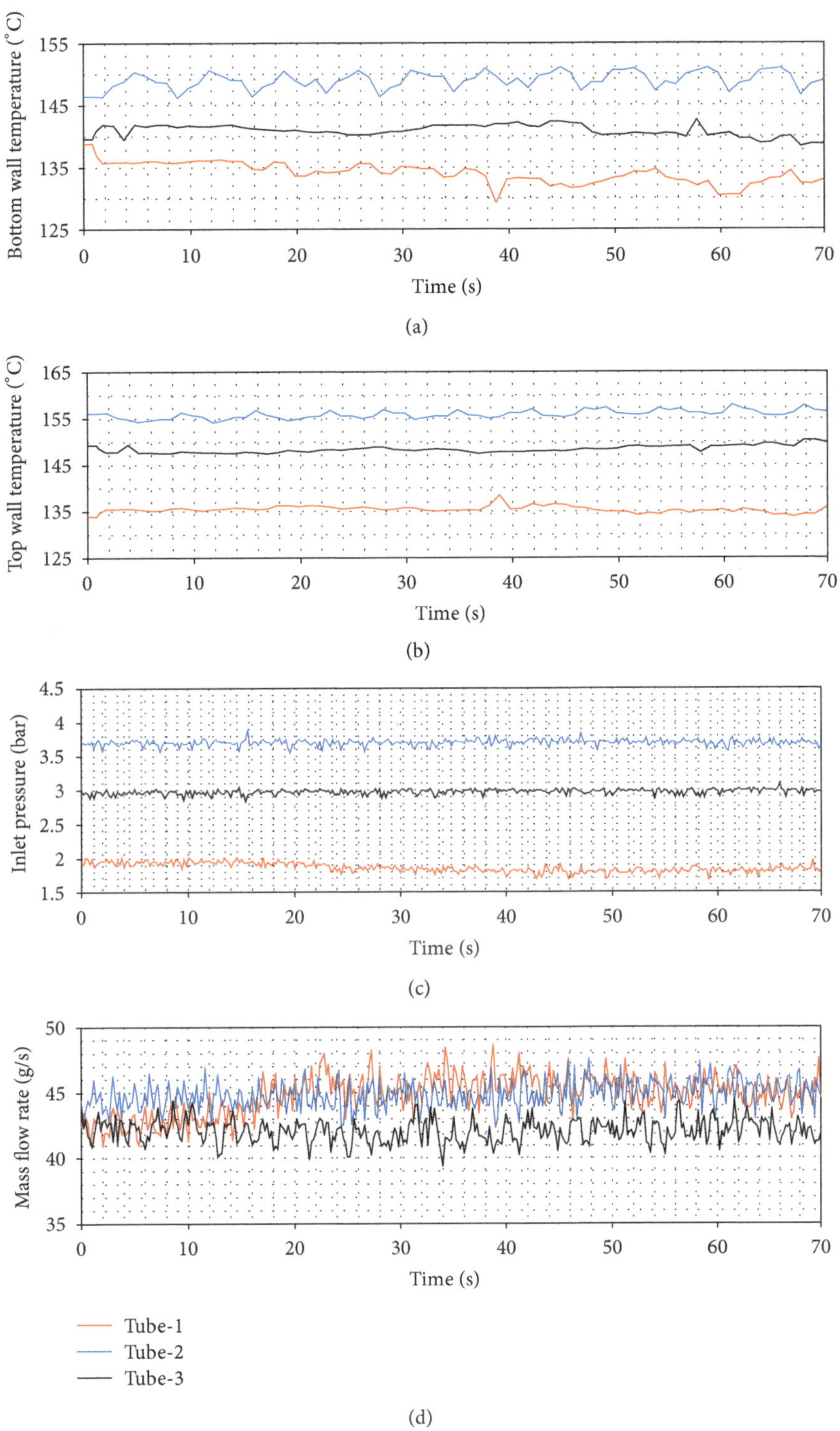

FIGURE 9: Pressure drop oscillations ($Q = 24\,\text{kW}$, $T_{in} = 15^\circ\text{C}$).

2.5. Experimental Measurements and Uncertainties. In the experimental facility, the temperatures were measured with T-type copper-constantan thermocouples with a 0.25 mm diameter. The reading measurement of the temperature taken by the thermocouples was within ±%0.5°C. The thermocouples were placed in a copper pipe for inlet and outlet temperature readings and fixed on the tube outer wall for defining two-phase flow regimes and oscillations. An Advantech data reading card was used for conversion of the signals taken from thermocouples, pressure transducer, and flow meters, separately. The total uncertainty in readings based on chosen acquisition level was in the range of 0.1–0.5°C.

The measured pressures in the experimental study are the pressure of the main tank, surge tank, and nitrogen tank. The

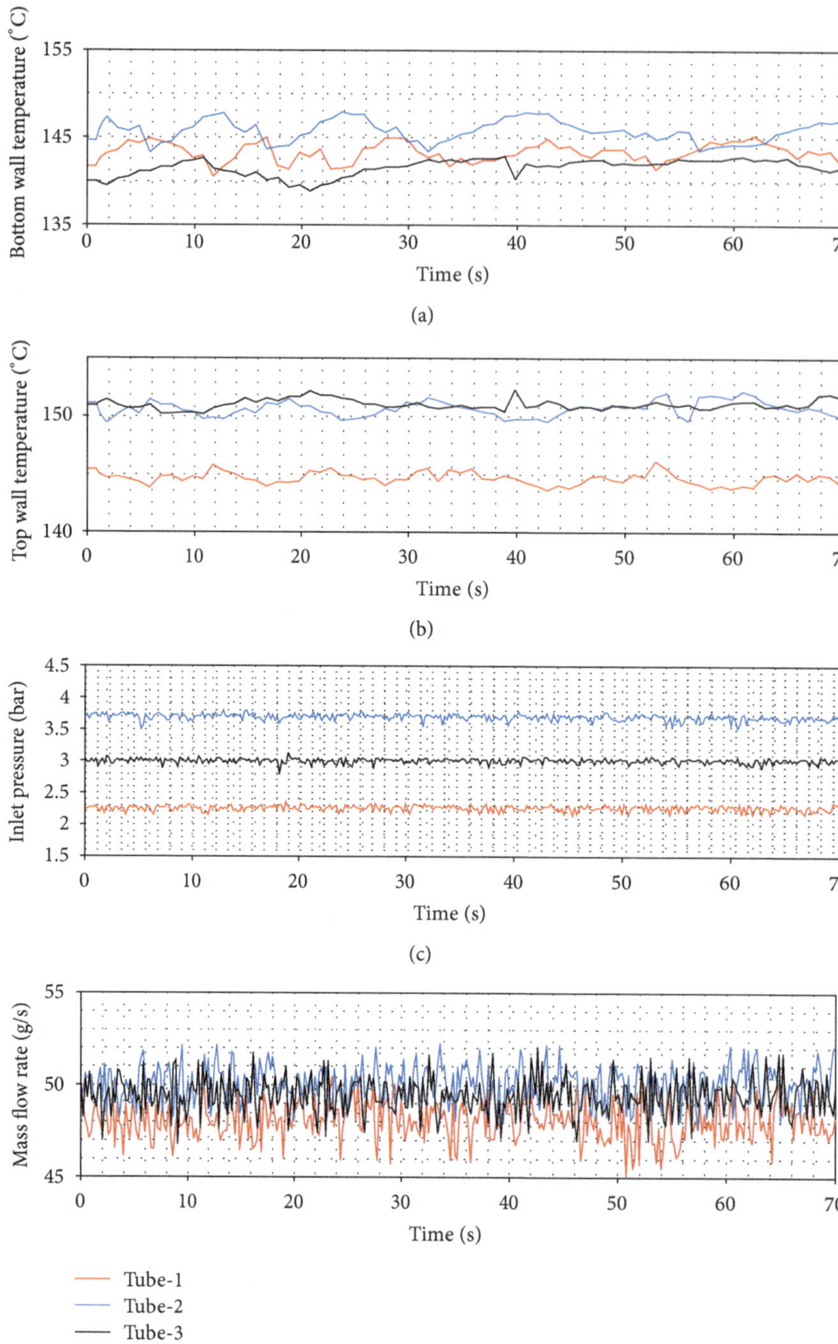

FIGURE 10: Pressure drop oscillations ($Q = 24\,\text{kW}$, $T_{in} = 25^{\circ}\text{C}$).

pressures at the orifice inlet and outlet were also measured. The uncertainty read on the analogue manometers was ±%0.1 bar. The outlet pressure of the orifice was measured with a digital manometer. The pressure transducer was only used to measure the oscillations occurring at the test tube inlet and outlet; the pressure uncertainty of the transducer is ±%0.1 bar.

On the determination of the oscillation and the stability boundaries, the mass flow rate measurements should be taken carefully. Two flow meters having flow rate setting on itself are used to measure and set the mass flow rate. The ranges of the flow meters are 0–400 and 0–1000 l/hour, respectively. By these flow meters, the uncertainty rate is ±%0.4 l/hour. The total uncertainty of the readings taken from the turbine

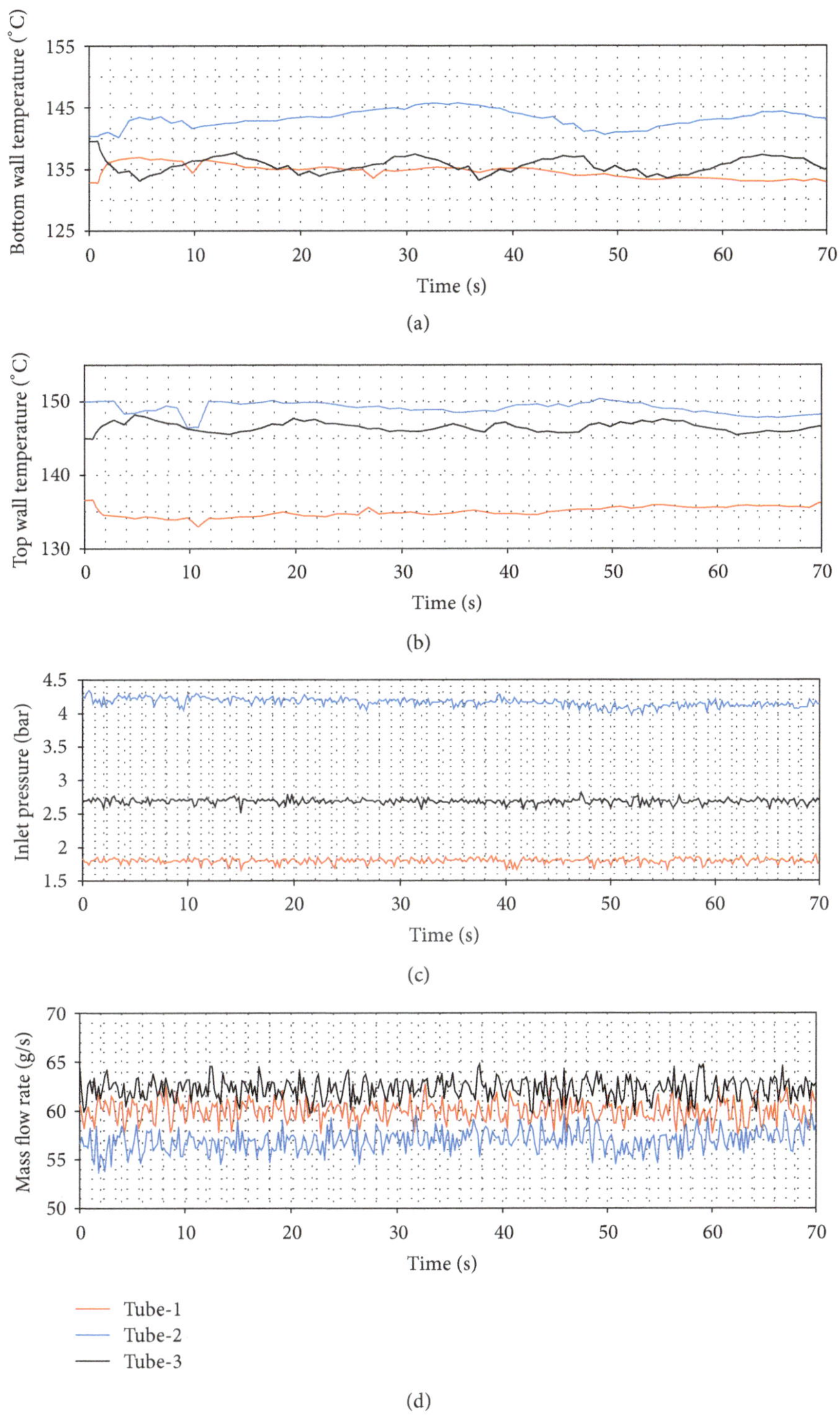

FIGURE 11: Pressure drop oscillations ($Q = 24$ kW, $T_{in} = 35^{\circ}$C).

type flow meter used to measure the flow rate oscillations is ±%0.05 l/hour. The power rate of the DC power supply was set with the setting button on the supplier, and the heat input values were read from the digital volt and current gauges. The DC power supply was controlled within ±%0.2.

3. Results and Discussion

In two-phase flow systems, the steady state characteristic curves that are used frequently are plotted by the pressure drop versus mass flow rate. These curves show the variations

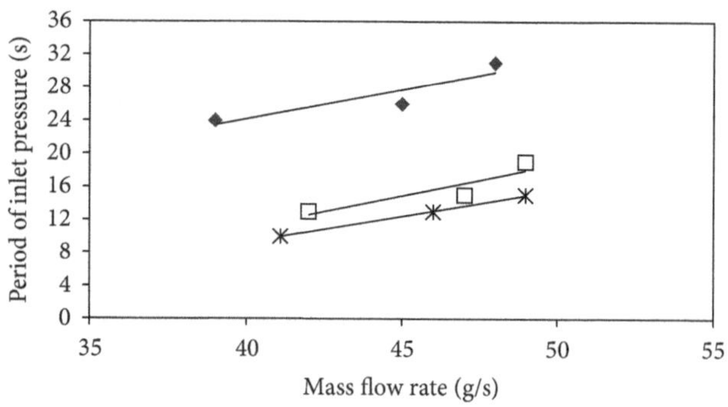

FIGURE 12: The variations in the period with the mass flow rate for all tubes (p. d. o.).

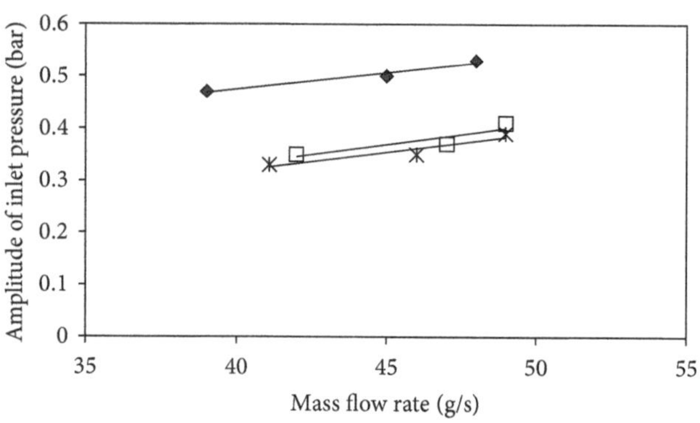

FIGURE 13: The variations in the amplitudes with the mass flow rate for all tubes (p. d. o.).

of pressure drop versus mass flow rate for each tube configuration. On measuring the pressure drop, the pressure drop between the surge tank and the test tube outlet is used. Also, the steady state curves depending on the inlet subcooling are plotted on defining the steady state characteristic curves. These curves are used for understanding the flow characteristics.

The positive sloped parts of the curves that correspond to high mass flow rate values represent the single phase flow region. The first bubbles are observed at the point on which the value of the pressure drop is at a minimum with a decreasing mass flow rate and the slope of the curve is negative. On the same point, two-phase flow starts by the observation of the first bubbles. Much more vapour flows with liquid phase as long as the number of the bubbles increases in two-phase flow. This increase in the number of the bubbles causes the density to decrease in terms of the liquid phase and thus the pressure drop to increase. As a result of the fact that the mass flow rate is decreased, the parts of the curves with a negative slope up to the saturated vapour boundary have a positive slope and the pressure drop starts to decrease. On the steady state characteristic curves, the region between the dashed lines represents the unstable region where the pressure drop type (p. d. o.) and the density wave type (d. w. o.) oscillations occur at the same time; this region is called the superimposed region. The left side of the superimposed region (the left side of the dashed line on the left hand side) indicates only the density wave type (d. w. o.) flow oscillations.

Using inner elements such as rings and twisted tapes in the flow field causes more friction than the smooth ones and also causes the pressure drop to increase. On the other hand, the creation of the turbulence and the removal or destruction of the boundary layer enhance the heat transfer rate. In Figures 4, 5, and 6, $Q = 0$ implies that no heat input is applied and this case presents the single phase steady state characteristics. The highest pressure drop is observed in tube-2, whereas the lowest pressure drop value is in tube-1. On the comparison of the pitch ratios, the pressure drop increases with the increase in the pitch ratio. Tubes are ordered from the highest pressure drop to the lowest value as tube-2, tube-3, and tube-1.

In Figure 7, the comparison of the steady state characteristic curves for all investigated tubes are presented at $T_i = 25°C$. As shown in the graph for five different test tubes, it is observed that the pressure drop between the maximum and the minimum points of the curves increases and the points where the boiling starts slip towards lower mass flow rates. Furthermore, the increase in the inlet subcooling increases the slope of the two-phase flow region and decreases the pressure drop values with the decreasing mass flow rates. It also can be said that the system becomes more unstable as long as the negative slope angle of the steady state characteristic curves increases. As the negative slope angle of the smooth tube is lower than the tubes with inner elements, it can be concluded that the smooth tube is more stable than the other tubes.

In two-phase flow systems, as the oscillations occurring under unsteady state conditions harm the systems, the fact that they are kept under control is highly important. Figure 10 presents the comparison of the oscillation boundaries for all tubes. On these curves, the region on the right side of the dashed lines implies the single phase flow region on high mass flow rates, and the space between the lines implies the unsteady region where the pressure drop and density wave type oscillations occur as imposed. The left side of the middle region represents the density wave oscillations only. In the single phase flow region corresponding to the high flow rates, no oscillation is observed. By the creation of the first bubbles, the oscillations begin occurring. These oscillations are the pressure drop type (p. d. o.) and the density wave type (d. w. o.) oscillations. The beginning and the end points of the p. d. o. slip towards lower flow rates as long as the inlet subcooling rate increases. By the decrease of the inlet temperature, the single phase flow region increases and thus the system becomes more stable.

The reason why the system is more stable is the fact that the oscillations start at lower mass flow rates. The oscillations in the tubes with insertions begin at lower flow rates than the

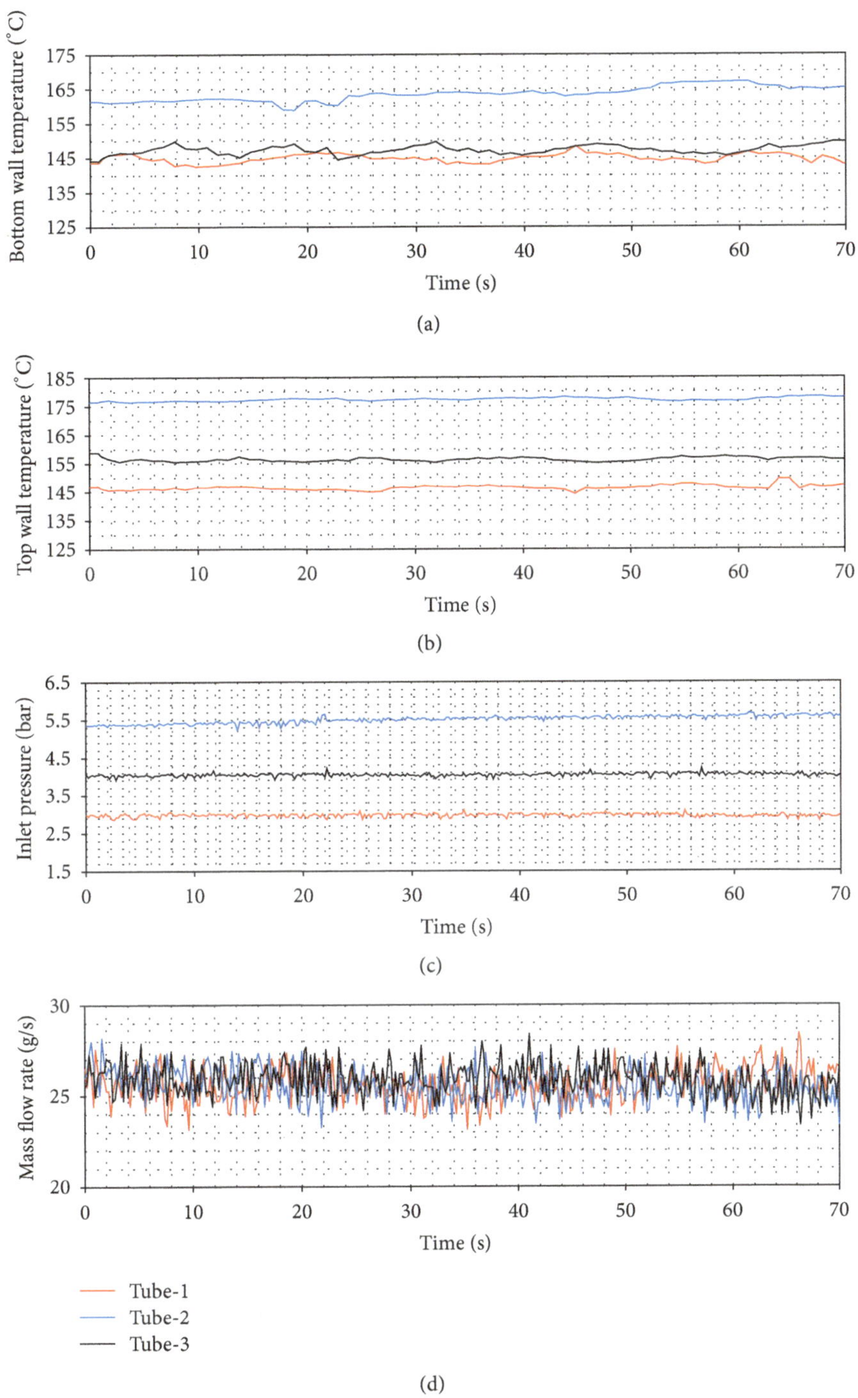

FIGURE 14: Density wave oscillations ($Q = 24\,\text{kW}$, $T_{in} = 15^{\circ}\text{C}$).

smooth tube and finish at higher flow rates. The insertions used in the horizontal tubes in which the phase separation occurs between the vapour and the liquid by the effect of the gravity disperse the vapour and cause the liquid bubbles to occur on the upper wall besides the vapour. Regarding the oscillation boundaries, the start point of the p. d. o. draws back towards the lower flow rates as long as the inlet temperature decreases. This case shows that increasing the inlet subcooling makes the system more stable for all tubes investigated. The stability boundary curves are in same manner for all tubes. The unsteady region of the flow expand, that is to say the imposed oscillation region of p. d. o. and d. w. o., grows by the increase of the gap between the boundaries.

In Figure 8, it is clearly seen that tube-1 has the widest boundaries. This means that the most unstable flow occurs in tube-1. The tubes with coiled wire insertions (tube-2 and tube-3) provide a more stable flow than the smooth tube. As for the pitch ratio, the fact that the pitch increases makes

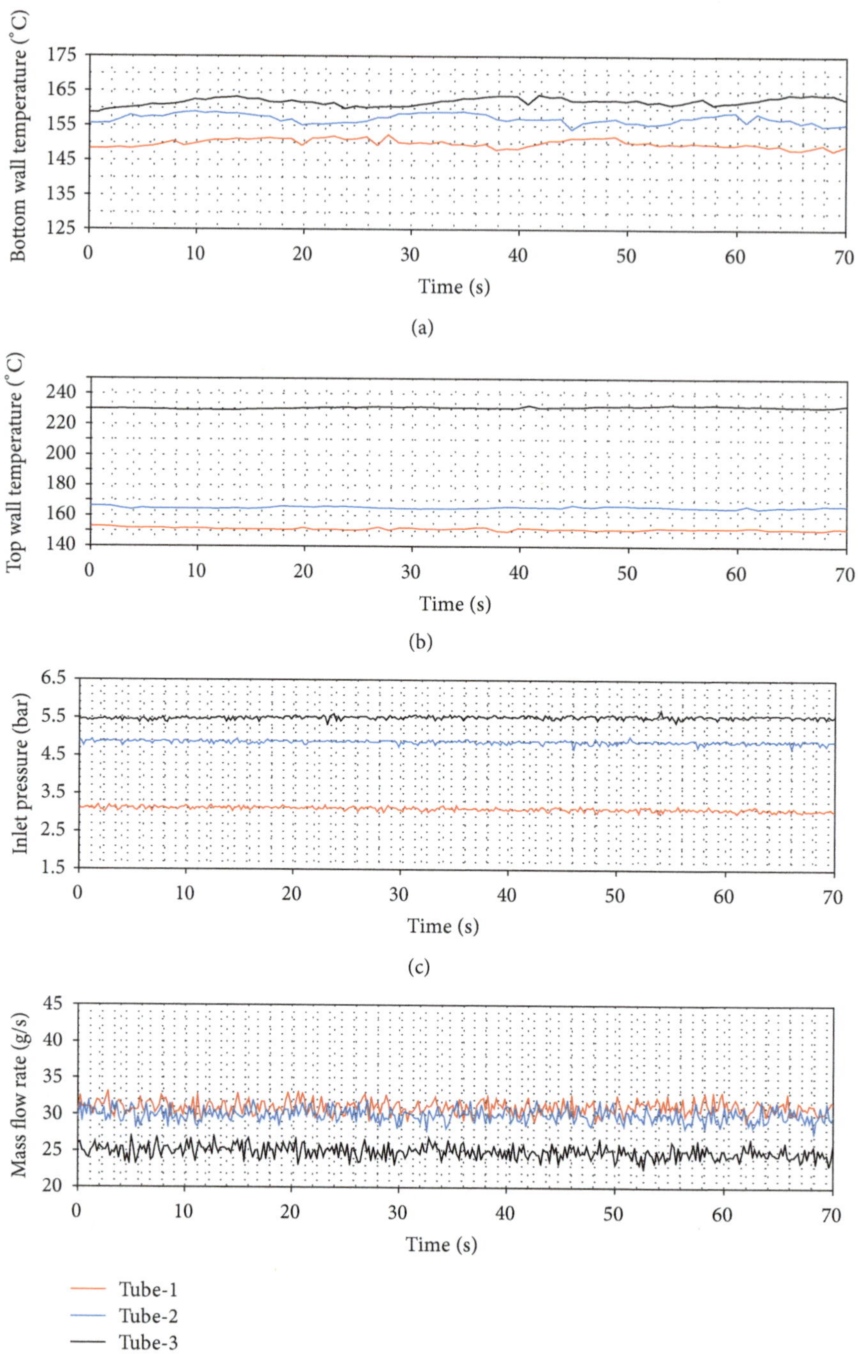

FIGURE 15: Density wave oscillations ($Q = 24$ kW, $T_{in} = 25$°C).

the system more unstable. In Figures 9, 10, and 11, the time dependent top and bottom wall temperatures, the inlet pressure, and the mass flow rate are shown schematically. The fact that the bubbles occurring because of the too small tube diameter cover both the top and the bottom walls prevents higher temperature differences to comprise between the walls. For each ring type, the p. d. o. oscillations are observed and these oscillations cause fluctuations with big amplitudes in the inlet pressure, top and bottom wall temperatures, and the mass flow rate. Also, oscillations in the mass flow rate are observed in addition to the inlet pressure. The amplitudes of the top wall are higher than at the bottom wall. When the thermal oscillations occur, the inlet pressure and the mass flow rate oscillate at lower frequency with large amplitude.

The pressure drop type flow oscillations are like dynamic instabilities occurring in unstable flows. The main effect causing pressure drop type flow oscillations to occur is the fact that a compressible volume exists. If the test section is

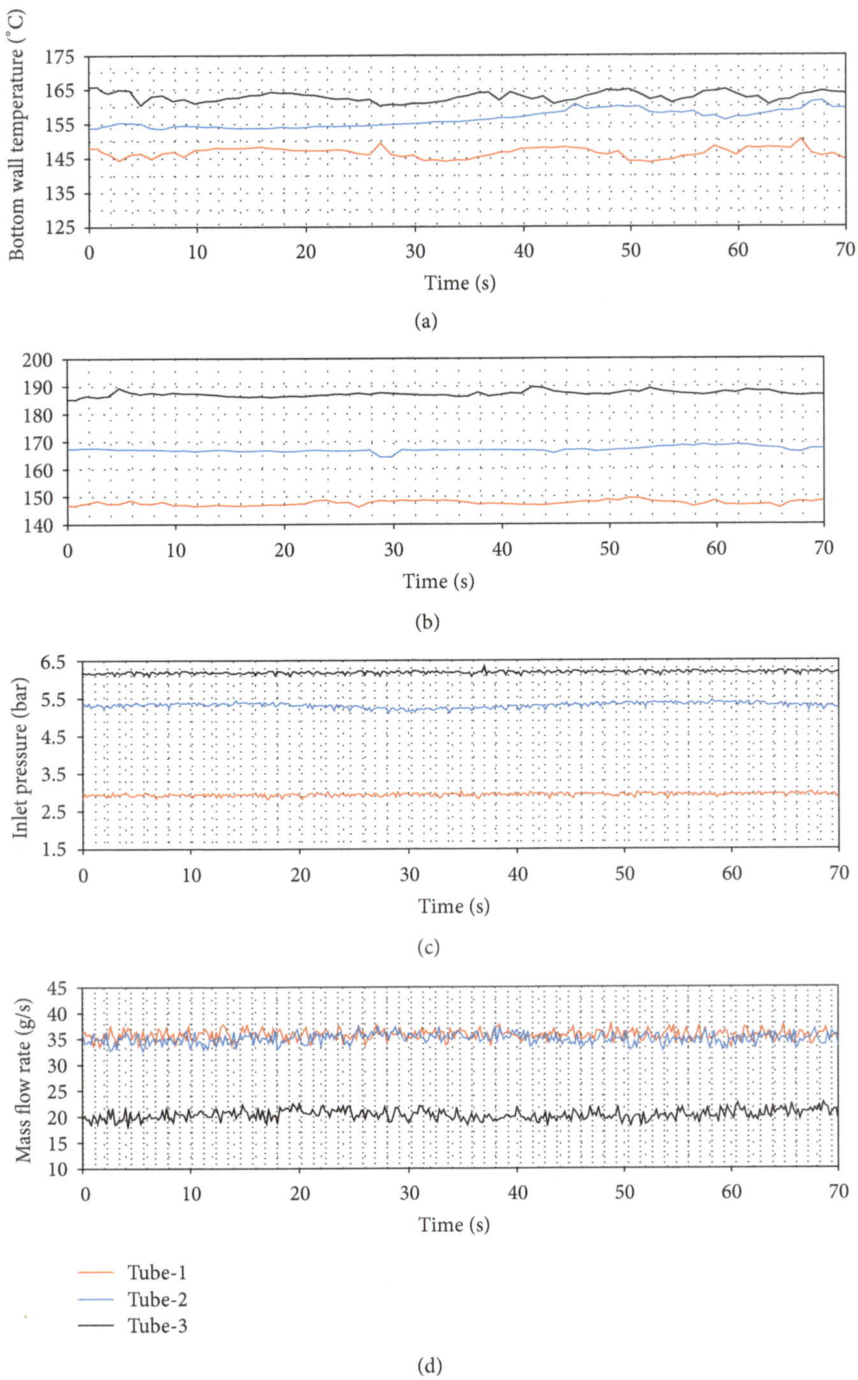

FIGURE 16: Density wave oscillations ($Q = 24\,\text{kW}$, $T_{\text{in}} = 35^{\circ}\text{C}$).

too long, the inner compressibility is enough for the pressure drop oscillations to occur; if not, the compressible volume is provided by a surge tank located at front side of the test section. In industry, the pressure drop oscillations for the systems including two-phase flows have a large oscillation period and high oscillation amplitude. This case causes the life of the system to decrease and makes it difficult to control the flow. Therefore, the fact that the pressure drop oscillations are determined and taken under control is vitally important.

In Figures 12 and 13, the effects of the mass flow rate on inlet pressure periods and amplitudes for all investigated tubes at $T_i = 15^{\circ}\text{C}$ are presented. For all tubes, the periods and the amplitudes belonging to the inlet pressure increase by the increase of the mass flow rate. If an order is made in terms of the periods and the amplitudes, it easily seen that the periods and the amplitudes in tube-1 (smooth tube) are higher than those of the tubes with insertions. The tubes are ordered from the highest period and amplitude to the lowest ones as tube-1,

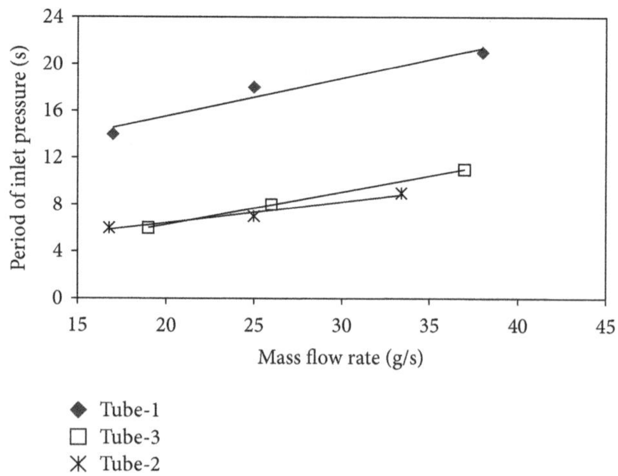

FIGURE 17: The variations of the periods with the mass flow rate for all tubes (d. w. o.).

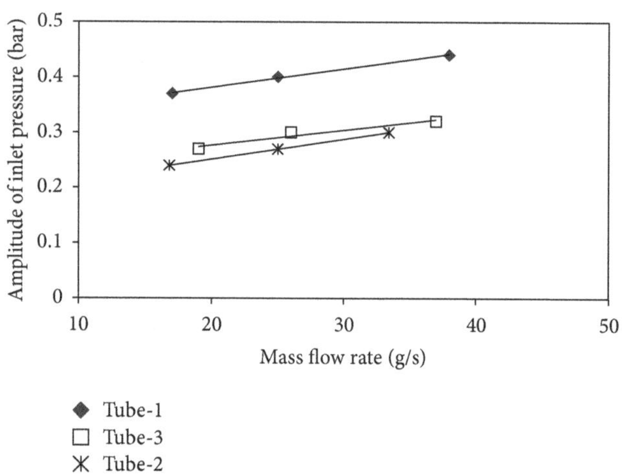

FIGURE 18: The variations in the amplitudes with the mass flow rate for all tubes (d. w. o.).

tube-3, and tube-2. According to this order, the smooth tube with the biggest effective diameter has the biggest period and amplitude, while tube-2 with the smallest effective diameter has the smallest period and amplitude. In the other words, it can be said that the period and the amplitude increase as long as the effective diameter increases. Furthermore, the periods and the amplitudes also increase with the increase in the pitch ratio.

The density wave flow oscillations occur as a result of boiling and the hydrodynamic behaviours of the flow type. The density wave flow oscillations take place together with the pressure drop oscillations as superimposed in the region with negative slope. The pure density wave oscillations are observed as long as they move to the upper points of the negative sloped region, towards the unstable region. The period and the amplitude values of the pure density wave oscillations are lower than the pressure drop type. Furthermore, the pure d. w. o. in smooth tubes occurs at higher flow rates relative to the tubes having insertions. In two-phase flows, the separation of the phases in the smooth tubes is observed more clearly than in the tubes with insertions.

In Figures 14, 15 and 16, the time dependent variations of both top wall and the bottom wall, the inlet pressure and mass flow rates are presented. For each ring type, the d. w. o. oscillations are observed and these oscillations cause fluctuations with big amplitudes in the inlet pressure, top and bottom wall temperatures, and the mass flow rate. Also, oscillations in the mass flow rate are observed in addition to the inlet pressure. The amplitudes of the bottom wall are higher than at the top wall. When the thermal oscillations occur, the inlet pressure and the mass flow rate oscillate at lower frequency with large amplitude.

Figures 17 and 18 show the variations of the periods and amplitudes with mass flow rates. For all investigated tubes, the periods and the amplitudes increase as long as the mass flow rate increases. The amplitude and period of the smooth tube are higher than the other configurations. In terms of both the amplitude and the period, the order from the biggest to the smallest is tube-1, tube-3, and tube-2. According to the order made for pure d. w. o., the smooth tube with the biggest effective diameter has the biggest period and amplitude, and the tube-2 with the smallest effective diameter has the smallest period and amplitude. Similar to the results of pressure drop oscillations, the period and the amplitude increase as long as the effective diameter increases. Also, the periods and the amplitudes also increase with the increase in the pitch ratio.

4. Conclusions

The aim of this study was to experimentally investigate the effect of the coiled wire insertions on dynamic instabilities and to compare the results with the smooth tube for forced convection boiling. The experiments were conducted in a circular tube, and water was used as the working fluid. Two different pitch ratios (H/D = 2.77 and 5.55) of coiled wire with circular a cross-section were utilised. The constant heat flux boundary condition was applied to the outer side of the test tube, and the constant exit restriction was used at the tube outlet. The remarkable conclusions are ordered as presented below.

(i) The rings had a higher pressure drop than the smooth tube and the pressure drop increased with the increase in the pitch ratio.

(ii) The system became more unstable as long as the negative slope angle of the steady state characteristic curves increased. As the negative slope angle of the smooth tube was lower than the tubes with inner elements, the smooth tube was more stable than the other tubes.

(iii) In the single phase flow region corresponding to the high flow rates, no oscillation was observed. The beginning and the end points of the p. d. o. slipped towards lower flow rates as long as the inlet subcooling rate increased. By the decrease of the inlet temperature, the single phase flow region increased and thus the system became more stable.

(iv) The reason why the system was more stable was the fact that the oscillations started at lower mass flow rates. The oscillations in the tubes with insertions began at lower flow rates than the smooth tube and finished at higher flow rates.

(v) The start point of the p. d. o. drew back towards the lower flow rates as long as the inlet temperature decreased. This case showed that increasing the inlet subcooling made the system more stable for all the tubes investigated.

(vi) The imposed oscillation region of p. d. o. and d. w. o. grew by the increase of the gap between the boundaries.

(vii) The fact that the bubbles occurred because of the too small tube diameter covered both the top and the bottom walls prevented higher temperature differences to comprise between the walls.

(viii) For each tube, the p. d. o. oscillations were observed and these oscillations caused the fluctuations with big amplitudes in the inlet pressure, top and bottom wall temperatures, and the mass flow rate.

(ix) The oscillations in the mass flow rate were observed in addition to the inlet pressure. The amplitudes of the top wall were higher than those of the bottom wall.

(x) When the thermal oscillations occurred, the inlet pressure and the mass flow rate oscillated at a lower frequency with a large amplitude.

Acknowledgment

This research was supported by BAP of the Research Fund of Ataturk University.

References

[1] S. Karagoz, *Investigation of the effects of heat transfer enhancement on two phase flow instabilities [Ph.D. thesis]*, Engineering Faculty, Ataturk University, Erzurum, Turkey, 2007.

[2] G. Omeroglu, *Investigation of the effects of the turbulators on two phase flow instabilities [Ph.D. thesis]*, Engineering Faculty, Ataturk University, Erzurum, Turkey, 2012.

[3] S. Karsli, M. Yilmaz, and O. Comakli, "The effect of internal surface modification on flow instabilities in forced convection boiling in a horizontal tube," *International Journal of Heat and Fluid Flow*, vol. 23, no. 6, pp. 776–791, 2002.

[4] O. Çomakli, S. Karsli, and M. Yilmaz, "Experimental investigation of two phase flow instabilities in a horizontal in-tube boiling system," *Energy Conversion and Management*, vol. 43, no. 2, pp. 249–268, 2002.

[5] A. Menteş, O. T. Yıldırım, S. Kakaç, and T. N. Veziroğlu, "Two-phase flow instabilities in a single channel up flow system," in *Proceedings of the Japan-U.S. Seminar on Two-Phase Flow Dynamics*, pp. 5–8, Lake Placid, NY, USA, 1984.

[6] S. Kakaç, C. O. Gavrilescu, and Ö. Çomaklı, "Two-phase flow instabilities with augmented surfaces in a horizontal in-tube boiling system," in *Proceedings of the Turkish National Conference on Thermal Sciences and Technologies*, pp. 1–26, Ankara, Turkey, September 1995.

[7] S. Kakaç, "A review of two-phase flow instabilities," in *Advances in Two-Phase Flow on Heat Transfer*, vol. 2, pp. 577–668, Martinus Nijhoff, Boston, Mass, USA, 1994.

[8] J. W. Coleman and S. Garimella, "Two-phase flow regimes in round, square and rectangular tubes during condensation of refrigerant R 134a," *International Journal of Refrigeration*, vol. 26, no. 1, pp. 117–128, 2003.

[9] L. K. H. Leung, D. C. Groeneveld, A. Teyssedou, and F. Aubé, "Pressure drops for steam and water flow in heated tubes," *Nuclear Engineering and Design*, vol. 235, no. 1, pp. 53–65, 2005.

[10] Ö. Çomaklı, M. Yılmaz, Ö. Bedir, and B. Şahin, "Isı Transfer İyileştirmesinin İki Fazlı Akış Kararsızlıklarına Etkisi," *Mühendis Ve Makina*, vol. 48, no. 565, pp. 9–17, 2007.

[11] L. Tadrist, "Review on two-phase flow instabilities in narrow spaces," *International Journal of Heat and Fluid Flow*, vol. 28, no. 1, pp. 54–62, 2007.

[12] Z. Y. Bao, D. F. Fletcher, and B. S. Haynes, "An experimental study of gas-liquid flow in a narrow conduit," *International Journal of Heat and Mass Transfer*, vol. 43, no. 13, pp. 2313–2324, 2000.

[13] Ö. Çomaklı, Ş. Karagöz, M. Yılmaz, and S. Karslı, "Yatay bir borudaki kaynamalı iki fazlı akış karasızlıkları ve akışkan tipinin etkileri," *Teknoloji*, vol. 11, pp. 15–24, 2008.

[14] S. Kakaç and L. Cao, "The effect of heat transfer enhancement on two-phase flow dynamic instabilities in a boiling system," in *Proceedings of the International Conference on Computational Heat and Mass Transfer (CHMT '99)*, A. A. Mohamad, S. I. Eastern, and G. Mağusa, Eds., pp. 448–462, Mediterranean University, April 1999.

[15] P. Poesio, "Experimental determination of pressure drop and statistical properties of oil-water intermittent flow through horizontal pipe," *Experimental Thermal and Fluid Science*, vol. 32, no. 8, pp. 1523–1529, 2008.

[16] G. Sotgia, P. Tartarini, and E. Stalio, "Experimental analysis of flow regimes and pressure drop reduction in oil-water mixtures," *International Journal of Multiphase Flow*, vol. 34, no. 12, pp. 1161–1174, 2008.

[17] J. Moreno Quibén and J. R. Thome, "Flow pattern based two-phase frictional pressure drop model for horizontal tubes, Part II: new phenomenological model," *International Journal of Heat and Fluid Flow*, vol. 28, no. 5, pp. 1060–1072, 2007.

[18] S. Kakac and B. Bon, "A review of two-phase flow dynamic instabilities in tube boiling systems," *International Journal of Heat and Mass Transfer*, vol. 51, no. 3-4, pp. 399–433, 2008.

[19] W. Yu, D. M. France, M. W. Wambsganss, and J. R. Hull, "Two-phase pressure drop, boiling heat transfer, and critical heat flux to water in a small-diameter horizontal tube," *International Journal of Multiphase Flow*, vol. 28, no. 6, pp. 927–941, 2002.

[20] M. Yılmaz, Ö. Çomaklı, Ş. Karagöz, and Ö. Bedir, *İki Fazlı Akışlarda Isı Transfer İyileştirmesi-1 Termodinamik*, vol. 176, 2007.

[21] M. Yılmaz, Ö. Çomaklı, Ş. Karagöz, and Ö. Bedir, *İki Fazlı Akışlarda Isı Transfer İyileştirmesi-2 Termodinamik*, vol. 176, 2007.

[22] F. Widmann, O. Comakli, C. O. Gavrilescu, Y. Ding, and S. Kakac, "The effect of augmented surfaces on two-phase flow instabilities in a horizontal system," *Journal of Enhanced Heat Transfer*, vol. 2, no. 4, pp. 263–271, 1995.

[23] S. Karagoz, S. Karsli, M. Yilmaz, and O. Comakli, "Density-wave flow oscillations in a water boiling horizontal tube with inserts," *Journal of Enhanced Heat Transfer*, vol. 16, no. 4, pp. 331–350, 2009.

Greenhouse Gas Inventory of a Typical High-End Industrial Park in China

Bin Chen,[1] Guoxuan He,[1] Jing Qi,[1] Meirong Su,[1] Shiyi Zhou,[2] and Meiming Jiang[2]

[1] *State Key Joint Laboratory of Environmental Simulation and Pollution Control, School of Environment, Beijing Normal University, Beijing 100875, China*
[2] *Beijing Development Area Co., Ltd., Beijing 100176, China*

Correspondence should be addressed to Bin Chen; chenb@bnu.edu.cn

Academic Editors: N. H. Afgan, A. Bahadori, M. Gambino, and A. L. Mayer

Global climate change caused by greenhouse gas (GHG) emissions, which severely limits the development of human society and threatens the survival of humanity, has drawn the international community's long-term attention. Gathering the most important production factors in the region, an industrial park usually represents the development level of specific industries in the region. Therefore, the industrial park should be regarded as the base unit for developing a low-carbon economy and reducing GHG emissions. Focusing on a typical high-end industrial park in Beijing, we analyze the carbon sources within the system boundary and probe into the emission structure in view of life-cycle analysis. A GHG inventory is thereby set up to calculate all GHG emissions from the concerned park. Based on the results, suggestions are presented to guide the low-carbon development of the high-end industrial park.

1. Introduction

Global climate change caused by greenhouse gas (GHG) emissions, which severely limits the development of human society and threatens the survival of humanity, has drawn the international society's attention. China has been considered as the largest GHG emitter in the world, and the Chinese government has been taking various measures to reduce GHG emission and mitigate climate change [1–6]. Among them, China is committed to cut the CO_2 emission per unit of GDP by 40–45% by 2020 against the 2005 level [7]. Within this context, low-carbon economy provides an efficient development path that maximizes the value in a low consumption, low pollution, and low GHG emission mode [8, 9].

Industrial parks are characterized as a clustering of industries designed to meet compatible demands of different organizations within one location [10]. Industrial agglomeration has proven to be vital to the economic growth of developed countries, as well as less-developed ones like China [11, 12]. According to the data from the Ministry of Land and Resources in China 2011, there were 208 nationally designated industrial parks and more than 3000 provincial ones in China. For the industrial land, the average fixed investment was 48.1 million RMB per hectare, and the output was 125.6 million RMB per hectare [13–15]. Taking Beijing as an example, all the industrial parks' contribution to the urban economy accounted for 24%. Collecting most important factors of production in the region, an industrial park usually represents the development level of specific industries in the region. Therefore, the industrial park should be regarded as the base unit for developing industrial economy and also a breakthrough for regional resource allocation and environmental management. With the emergence of low-carbon economy, more and more industrial parks are seeking a low-carbon mode incorporating production, consumption, and resource allocation issues.

In the industrial park, enterprises of identical or similar industries are not simply spatially concentrated. The administrator must plan the layout, build the public infrastructure, and handle the daily maintenance. Therefore, GHG emissions from industrial parks could be divided into two parts, and the responsibility for emission cutting could be shared by different parties including the enterprises and the administrator of the park. Currently, there have been increasing research interests focused on the GHG emissions from enterprises and industrial sectors, shaping the GHG inventories in expanding scopes, from the combustion of fossil fuel to the whole production chain [16–21]. This means, in spite of the direct emission from energy consumption and chemical process, the inventories also include indirect emission owing to the total supply chain extend to the production gate [22]. In fact, it is necessary to calculate the GHG emission from the perspective of administrators and put forward useful suggestions on carbon reduction, which may supplement the existing literatures and help industrial parks achieve the low-carbon mode.

High-end industrial park is a type of industrial park that usually agglomerates the head offices or research and development centers of high-tech industries. There are few manufacturing productive processes in such industrial parks, so less fossil fuel and fewer raw materials are consumed by the enterprises. As a result, the administrator would take more responsibilities compared to those of the other types of industrial parks. In this paper, a GHG inventory is set up to analyze the life-cycle GHG emissions of a high-end industrial park in Beijing. Based on the results, suggestions are given to guide the low-carbon development of the high-end industrial park.

2. Methodology

The life cycle of a high-end industrial park is divided into three stages, that is, construction stage, operation stage, and demolition stage. In the construction stage, the concerned buildings and affiliated municipal facilities are constructed, as well as the garden landscape. The operation stage usually lasts for 40 to 50 years. During this stage, the industrial parks need to consume a large amount of electricity, heat power, and water. Meanwhile, wastewater and solid waste are discharged daily from the buildings and public space. As a general rule, a high-end industrial park has a 50-year term for the land use rights, and then it goes into the demolition stage. In this study, the GHG emission in the demolition stage is estimated based on the size and construction structure of the industrial park.

We consider three types of GHGs, that is, carbon dioxide (CO_2), methane (CH_4), and nitrogen oxide (N_2O). The inventory includes five sorts of GHG emissions sources: (1) direct and indirect emission from the consumption of primary energy and second energy; (2) direct emission from industrial production process; (3)-(4) emission from the production and transportation processes of all the materials and equipment used in the high-end industrial park; (5) emission from the sewage treatment and solid waste disposal processes.

2.1. Energy Consumption. Energy consumption is an important source of GHG emissions in the industrial park. As defined in this study, energy consumption processes involved mainly includes fossil fuel combustion, electricity and heat energy production, and transportation vehicles. The estimation of the GHG emission from above processes refers to the 2006 IPCC Guidelines for National Greenhouse Gas Inventories [22].

For the fossil fuel combustion, (1) is given as

$$E^{c}_{\mathrm{GHG},i} = Q^{c}_{i} \times \mathrm{EF}^{c}_{\mathrm{GHG},i}, \tag{1}$$

where $E^{c}_{(\mathrm{GHG},i)}$ is the emissions of a given GHG by type of fuel (kg), Q^{c}_{i} is the amount of fuel combusted (TJ), and $\mathrm{EF}^{c}_{\mathrm{GHG},i}$ is the emission factor of a given GHG by type of fuel (kg GHG/TJ).

For the electricity and heat energy production processes, if the power comes from the sectors outside the industrial park, the concerned GHG emission can be calculated according to the average energy consumption level of the local sector. If the power is produced by the enterprise just in the industrial park, the concerned GHG emission can be estimated as zero to avoid the repeated calculation based on the fossil fuel of combustion process. Equation (2) is presented to calculate the GHG emission from electricity and heat energy production processes as

$$E^{e}_{\mathrm{GHG}} = Q^{e} \times \mathrm{EF}^{e}_{\mathrm{GHG}}, \tag{2}$$

where E^{e}_{GHG} is the emissions of a given GHG by electricity or heat (kg), Q^{e} is the amount of electricity or heat produced (kWh or TJ), and $\mathrm{EF}^{e}_{\mathrm{GHG}}$ is the emission factor of a given GHG by electricity or heat (kg GHG/kwh or kg GHG/TJ).

As for the transportation process, it refers to the transportation of the energy resources, raw materials, instruments and equipment from the origin to the industrial park. The GHG emissions from the transportation process can be estimated from the fuel consumed:

$$E^{t}_{\mathrm{CO}_2} = \sum_{i} \left(Q^{t}_{i} \times \mathrm{EF}^{t}_{\mathrm{CO}_2,i} \right), \tag{3}$$

where $E^{t}_{\mathrm{CO}_2}$ is the emissions of CO_2 (kg), Q^{t}_{i} the fuel consumed (TJ), and $\mathrm{EF}^{t}_{\mathrm{CO}_2,i}$ is the emission factors (kg CO_2/TJ), or calculated by the distance covered by the vehicles as

$$E^{t}_{\mathrm{NH}_4,\mathrm{N}_2\mathrm{O}} = \sum_{a,b,c,d} \left(L_{a,b,c,d} \times \mathrm{EF}^{t}_{a,b,c,d} \right) + \sum_{a,b,c,d} C_{a,b,c,d}, \tag{4}$$

where $E^{t}_{\mathrm{NH}_4,\mathrm{N}_2\mathrm{O}}$ is the emissions of CH_4 or N_2O (kg), $L_{a,b,c,d}$ is the distance travelled during thermally stabilized engine operation phase for a given mobile source activity (km), $\mathrm{EF}^{t}_{a,b,c,d}$ is the emission factor (kg/km), $C_{a,b,c,d}$ is the emissions during warm-up phase (cold start)(kg), a is the fuel type (e.g., diesel, gasoline, natural gas, and LPG), b is the vehicle type, c is the emission control technology (such as uncontrolled, catalytic converter, etc.), and d is the operating conditions (e.g., urban or rural road type, climate, or other environmental factors).

In general, the first approach (fuel sold) is appropriate for CO_2 while the second (distance travelled by vehicle type and road type) is suitable for CH_4 and N_2O.

2.2. Industrial Production. The GHG emissions from industrial production processes of specific products are mainly due to the physical or chemical reaction rather than the combustion of fossil fuels. In an industrial park, the GHG type and amount depend on the leading industry type, and the accounting method varies accordingly. According to the activity data obtained and the applicable scope of emission factors, there are two scales to calculate the emission from the industrial production process: an industry or a product. If it is an industry, the calculation is less precise. The activity data can be the economic scale (such as the GDP) or the production scale (such as the production output), and the emission factors can be the recommended ones by IPCC or the average level of the specific countries or territories. Regarding a product, the calculation must focus on the supply chain of the product, even narrowed down to production device and working conditions. For a high-end industrial park in this study, there is almost no real supply chain of product.

2.3. Material Input. Similar to the electricity and heat energy production processes, repeated calculations should also be avoided when studying the GHG emission due to the material input. There is a development trend of the industrial parks that more and more industrial parks will generate internal logistics networks, and then the products, by-products, or waste of one enterprise will soon be used by another enterprise as raw material or energy. Therefore, in this study, the objects are the materials coming from the outside, and the calculation will cover all the GHG emission during the production and transportation processes that is usually defined as "from cradle to gate." GHG emission of this part could be calculated by

$$E^{m}_{\mathrm{GHG},i} = Q^{m}_{i} \times \mathrm{EF}^{m}_{\mathrm{GHG},i}, \quad (5)$$

where $E^{m}_{\mathrm{GHG},i}$ is the emissions of GHG during the production and transportation processes of type of material (kg), Q^{m}_{i} is the amount of material input, and $\mathrm{EF}^{m}_{\mathrm{GHG},i}$ is the emission factor.

2.4. Equipment Employment. Some equipment will be applied directly by enterprises for industrial production activities, such as blast furnace and converter that are essential to iron and steel industry; and the other equipment will be used to maintain the daily operation of the industrial parks such as water pump, draft apparatus, and so forth. Similar to the material input process, the calculation of equipment employment will also cover the "from cradle to gate" GHG emission. There are two methods to obtain the emission factors of equipment: (1) one is based on the input-output between the equipment production sector and other sectors, and the result will be in the form of GHG emission per unit currency, as described in [23, 24]; (2) the other one is to analyze the raw materials dosage and processing energy consumption of a specific equipment, and the result will be in the form of GHG emission per piece of equipment, as described in references [25–27].

2.5. Sewage Treatment and Solid Waste Disposal. The conventional disposal of solid waste is landfill, composting, and burning. The landfill and composting will mainly generate CH_4, and burning produces CO_2. Currently, landfill is the most popular method in the Chinese cities to dispose of the solid waste, and the technology of GHG collection under such condition is developing rapidly and being promoted widely. In this study, the GHG emission from solid waste disposal process will be calculated using the first-order attenuation method as recommended by IPCC as

$$E^{s}_{\mathrm{CH}_4} = \left(\sum_{x} E^{s}_{\mathrm{CH}_4,x,T} - R_T \right) \times \left(1 - \mathrm{OX}_T\right), \quad (6)$$

where $E^{s}_{\mathrm{CH}_4}$ is the emission of CH_4 from solid waste disposal (kg), $E^{s}_{\mathrm{CH}_4,x,T}$ is the emission of CH_4 from a type of solid waste disposal (kg), R_T is the recovery of CH_4 (kg), and OX_T is the oxidation factor (%).

The usual sewage anaerobic treatment will discharge CH_4 and N_2O, where the amount depends on the contents of biodegradable organics and nitrogenous substance in the sewage water. The calculation method recommended by IPCC is chosen to estimate the GHG emission from the sewage treatment process as

$$E^{w}_{\mathrm{CH}_4} = \mathrm{TOW} \times \mathrm{EF}^{w}_{\mathrm{CH}_4} - R, \quad (7)$$

where $E^{w}_{\mathrm{CH}_4}$ is the emission of CH_4 from sewage treatment (kg), TOW the total organic degradable material in wastewater (kg), $\mathrm{EF}^{w}_{\mathrm{CH}_4}$ the emission factor (kg/kg), R is the recovered CH_4 (kg) and

$$E^{w}_{\mathrm{N}_2\mathrm{O}} = \mathrm{N}^{w} \times \mathrm{EF}^{w}_{\mathrm{N}_2\mathrm{O}} \times \frac{44}{28}, \quad (8)$$

where $E^{w}_{\mathrm{N}_2\mathrm{O}}$ is the emission of N_2O from sewage treatment (kg), N^{w} is the nitrogen in effluent (kg), $\mathrm{EF}^{w}_{\mathrm{N}_2\mathrm{O}}$ is the emission factor (kg/kg), and44/28 is the conversion factor of kg N_2O-N into kg N_2O.

3. Case Study

3.1. Data Sources. The concerned industrial park is located in the southwest of Beijing. It is a typical high-end industrial park known for its good environmental quality and high-end industry concentration. The park has an area of 119235.6 m^2. The building density is 31.62%, while the landscaping ratio is 41%, which means 1 square meter of park area is covered by 0.3162 square meters of building and 0.41 square meters of landscape. All the data in our case industrial park are obtained by field research, while the emission factors from the public data sources. In order to compare the results, all the GHG emissions in this study are converted to the form of CO_2-eq based on the global warming potential (GWP) recommended by IPCC [28].

3.1.1. Energy Consumption. The emission factors of primary energy are specific for China, which are referenced to the default emission factors by IPCC and the average lower heat values in China's energy statistics yearbook [29]. The emission factors of the secondary energy (such as the electricity and heat) are obtained based on the amount of primary energy consumed by the electricity and heat production sector in Beijing. The basic data can be found in the reports of the National Development and Reform Commission and *Beijing Statistical Yearbook* [30].

3.1.2. Industrial Production. The industrial park is designed with high-end positioning that is a well-operated cluster of the head offices or research and development centers of high-tech industries. Thus, the GHG emission of industrial production process in our case is neglected.

3.1.3. Material Input. As discussed in Section 3.1.2, there are few industrial production processes in such high-end industrial parks. We thus have the hypothesis that the raw material input necessary for industrial production can be ignored. As a result, this study focuses on the GHG emission from the construction materials input during the construction stage. The emission factors of the construction materials can be divided into 3 types [31–43].

Type 1. The products are easy to decide their boundaries and technological processes to choose the precise matching emission factors.

Type 2. The products are not well studied by life-cycle analysis but the production technological processes are clear. We can obtain the emission factors based on the amount of energy consumption and raw material input, on the condition that these energy consumption and raw material input are of Type 1.

Type 3. The products have never been studied. We have to select an alternative that has similar production process, raw material, and function, with an assumption that the emission factors of them are the same.

3.1.4. Equipment Employment. The method with which we obtained the emission factors of the equipment is the same with the material input. However, the equipment has a more complicated production process and various specifications and types, so it is difficult to make clear the exact emission factors of each piece of equipment [44–47]. As a result, the majority of equipment belongs to Type 3, and the uncertainty of the calculation results is increased.

3.1.5. Sewage Treatment and Solid Waste Disposal. According to the investigation, all the sewage generated by the case industrial park is discharged into the municipal drainage sewage pipe network, so we assume that the emission factors of the sewage from the industrial park are the same as the average of the municipal sewage in Beijing. With reference to the literatures [48, 49], the emission factors of the sewage treatment process are $(5.61E-03)$ kg CO_2/kg, $(8.89E-08)$ kg N_2O/kg, and $(1.59E-02)$ kg CH_4/kg.

Similarly, the emission factors of the solid waste disposal process are assumed to be the same as the average level in Beijing. With reference to the literatures [50], the CH_4 correction factor (MCF) is 1.0, the dissolved organic carbon (DOC) 6.5%, the fraction of DOC dissolved (DOC_f) 0.5, the CH_4 volume fraction of the landfill gas 0.5, and the oxidation factor (OX_T) 0.1.

3.2. Results and Discussion. The GHG emission of each stage in the life cycle of the industrial park is calculated based on the proposed method. The overall GHG emission of the lifecycle is 1872177 t CO_2-eq. The construction stage takes up 4.546%, which amounts to 85105.82 t CO_2-eq GHG emission with an intensity of 801.69 kg CO_2-eq/m^2; the demolition stage takes up 0.102%, contributing 1917.3 t CO_2-eq GHG emission with an intensity of 18.06 kg CO_2-eq/m^2. As can be seen, the operation stage contributes the majority of GHG emission, which achieves a proportion of 95.352%. The GHG emission amount of operation stage is 37717.18 t CO_2-eq, with the intensity being 355.29 kg CO_2-eq/m^2.

The construction stage of the industrial park is decomposed into 12 steps with different functions, which are structure (S), indoor decoration (ID), outdoor decoration (OD), building electric (BE), building water supply and drainage (BWSD), heating (H), ventilation (V), fire protection (FP), road (R), municipal electric (ME), municipal water supply and drainage (MWSD), and landscaping (L). During the construction stage, GHG emission from construction material input process is accounted to be 82509 t CO_2-eq, which takes the 96.95% part of the stage. The contribution of each step to the amount can be found in Figure 1. As it is shown, the top three emission sources are S (59.71%), ID (20.33%), and OD (11.40%) and then L (3.74%), V (1.78%), and R (1.09%). The other six steps only take up the proportion of less than 1%.

The operation stage should be of the focus of GHG emission reduction in the industrial park for its significant contribution. In order to get more specific and meaningful results, the overall GHG emission of this stage is further decomposed into seven processes. As shown in Figure 2, the processes of sewage treatment, heat energy consumption and electricity consumption should be paid more attention, which contributes to 98.69% of the operation stage emission.

For the operation stage, some strategies are adopted to change the energy consumption and sewage treatment as shown in Table 1. The GHG emission of operation stage will decline as shown in Figure 3. The overall emission of operation stage will be reduced to 27443.58 t in 2020 and 17711.66 t in 2050, which are just 72.76% and 46.96% of those in 2011.

4. Conclusions

For a high-end industrial park, the GHG emission of the construction stage is very intensive with significant environmental impacts, which is a key point to be reckoned with by the administrators. Based on the calculation results, we

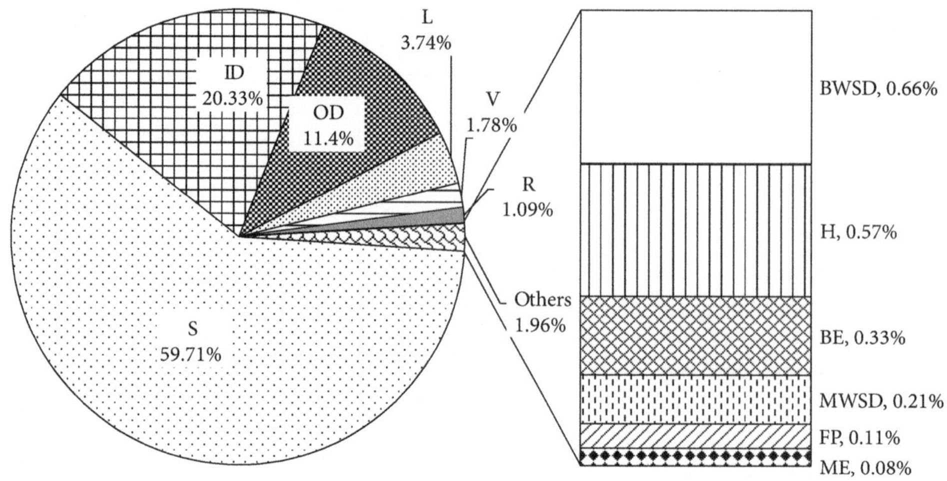

FIGURE 1: GHG emission from construction material of construction stage.

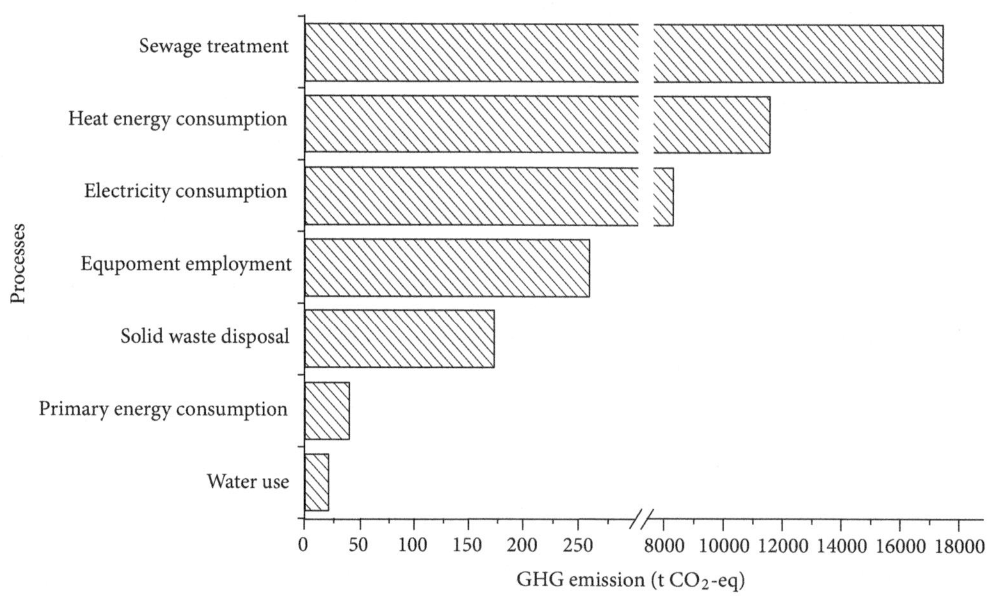

FIGURE 2: GHG emission of operation stage.

can control the GHG emission of construction stage in the following ways.

(a) *Using local construction materials to reduce the GHG emission from the transportation processes of the construction materials.* Most construction materials are of a large size and usually in high demand. Thus, there will be a considerable quantity variance of GHG emission between different transport distances. Employing the local construction materials can not only save the transportation cost and transportation time, but also reduce the GHG emissions, which may achieve much more benefits in economy and environment.

(b) *Employing the low-carbon and regeneration construction materials instead of the traditional ones to reduce the GHG emission from the upstream production process and downstream disposal.* After the quantitative evaluation of performance and cost, decision makers may prefer the low-carbon and regeneration construction materials. The low carbon and regeneration characteristics imply that these materials consume less energy and fewer resources during the production processes compared to the traditional materials, thereby having a better performance and lower price.

(c) *Optimizing the construction progress to promote a safe and low-carbon form of construction engineering.* The arrangement of the construction schedule can be optimized to ensure the project to be finished on time. For example, less night work can

Table 1: Scenario analysis of operation stages.

Emission source	Parameter	2010	2020	2030	2040	2050
Electricity consumption	Emission factor (g CO_2-eq/kWh)	996.36	862.52	759.91	654.78	552.17
Heat energy consumption	Emission factor (kg CO_2-eq/GJ)	101.43	44.88	39.27	33.66	28.05
Sewage treatment	Emission factor (g CO_2-eq/kg sewage)	435.45	413.20	392.21	372.22	353.21

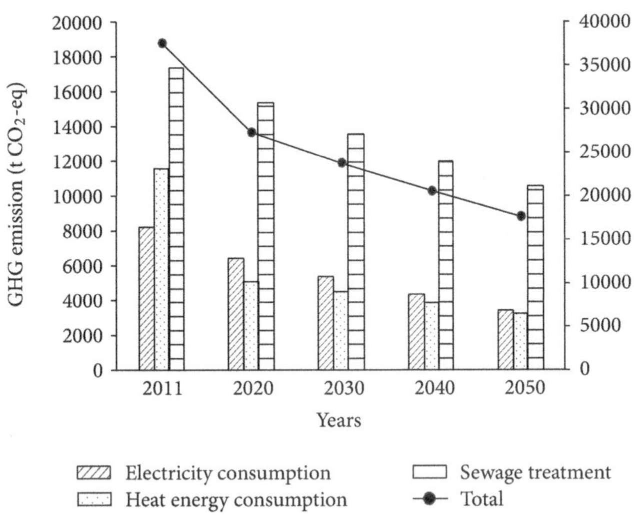

Figure 3: GHG emission of operation scenarios of different strategies.

avoid unnecessary power consumption and noise pollution.

The administrators of industrial parks should take the responsibilities to reduce the GHG emission during the operation stage, which normally lasts for about 40 to 50 years. As implied by the accounting result, more attention should be paid to the processes of sewage treatment, heat energy consumption, and electricity consumption when controlling the GHG emission of this stage. For example, the industrial park in our case produces 40,000 t sewages every year, and the treatment process of the sewage will result in a GHG emission of 17373.46 t CO_2-eq. Therefore, the administrator can control the emission by bringing in water saving technology and building the recovery system of waste water.

Acknowledgments

This study was supported by the Fund for Creative Research Groups of the National Natural Science Foundation of China (no. 51121003), Key Program of National Natural Science Foundation (nos. 50939001, 41271543), National Key Technology Research and Development Program (no. 2012BAK30B03), and Program for New Century Excellent Talents in University (NCET-09-0226). The authors are especially grateful for the financial support from Beijing Development Area Co., Ltd.

References

[1] The State Council Information Office, China's Climate Change Policies and Actions, 2011, http://www.gov.cn/jrzg/2011-11/22/content_2000047.htm.

[2] The State Council Information Office of the People's Republic in China, "China's climate change policies and actions," White paper, 2011, http://www.gov.cn/jrzg/2011-11/22/content_2000047.htm.

[3] Ministry of Science and Technology Web Portal, "China's response to climate change science and special action," Published by 14 Ministries, 2007, http://www.most.gov.cn/shfzs/sfqqhj/200708/t20070812_52141.htm.

[4] Information office of the state council, China's climate change policies and actions, 2008, http://www.gov.cn/zwgk/2008-10/29/content_1134378.htm.

[5] The central government portal website, The state council board of managing directors research in our country to control greenhouse gas emissions targets, 2009, http://www.gov.cn/ldhd/2009-11/26/content_1474016.htm.

[6] China' news network, The 12th five year plan: China will greatly reduce the energy consumption intensity, 2011, http://www.chinanews.com/ny/2011/03-05/2886184.shtml.

[7] The central government portal website, The state council board of managing directors research in our country to control greenhouse gas emissions targets, 2009, http://www.gov.cn/ldhd/2009-11/26/content_1474016.htm.

[8] Chinese Academy of Sciences, the Sustainable Development Strategy, *2009 China Strategy of SuStainable Development Report—Explore China's characteristics of Low Carbon Road*, Science press, Beijing, China, 2009.

[9] H. X. Gao, *Philosophical Thinking of Low Carbon Society*, Party School of the CPC Central Committee, Beijing, China, 2011.

[10] Y. Geng and Z. Hengxin, "Industrial park management in the Chinese environment," *Journal of Cleaner Production*, vol. 17, no. 14, pp. 1289–1294, 2009.

[11] M. Fujita and J. F. Thisse, *Economics of Agglomeration: Cities, Industrial Location, and Regional Growth*, Cambridge University Press, Cambridge, UK, 2002.

[12] C. C. Fan and A. J. Scott, "Industrial agglomeration and development: a survey of spatial economic issues in East Asia and a statistical analysis of Chinese Regions," *Economic Geography*, vol. 79, no. 3, pp. 295–319, 2003.

[13] The territory of the People's Republic of the national development and reform commission news center, The national development and reform commission officials answers to the rectification issues in development zone, 2007, http://xwzx.ndrc.gov.cn/zcjd/t20070429_133207.htm.

[14] "The territory of the People's Republic of the national development and reform commission news center, land intensive utilization about the development zone in 2010," The

ministry of land and resources, 2011, http://xwzx.ndrc.gov.cn/ywdt/t20111116_444930.htm.

[15] The ministry of land and resources of the People's Republic of China, distribution chart of Chinese development zone, 2008, http://www.mlr.gov.cn/tdzt/tdgl/direction/200801/t20080103_667861.htm.

[16] H. M. Chen, "Analysis on embodied CO_2 emissions including industrial process emissions," *Chinese Journal of Population Resources and Environment*, vol. 19, no. 3, pp. 25–30, 2009.

[17] H. S. Matthews, C. T. Hendrickson, and C. L. Weber, "The importance of carbon footprint estimation boundaries," *Environmental Science and Technology*, vol. 42, no. 16, pp. 5839–5842, 2008.

[18] Y. A. Huang, C. L. Weber, and H. S. Matthews, "Categorization of scope 3 emissions for streamlined enterprise carbon footprinting," *Environmental Science and Technology*, vol. 43, no. 22, pp. 8509–8515, 2009.

[19] Y. A. Huang, C. L. Weber, and H. S. Matthews, "Carbon footprinting upstream supply chain for electronics manufacturing and computer services," in *Proceedings of the IEEE International Symposium on Sustainable Systems and Technology (ISSST '09)*, pp. 1–6, May 2009.

[20] L. P. Ju, *Evaluation and Prediction of City Greenhouse Gas Emission-a Case Study of Chongqing*, Beiing Normal University, Beijing, China, 2011.

[21] World Resoures Institute (WRI) & World Business Council for Sustainable Development (WBCSD), Corporate Value Chain (Scope 3) Accounting and Reporting Standard, USA, 2011.

[22] IPCC, *2006 IPCC Guidelines For National Greenhouse Gas Inventories*, Institute for Global Environmental Strategies, Hayama, Japan, 2006, http://www.ipcc-nggip.iges.or.jp/public/2006gl/index.html.

[23] Guillaume FABRE, LCB Method 3.0., 2011.

[24] G. Q. Chen, *Systematic Caculation on Energy Consumption Asn Carbon Emission for Buildings*, Xinhua Press, Beijing, China, 2010.

[25] L. J. Gu, *Impact Study of Chinese Building Industry Environmental Based on Life Cycle Assessment*, Tsinghua University, Beijing, China, 2009.

[26] Z. J. Li, *Study on the Life Cycle Consumption of Energy and Resource of Air Conditioning in Urban Residential Buildings in China*, Tsinghua University, Beijing, China, 2007.

[27] Y. B. Liu, X. T. Wang, B. R. Lin, and B. Peng, "Research progress in life cycle energy consumption data of building subassembly and equipment," *Building Science*, vol. 27, no. 2, pp. 255–262, 2011.

[28] IPCC, *2006 IPCC Guidelines For National Greenhouse Gas Inventories*, Institute for Global Environmental Strategies, Hayama, Japan, 2006.

[29] National Bureau of Statistics of China, *China's Energy Statistics Yearbook 2010*, China Stastices Press, Beijiing, China, 2011.

[30] Beijing Buoeau of Statistics, Beijing statistical yearbook 2011, http://www.bjstats.gov.cn/nj/main/2011-tjnj/index.htm.

[31] G. Hammond and C. Jones, *Inventory of Carbon & Energy 2.0*, University of Bath, Bath, UK, 2011.

[32] M. Nisbet, M. Marceau, and M. VanGeem, *Environmental Life Cycle Inventory of Portland Cement Concrete*, Portland Cement Association, Skokie, Ill, USA, 2002.

[33] Concrete Pipeline Systems Association, Cradle-to-Gate Analysis for Concrete Pipeline, Manhole Ring, and Cover Slabs, 2010.

[34] Quarry Products Association, Sustainable Development Report, 2008.

[35] Department for Environment Food and Rural Affairs/Department of Energy & Climate Change, Guidelines to Defra/DECC's GHG Conversion Factors for Company Reporting: Methodology Paper for Emissions Factors, 2010.

[36] U.S. Energy Information Administration, Manufacturing Energy Consumption Survey, 2002.

[37] G. M. Nicoletti, B. Notarnicola, and G. Tassielli, "Comparative Life Cycle Assessment of flooring materials: ceramic versus marble tiles," *Journal of Cleaner Production*, vol. 10, no. 3, pp. 283–296, 2002.

[38] Enviros Consulting, Glass Recycling-Life Cycle Carbon Dioxide Emissions, 2003.

[39] M. McGill, *Carbon Footprint Analysis of Textile Reuse and Recycling*, Imperial College London, London, UK, 2008.

[40] R. J. Murphy and A. Norton, *Life Cycle Assessment of Natural Fiber Insulation Materials*, 2008.

[41] A. C. Schmidt, A. A. Jensen, A. U. Clausen, O. Kamstrup, and D. Postlethwaite, "A comparative life cycle assessment of building insulation products made of stone wool, paper wool and flax: part 1: background, goal and scope, life cycle inventory, impact assessment and interpretation," *International Journal of Life Cycle Assessment*, vol. 9, no. 1, pp. 53–66, 2004.

[42] J. B. Wilson, "Life-cycle inventory of particleboard in terms of resources, emissions, energy and carbon," *Wood and Fiber Science*, vol. 42, no. 1, pp. 90–106, 2010.

[43] Natural Stone Council, Limestone Quarrying and Processing: A Life Cycle Inventory, 2008.

[44] Z. J. Li, *Study on the Life Cycle Consumption of Energy and Resource of Air Conditioning in Urban Residential Buildings in China*, Tsinghua University, Beijing, China, 2007.

[45] J. Deng, W. Li, and Z. Zhang, "Gas engine driven heat pump system life cycle assessment," *Journal of Xi'an Jiaotong University*, vol. 40, no. 11, pp. 1258–1262, 2006.

[46] Y. B. Liu, X. T. Wang, B. R. Lin, and B. Peng, "Research progress in life cycle energy consumption data of building subassembly and equipment," *Building Science*, vol. 27, no. 2, pp. 255–262, 2011.

[47] L. J. Gu, *Impact Study of Chinese Building Industry Environmental Based on Life Cycle Assessment*, Tsinghua University, Beijing, China, 2009.

[48] X. Ma, *Greenhouse Gas Emission Analysis from Municipal Sewage Treatment Plants of China*, Beijing Forestry University, Beijing, China, 2011.

[49] C. Zhang, Y. Chen, and X. J. Tan, "Discussion on greenhouse gas generation and reduction strategies in urban wastewater system," *Water Supply and Drain*, vol. 36, no. 9, pp. 29–33, 2010.

[50] W. P. Du, *Primary Study on Greenhouse Gas of MSW*, Nanjing University of Information Science &Technology, Nanjing, China, 2006.

19

An Assessment of Japanese Carbon Tax Reform Using the E3MG Econometric Model

Soocheol Lee,[1] Hector Pollitt,[2] and Kazuhiro Ueta[3]

[1] *Faculty of Economics, Meijo University, 1-501 Shiogamaguchi Tenpak-ku, Nagoya 468-8502, Japan*
[2] *International Modelling, Cambridge Econometrics, Covent Garden, Cambridge CB1 2HT, UK*
[3] *Graduate School of Economics, Kyoto University, Yoshida-Honmachi Sakyo-ku, Kyoto 606-8501, Japan*

Correspondence should be addressed to Soocheol Lee, slee@meijo-u.ac.jp

Academic Editors: B. Chen, Z.-M. Chen, and H.-S. Tang

This paper analyses the potential economic and environmental effects of carbon taxation in Japan using the E3MG model, a global macroeconometric model constructed by the University of Cambridge and Cambridge Econometrics. The paper approaches the issues by considering first the impacts of the carbon tax in Japan introduced in 2012 and then the measures necessary to reduce Japan's emissions in line with its Copenhagen pledge of −25% compared to 1990 levels. The results from the model suggest that FY2012 Tax Reform has only a small impact on emission levels and no significant impact on GDP and employment. The potential costs of reducing emissions to meet the 25% reduction target for 2020 are quite modest, but noticeable. GDP falls by around 1.2% compared to the baseline and employment by 0.4% compared to the baseline. But this could be offset, with some potential economic benefits, if revenues are recycled efficiently. This paper considers two revenue recycling scenarios. The most positive outcome is if revenues are used both to reduce income tax rates and to increase investment in energy efficiency. This paper shows there could be double dividend effects, if Carbon Tax Reform is properly designed.

1. Introduction

In recent years, the Japanese Government has proposed to introduce low-carbon policy instruments such as a carbon tax and ETS by introducing the bill of the Basic Act on Global Warming Countermeasures in 2009. The carbon tax plan was approved at the cabinet meeting in December 2011 (instead of ETS) due to strong ETS opposition from business circles. The bill has now been passed by the Japanese diet (March, 2012), making the carbon tax the first one to be introduced in Asia. The effects of carbon taxes on GHG emissions and the wider economy are now attracting the notice of many researchers and policy makers in this area.

The aim of this paper is to analyse the potential economic and environmental effects of implementing low-carbon policies, notably carbon taxation, in Japan. The analysis provides an input to the discussion by presenting a quantitative assessment of carbon taxation in Japan. The approach is model based and uses E3MG, a global macroeconometric model that links the world's economies to their energy systems and associated emissions. The assessment approaches the issue from two angles by considering first the impacts of the tax increases put forward in 2010 and then the measures that would be needed to reduce Japan's emissions in line with its Copenhagen pledge of −25% compared to 1990 levels.

This analysis builds on previous work by, among others, Park [1], Kawase et al. [2], and Takeda [3], who examined the effects of ETR using CGE (Computable General Equilibrium) models. E3MG shares many of the features of such models, for example, GTAP [4], the Monash model, and GEM-E3 [5], but relaxes some of the common assumptions to the CGE approach (such as fully rational behaviour), instead using an empirical approach (see Section 3). The structure of E3MG makes it more similar to the approach used in Sugawara [6], but E3MG is much larger in scale and with a much larger disaggregation of sectors.

The paper is structured as follows: Section 2 summarises the policy environment in Japan that provides the basis for this analysis; Section 3 describes the E3MG model, while Sections 4 and 5 describe the scenarios that were assessed and the results from these scenarios, respectively; Section 6 concludes.

2. Background: Japanese Climate Change Policy

2.1. A Brief Description of Japanese Climate Change Policy. Japan has been pushing for the adoption of measures to combat climate change since 1997, just after the agreement of Kyoto Protocol. The Law Concerning the Promotion of the Measures to Cope with Global Warming, for example, became effective in 1998. However, this law required business circles and householders to try to reduce GHG emissions without any imperative policy instruments like carbon taxes or ETS. Emission reduction efforts from business have depended on the Voluntary Action Plan led by Keidanren (Keidanren is the most influential general business association in Japan to which 1,281 leading companies and 127 industrial associations belong). Reductions from households have depended on government-led campaigns such as Cool Business, which requests people to control their office and room air conditioner over 28°C with light dress (e.g., no necktie and no jacket, etc.) during the summer season.

In 2009, the Japanese Government submitted the bill of the Basic Act on Global Warming Countermeasures to parliament. The bill outlined a mid-term goal to reduce GHG emissions by 25% below the 1990 level in 2020 and a long-term goal of 80% below the 1990 level in 2050. Further goals include raising the share of renewable energy within total primary energy supply to 10% by 2020. Measures to achieve these targets have included proposals for carbon taxes and also an emissions trading scheme (ETS). The Committee on Institutional Design for Emissions Trading was established in 2000 by the Environmental Agency and began examining the introduction of an ETS at national level. The committee investigated an ETS as a domestic measure to achieve Japan's greenhouse gas reduction target under the Kyoto Protocol (see Committee on Institutional Design for Emissions Trading [7]).

In 2010, the Ministry of the Environment in Japan proposed an ETS that would be implemented in 2013. However, despite strong requests to implement an ETS from environmental NGOs and academia, the ruling party (i.e., the Democratic Party of Japan) proposed a postponement of its implementation on December 17, 2010, due to opposition from Keidanren and industry as well as potential negative economic impacts (advocates for the implementation of a cap-and-trade ETS include the Kiko Network [8], an environmental NGO, and Morotomi and Ayukawa [9]). At the end of 2010, the Japanese government opted to introduce a carbon tax in 2011 instead of an ETS. However, the implementation of this tax was put off until 2012 by the influence of the 2011 tsunami in Japan.

Nevertheless, from the perspective of promoting global warming mitigation measures and pushing for energy conservation after the 2011 tsunami, there was a determined push in Japan to introduce a carbon tax (Special Provisions for Carbon Dioxide Tax of Global Warming Measures), and the Japanese FY2012 Tax Reform Revision passed the House of Councilors on March 30, 2012. It will be implemented from October 1, 2012. As a result, the government will increase the rates, depending on carbon content, of the present Petroleum and Coal Tax, which is imposed on all fossil fuels.

Table 1: FY2012 proposed tax rates.

	Crude petroleum/petroleum products (per kilo liter)	Gaseous hydrocarbon (per ton)	Coal (per ton)
Present	JPY 2,040	JPY 1,080	JPY 700
October 1, 2012	JPY 2,290	JPY 1,340	JPY 920
April 1, 2014	JPY 2,540	JPY 1,600	JPY 1,140
April 1, 2016	JPY 2,800	JPY 1,860	JPY 1,370

The first scenario in this paper simulates the potential effects of FY2012 Tax Reform.

2.2. Carbon Tax Plan Details. The FY2012 Tax Reform (see http://www.mof.go.jp/english/tax_policy/tax_reform/fy2012/tax2012a.pdf) specifies that the Japanese Government will introduce a Carbon Dioxide Tax with the aim of controlling energy-originated CO_2 emissions, which currently account for around 90% of GHG emissions.

The government will add the following tax rates, corresponding to the amount of CO_2 emissions, to existing fossil fuel prices (including existing taxes):

(i) crude oil, petroleum products: JPY 760/kl,

(ii) gaseous hydrocarbons: JPY 780/t,

(iii) coal: JPY 670/t.

The additional taxes are set to be introduced, in part, on October 1, 2012 and increased progressively such that they are fully implemented by 2016. Table 1 specifies the interim measures in further detail. These figures are used to define the first scenario in this paper.

3. The E3MG Model

This section briefly describes the E3MG model that was used to carry out the analysis. For further information about the model, the reader is referred to Barker et al. [10] and the website http://www.e3mgmodel.com

3.1. Basic Model Structure. The E3MG model (energy-environment-economy model at the global level) is a computer-based tool that has been constructed by international teams at the University of Cambridge and Cambridge Econometrics. The model is econometric in design and is capable of addressing issues that link developments and policies in the areas of energy, the environment, and the economy. The essential purpose of the model is to provide a framework for policy evaluation, particularly policies aimed at achieving sustainable energy use over the long term. However, the econometric specification that the model uses also allows for an assessment of short-term transition effects.

The current version of E3MG consists of 22 world regions, although in this analysis we focus solely on Japan. The basic structure of E3MG is presented in Figure 1. The model integrates energy demand and emissions with the economy; fuel demand is determined by prices and economic activity with feedback through the energy supply sectors. Energy combustion results in greenhouse gas emissions.

The economic module in E3MG contains a full representation of the National Accounts, as formulated in Cambridge by Richard Stone, and formally presented in European Communities et al. [11]. A key feature of E3MG is its sectoral disaggregation, with 42 economic sectors, linked by input-output relationships; this aspect is particularly important in modelling carbon taxes as the different sectors use different fuels in varying degrees of intensity and have different technological options for changing consumption patterns.

Exogenous inputs to the model include population, government tax and spending rates, and international energy prices. The outputs include a range of economic and labour market indicators, defined at sectoral level, plus indicators for energy consumption and emissions.

Figure 2 shows the mechanism through which a carbon tax could affect macroeconomic outcomes. The taxes are levied on consumption of fuel use, leading to reductions in fuel demand but also higher costs for industries and households. Higher industry costs may be absorbed as loss of profits or passed on to final consumers. Higher prices mean losses of real output for domestic consumers and for exporters.

However, the revenues from carbon taxes may also be used to reduce other tax rates, with positive economic benefits. In the scenarios in this paper, a large share is used to reduce income taxes. The effects this has on the economy are shown in Figure 3; reduced income taxes lead to higher incomes, which are spent on consumer goods and lead to increases in domestic production, creation of jobs, and further income rises (i.e., a multiplier effect).

E3MG's treatment of energy demand is largely top-down in nature. Econometric equations are estimated for aggregate energy demand and demand for the four main fuel types (coal, fuel oil, natural gas, electricity). Energy demand, for 19 different user groups, is a function of economic activity, relative prices, and measures of technology. The model solves all equations simultaneously and adjusts the individual fuels to sum to the total for each user. Feedbacks to the economy are provided by adjusting input-output coefficients and household energy demand.

The following equations provide an example of E3MG's econometric, error-correction equations for aggregate energy consumption (*EnCon*) at time t. First a long-run equation is estimated based on levels of economic activity (*Act*), energy prices (*EnPrice*), investment (*Inv*), and *R&D*. The lagged errors from this equation (e) are then used in the short-run equation which uses differences of the same independent variables, plus the lagged dependent variable.

Long Run:

$$\begin{aligned} EnCon_t &= a_1 + (b_1 * Act_t) + (b_2 * EnPrice_t) \\ &\quad + (b_3 * Inv_t) + (b_4 * R\&D_t) + e_t. \end{aligned} \tag{1}$$

Short Run:

$$\begin{aligned} \Delta EnCon_t &= a_2 + (b_1 * \Delta Act_t) + (b_2 * \Delta EnPrice_t) \\ &\quad + (b_3 * \Delta Inv_t) + (b_4 * \Delta R\&D_t) + (b_5 * e_{t-1}) \\ &\quad + (b_6 * \Delta EnCon_{t-1}) + \in_t. \end{aligned} \tag{2}$$

The exception to this top-down treatment is in power generation, as the historical data do not provide the basis to estimate econometric equations in new technologies. In this sector E3MG includes a bottom-up representation with 28 specific generation technologies, made up of both conventional and renewable supplies. The model bases future investments on the relative prices of each technology, including the effects of carbon taxation. This part of the model is described in Barker et al. [12].

Emissions are estimated using a fixed coefficient to fuel demand. Nonenergy emissions are included in the model so that global totals are met but are treated as exogenous in this paper.

E3MG also includes endogenous measures of sectoral technological progress. The indices used in the model are functions of accumulated capital, enhanced by R&D, adapted from Lee et al. [13]. Endogenous technological progress is allowed to influence several of the model's equation sets, including energy demand, international trade, price formation, and the labour market.

3.2. Data Sources and Equation Estimation. As an econometric model with sectoral detail, E3MG requires extensive data inputs. A large time-series database covering 1970–2008 annually (with more recent aggregate figures where available) has been constructed, in the main based on international datasets. For Japan the main data source for economic data is the OECD Structural Analysis database, with other macro-level indicators being obtained from the IMF and the World Bank. If there are gaps in the data these are filled using national figures. The main cross-sectional data (the input-output table and bilateral trade flows) are sourced from the OECD.

The main source for energy data is the IEA. CO_2 emissions have also been made consistent with IEA figures.

E3MG consists of 22 estimated sets of equations (each disaggregated by sector and by country). These cover the components of GDP, prices, the labour market, and energy demand.

The estimation method utilises developments in time-series econometrics, in which dynamic relationships are specified in terms of error correction models (ECM) that allow dynamic convergence to a long-term outcome.

The specific functional form of the equations is based on the econometric techniques of cointegration and error-correction, particularly as promoted by Engle and Granger

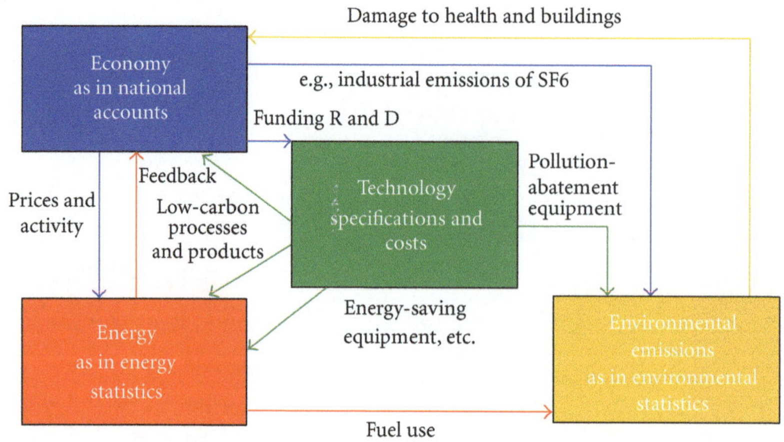

FIGURE 1: E3 interactions within E3MG.

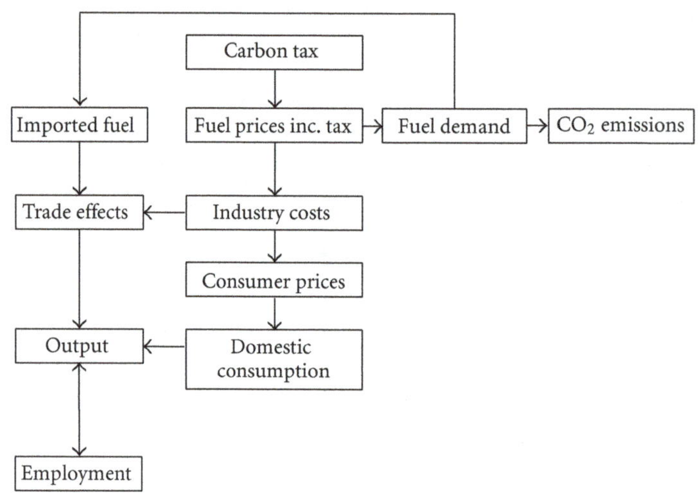

FIGURE 2: Potential effects of a carbon tax.

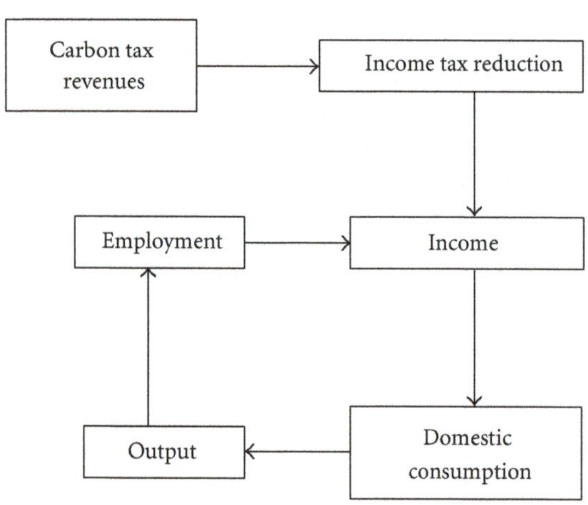

FIGURE 3: Potential effects of revenue recycling.

[14] and Hendry et al. [15]. In brief, the process involves two stages. The first-stage is a levels relationship, whereby an attempt is made to identify the existence of a cointegrating relationship between the chosen variables, selected on the basis of economic theory and a priori reasoning. For example, for employment demand the list of variables contains real output, real wage costs, hours worked, energy prices, and a measure of technological progress. If a cointegrating relationship exists, then the second stage regression is known as the error-correction representation and involves a dynamic, first-difference, regression of all the variables from the first stage, along with the lagged difference in the dependent variable, and the error-correction term (the lagged residual from the first stage regression).

3.3. Previous Analysis with E3MG. The E3MG model has been under development for much of the past decade. It is now used for policy analysis at European level, including the 2010 European Commission communication on the impacts of moving to a 30% GHG target. The model has also been used repeatedly for assessing decarbonisation pathways at different international levels [10] and in the UK [16]. Most recently E3MG was applied in Barker et al. [17] to provide an economic assessment of the IEA's 450 ppm scenario [18].

Also of potential application to these scenarios and their underlying policy context is the model's assessment of rebound effects [23]. In this paper the E3MG model was used to show that long-run rebound effects can cancel out up to 50% of the environmental gains from efficiency measures; the analysis goes on to recommend carbon pricing as a means to reduce the rebound effect.

3.4. Comparison to CGE Modelling. In terms of basic structure, purpose, and coverage, there are many similarities between E3MG and comparable CGE models, such as GTAP [4], the Monash model, and GEM-E3 [5]. Each is a computer-based economic model that considers energy-environment-economy interactions at the global level, broken down into sectors and world regions. In addition the regional and sectoral disaggregations are broadly similar. Both modelling approaches are based on a consistent national accounting framework and make use of similar national accounts data.

However, beneath the surface there are substantial differences in modelling approach and it is important for the reader to be aware of this when interpreting results. The two types of model come from distinct economic backgrounds. While the models are quite consistent in their accounting, identity balances, they differ substantially in their treatment of unobservable behavioural relationships. The CGE model favours setting these in line with economic theory, for example, by assuming that individuals act rationally in their own self-interest. In contrast, the econometric model interrogates historical datasets to try to determine these factors on an empirical basis.

Both approaches have their relative strengths and weaknesses; for example, the assumption of optimising rational behaviour in CGE models has been increasingly questioned since the recession, while econometric models are reliant on having high-quality time-series data. Although subtle, these differences in theoretical approach can lead to different conclusions being drawn from the model results; for example, the econometric model does not assume optimal behaviour in the baseline, implying that negative-cost emission reductions are available. Jansen and Klaassen [24] and Bosetti et al. [25] describe some of the differences in the context of ETR, including revenue recycling options.

This distinction is important when comparing the analysis in this paper to previous model-based assessments in Japan, which have almost exclusively used a CGE approach, as discussed in Section 1. In Europe it is now common for CGE and macroeconometric models to be run in tandem so that results are not dependent on a single set of modelling assumptions (e.g., [19]).

4. Scenarios

4.1. Baseline. The baseline that has been used for this analysis has been scaled to be consistent with the current policies scenario in *World Energy Outlook*, 2010 ([18], henceforth referred to as WEO). E3MG's equation results are set to match WEO figures for energy demand and emissions, but also to use the same figures for economic drivers in order to retain consistency throughout the model's internal relationships. In summary, by 2020,

(i) total energy-related CO_2 emissions are expected to fall to 998 mtCO_2, from 1,147 mtCO_2 in 2008 and 1,063 in 1990;

(ii) total primary energy demand is expected to be stable at current levels;

(iii) final consumption of oil is expected to slowly decrease, but consumption of electricity to increase;

(iv) additional electricity will be generated from nuclear power.

The 2010 edition of WEO does not include the latest data regarding the recession or the impacts of the Fukushima nuclear accident (which clearly raises questions on the suitability of the final bullet point) but, while this may impact on the magnitude of results, in our view it does not change the direction of results, or the overall conclusions from this paper. Nevertheless, one of our recommendations is to repeat this exercise once there is more certainty about the global economy and Japanese energy and climate policy.

4.2. Policy Scenarios. We consider four policy scenarios. The first scenario assesses the economic and environmental impacts of the tax increases that were announced for FY2012. The other scenarios consider the measures that would be necessary to reduce GHG emissions by 25% from 1990 levels, in line with Japan's Copenhagen pledge. There are three variants of this scenario, one with the carbon tax levied on its own and two options with different methods of revenue recycling (where revenues from carbon taxes are used to reduce revenues from other taxes). In scenario 2b, some of the revenues are used to fund a public investment programme in energy efficiency (the investment in energy efficiency is assumed to lead to reductions in energy demand using the ratios of investment and energy savings for OECD countries published in IEA [18]) in buildings, while in S2c they are used to reduce labour costs.

The scenarios are summarised in Table 2, while Table 3 illustrates the tax inputs for each scenario. Inputs for S2a, S2b, and S2c are not derived from Japanese policy but are estimated by E3MG based on achieving the CO_2 emissions reduction target of 25% in 2020 compared to 1990.

5. Results

5.1. Environmental Impacts. Figure 4 shows the impact on energy-related CO_2 emissions, compared to baseline, in each of the scenarios.

The first result is that the FY2012 measures that are modelled in S1 have only a small impact on emissions levels; this is because their impact on fuel prices (i.e., the relative effect once the taxes are added to fuel costs and existing taxes) is very small, in the range of 1–3%. This provides little incentive for behavioural change, and there are only small reductions in fuel consumption. In the other scenarios, emissions fall by slightly more than 20% compared to baseline (which is 25% below 1990 levels). However, quite

TABLE 2: Summary of scenarios.

	Carbon tax rates	Revenue recycling
S1	FY2012 reform	None
S2a	To reach 25% GHG reduction	None
S2b	To reach 25% GHG reduction	95% of revenues used to reduce income taxes, 5% used for investment in energy efficiency
S2c	To reach 25% GHG reduction	75% of revenues used to reduce income taxes, 25% used to reduce employers' social security contributions

TABLE 3: E3MG carbon tax inputs (JPY/toe).

	S1			S2a	S2b	S2c
	Oil	Gas	Coal	All fuel types		
2012	325	305	371	17,721	20,529	19,812
2013	325	305	371	26,982	25,843	29,316
2014	651	610	743	29,508	29,215	32,487
2015	651	610	743	30,709	28,652	32,468
2016	989	915	1,131	32,500	27,460	33,318
2017	989	915	1,131	35,257	28,750	35,868
2018	989	915	1,131	39,749	33,140	41,531
2019	989	915	1,131	43,192	36,528	45,426
2020	989	915	1,131	44,240	35,615	45,811

Notes: Figures based on 1 USD = 76.8 JPY.
S1 values derived from FY2012 tax rates.
S2 values required to achieve 25% reduction in CO_2 emissions between 1990 and 2020.
Source(s): E3MG, Cambridge Econometrics.

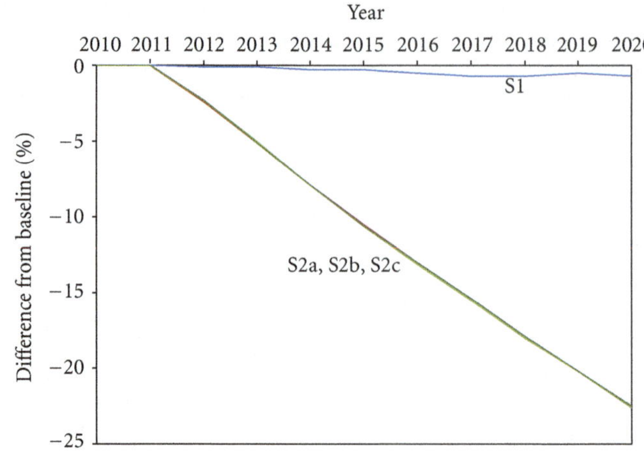

Source(s): E3MG, Cambridge Econometrics

FIGURE 4: Energy-based CO_2 emissions, Japan.

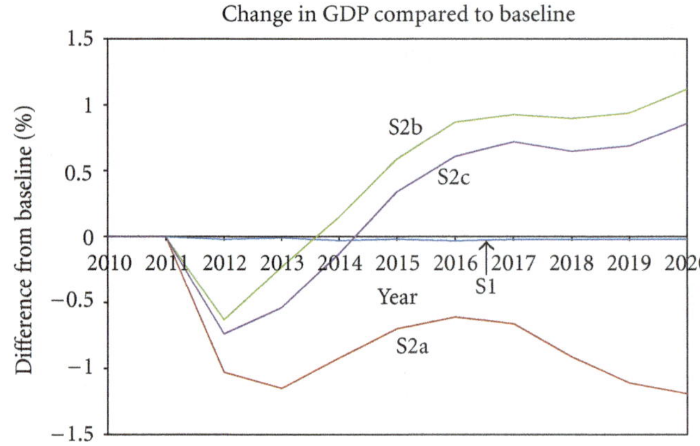

Source(s): E3MG, Cambridge Econometrics

FIGURE 5: Japan GDP.

a high carbon price is required to do this in such a short period of time; the model results suggest that electricity prices would need to increase by up to 50% and motor fuel prices increase by around 40%.

5.2. Macroeconomic Impacts. Our results suggest that the average annual revenues raised during 2012–2020 would be approximately ¥310 bn ($4 bn) in S1 and ¥11,500 bn ($150 bn) per year in the variants of S2. The latter equates to around 3% of GDP and is clearly enough to have an impact on macroeconomic indicators.

Previous model-based studies, including Andersen and Ekins [26] and Ekins and Speck [27], have found that it is possible to reduce CO_2 emissions while simultaneously increasing GDP through carefully designed ETR. The results in this paper are consistent with this finding, on the condition that the revenues generated by carbon taxes are recycled effectively. As Figure 5 shows, modest GDP gains of up to 1.2% in 2020 (compared to baseline) are possible under these scenarios. It is also notable that the potential costs to GDP in S2a are also quite small.

The reason that GDP increases in the scenarios with revenue recycling is that the positive effects of reducing income taxes outweigh the negative effects of higher energy costs. The recycled revenues (including the share allocated to investment or reducing social security contributions) sum to the revenues from the carbon taxes, so the same amount of demand is in the system. However, there is a reallocation within the system that produces economic benefits for two main reasons.

(i) As Japan imports such a large share of its energy, any measures that reduce energy consumption are likely to boost the trade balance and hence GDP.

(ii) Much of the higher energy costs fall on business which may not pass these costs on to final consumers (e.g., due to international competition); there is therefore a redistribution from companies (lower profits) to workers (higher incomes), but workers have a lower savings ratio meaning there is additional spending overall.

The second of these effects is dependent on cost pass-through rates which are largely determined by econometric estimates. Our results suggest that pass-through rates in

TABLE 4: Macroeconomic impacts, Japan.

	S1	S2a	S2b	S2c
GDP	0.0	−1.2	1.1	0.9
Employment	0.0	−0.1	0.4	0.4
H'hold consumption	0.0	−1.6	2.0	1.7
Investment	0.0	−0.6	0.9	0.7
Exports	0.0	−0.5	−0.4	−0.5
Imports	0.0	−0.3	1.1	1.1
Price level	0.1	2.5	1.4	2.0

Values are % difference from the Baseline.
Source(s): E3MG, Cambridge Econometrics.

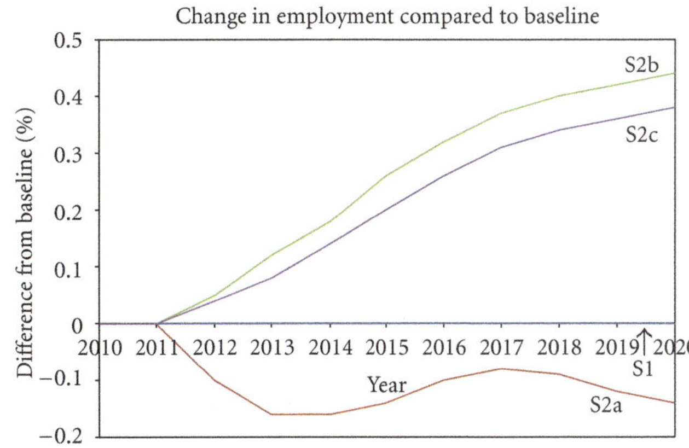

Source(s): E3MG, Cambridge Econometrics

FIGURE 6: Japan employment.

Japan are often low, which contributes to the GDP benefits. (The reasons for this are not clear but could be linked to the period of deflation. To test the impact of this we carried out a sensitivity analysis with 100% pass-through, as is common in CGE models. The results showed that GDP still increased in scenarios 2b and 2c but by less than the amount reported in this paper.) Table 4 shows the impacts on the price level and other macroeconomic indicators.

Another reason that the measures with revenue recycling have a positive impact on GDP is the fact that Japan is relatively less exposed to international trade. This means that the competitiveness effects of carbon taxes (higher domestic costs leading to higher imports and lower exports) are less, while the impacts of the revenue recycling are higher (consumers spend less of the extra income on imports). This means that, despite a worsening trade balance, increases in household expenditure lead to quite positive results to GDP overall.

Impacts on investment are quite small, except when a share of the revenues is used to fund investment programmes. The impacts on employment are smaller in scale than the impacts on GDP (see Figure 6) but generally follow the same pattern. In S2c additional jobs are created as a result of a share of the revenues being used to reduce labour costs.

Additionally, the model results enable us to compare the macroeconomic effects between different approaches to revenue recycling. The revenue recycling scenarios both cause GDP and employment to increase in comparison to the scenario without revenue recycling (S2a). S2b, which considers using revenues to both reduce income tax and increase investment in energy efficiency, has slightly greater positive effects than S2c, which also uses revenues to reduce income tax but combined with reducing employers' social security contributions.

Our model results are consistent with the "double dividend hypothesis." (This is explained in Goulder [28].) Apart from increasing welfare due to lower pollution externalities (a green dividend), environmental taxes raise revenue that can be used to lower other pre-existing tax distortions, resulting in economic welfare gains from a smaller deadweight loss of the tax system, or "efficiency" dividend.

5.3. Sectoral Impacts. As with any new policy there will be winners and losers created. Table 5 summarises the main sectoral impacts from the model results in terms of changes in real output. It is important to note that the level of sectoral detail in the model is limited by the available data and that there will be subsectors and specific firms that will be affected by more than the sectoral figures shown here.

The patterns in outcomes are as would be expected from such a modelling exercise. The energy sectors suffer a loss in demand from their output, although much of this is met by lower imports. Electricity sees the greatest reduction in output in all of the scenarios. Other sectors that stand to lose out are those that are intensive users of energy and are exposed to international competition.

The sectors that stand to benefit are typically those that are not carbon intensive and supply products to consumers (see Table 5). Within S2a, some consumer sectors, such as hotels and catering, benefit from a shift in spending away from energy, but the overall outlook is negative. As household incomes increase in S2b and S2c, the same sectors benefit from additional spending that is directed towards them. In addition, the investment sectors benefit in S2b as a result of the spending programme. Output in textiles and clothing is quite sensitive, with strong negative effects in S2a but notable positive effects in S2b and S2c. This suggests a high income elasticity of demand in the sector.

The patterns for sectoral employment are similar to the ones for sectoral output. Figure 7 also illustrates a more detailed time-series representation of the most positively and negatively affected sectors in S2a.

5.4. Comparison to Previous Results in Japan. While the modelling results are consistent to similar exercises carried out in Europe using a similar modelling approach (e.g., [26]), they should also be compared to previous analysis carried out in Japan.

In Japan, the CGE model has been the main tool to analyse the economic and environmental effects of carbon taxes. Using a CGE model, Park [1] shows double dividend effects of a carbon tax for a 20% emission reduction target in 2010, using the revenues to cut employers' social contributions (as has been done in S2c). This carbon tax

Table 5: Real output effects, selected sectors.

S1		S2a		S2b and S2c	(S2b)	(S2c)
Gas supply	−0.3	Basic metals	−2.0	Printing and publishing	6.1	5.6
Electricity	−1.0	Textiles and clothing	−2.6	Hotels and catering	5.3	5.1
		Gas supply	−6.6	Food/drink and tobacco	5.3	4.8
		Electricity	−28.5	Textiles and clothing	4.8	4.1

Notes: values are percent of difference from the baseline.
Source(s): E3MG, Cambridge Econometrics.

scenario resulted in a GDP increase between 0.16 and 0.49% in 2010 compared to baseline. Also based on CGE modelling, Park [29] introduces a 30,000 yen/tCO_2 carbon tax to achieve the Japanese Kyoto Protocol target of 6% GHG reductions in 2010 compared to 1990. The results show double dividend effects for a carbon tax with revenue recycling.

Kawase et al. [2] analysed the effects of a 3,000 yen/tCO_2 carbon tax using CGE and industrial I-O analysis. The results showed a reduction in CO_2 between 0.11 and 0.27% and an increase in GDP between 0.01 and 0.09% compared to baseline, with the revenues recycled through cuts in consumption or income taxes.

Sugawara [6] suggested that Japanese GDP will increase by 1-2% annually without introducing low-carbon policy. But, based on analysis using a macroeconometric model, this would be reduced to zero (and the unemployment rate would increase by 3 percentage points) if Japan introduced a carbon tax to meet its 25% GHG 2020 target. However, if the tax revenues are recycled to public investment, the negative impacts would be lessened.

For the main part, the studies above show results that are consistent with the analysis presented in this paper. A carbon tax with revenue recycling could have a positive impact on GDP and seems very likely to lead to reduced CO_2 emissions. But the results for achieving a 25% reduction target in this paper suggest a larger economic effect than previous studies. This partly reflects the structure of the E3MG model, with many of the assumptions that are common to CGE models being relaxed. However, it also reflects the ambitious nature of the 25% reduction target, high international energy prices (meaning that reductions in imported fuels have greater economic benefit), and the spare economic capacity that is available in the Japanese and global economies following the financial and economic crisis.

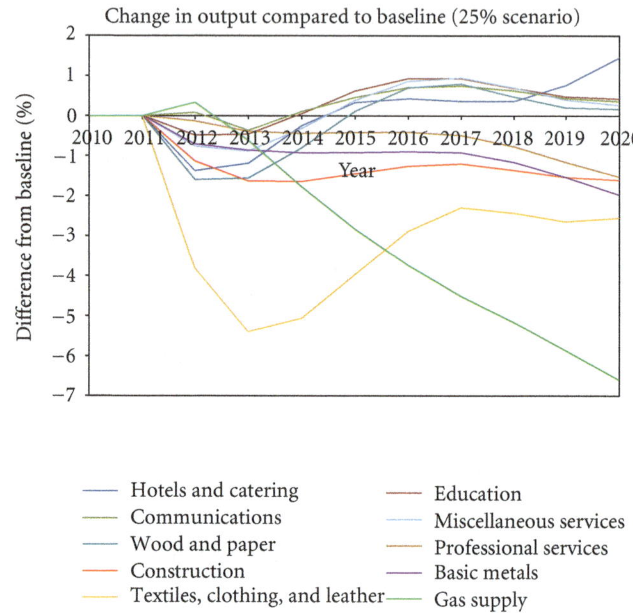

Note(s): values are based on the 5 most positively and negatively affected sectors in S2a
The graph does not include the electricity sector (−28.5% in 2020)
Source(s): E3MG, Cambridge Econometrics

Figure 7: Selected output effects under 25% reduction target.

6. Conclusions

This paper analyses the effects of carbon taxes on GHG emissions and the wider economy for the Japanese FY2012 Tax Reform Revision. It also includes three scenarios of carbon taxes that are simulated to reduce GHG emissions by 25% in 2020 compared to 1990 levels. The analysis uses the E3MG model developed by the University of Cambridge and Cambridge Econometrics.

The results from the model suggest that the FY2012 Tax Reform Revision has only a small impact on emission levels and no significant impact on GDP and employment.

The potential costs of reducing emissions by a large percentage (to meet the Copenhagen target for 2020) are quite modest, but noticeable (GDP falls by around 1.2% compared to baseline and employment by 0.4% compared to baseline). But this could be offset, possibly with some economic benefits, if revenues are recycled efficiently. The results suggest that there could be double dividend effects, if the revenues from carbon taxes are recycled efficiently.

References

[1] S. J. Park, "A CGE analysis of environmental tax reform," *Journal of Political Economy and Commercial Science (Kokumin-KeizaiZasshi)*, vol. 186, no. 2, pp. 1–16, 2002.

[2] A. Kawase, Y. Kitaura, and K. Hashimoto, "Environmental tax and the double dividend: a computational general equilibrium analysi," *Public Choice Studies (Kokyo Sentaku No Kenkyu)*, vol. 41, pp. 5–23, 2003.

[3] S. Takeda, "The double dividend from carbon regulations in Japan," *Journal of the Japanese and International Economies*, vol. 21, no. 3, pp. 336–364, 2007.

[4] T. Hertel, *Global Trade Analysis: Modeling and Applications*, Cambridge University Press, 1999.

[5] P. Capros, D. Van Regemorter, L. Paroussos, and P. Karkatsoulis, "The GEM-E3 model," IPTS Scientific and Technical Report, 2012.

[6] H. Sugawara, "Greenhouse gas emission policies and impacts on the Japanese economies," *Journal of International Business Management of Kanagawa University*, vol. 39, pp. 17–32, 2010.

[7] Committee on Institutional Design for Emissions Trading, *A Proposal for National Emissions Trading*, Ministry of the Environment, 2000.

[8] Kiko Network, *The Opinion Expressed at the Subcommittee on the National Emissions Trading System*, Kiko Network, 2010.

[9] T. Morotomi and Y. Ayukawa, *A Low-Carbon Society and Emissions Trading: Policy Mix Centering on National Emissions Trading*, Nippon Hyoronsha, Tokyo, Japan, 2007.

[10] T. Barker, H. Pan, J. Köhler, R. Warren, and S. Winne, "Avoiding dangerous climate change by inducing technological progress: scenarios using a large-scale econometric model," in *Avoiding Dangerous Climate Change*, H. J. Schellnhuber, W. Cramer, N. Nakicenovic, T. Wigley, and G. Yohe, Eds., chapter 38, Cambridge University Press, 2005.

[11] *System of National Accounts 2008*, European Communities, International Monetary Fund, Organisation for Economic Co-operation and Development, United Nations and World Bank, New York, NY, USA, 2009.

[12] T. Barker, H. Pollitt, and O. Lofsnaes, "The ETM in E3ME43," Tech. Rep., Cambridge Econometrics Working Paper, 2007.

[13] K. Lee, M. H. Pesaran, and R. G. Pierse, "Aggregation Bias in Labour Demand Equations for the UK Economy," in *Disaggregation in Econometric Modelling*, T. Barker and M. H. Pesaran, Eds., chapter 6, Routledge, 1990.

[14] R. F. Engle and C. W. J. Granger, "Cointegration and error correction: representation, estimation and testing," *Econometrica*, vol. 55, pp. 251–276, 1987.

[15] D. F. Hendry, A. Pagan, and J. D. Sargan, "Dynamic specification," in *Handbook of Econometrics*, Z. Griliches and M. D. Intriligator, Eds., vol. 2, North Holland, Amsterdam, The Netherlands, 1984.

[16] A. S. Dagoumas and T. S. Barker, "Pathways to a low-carbon economy for the UK with the macro-econometric E3MG model," *Energy Policy*, vol. 38, no. 6, pp. 3067–3077, 2010.

[17] T. Barker, A. Anger, U. Chewpreecha, and H. Pollitt, "A new economics approach to modelling policies to achieve global 2020 targets for climate stabilisation," *International Review of Applied Economics*, vol. 26, no. 2, pp. 205–221, 2012.

[18] IEA, International Energy Agency, *World Energy Outlook 2010*, OECD/IEA, Paris, Francs, 2010.

[19] European Commission, "Analysis of options to move beyond 20% greenhouse gas emission reductions and assessing the risk of carbon leakage," COM(2010) 265 final, Brussels, 2010.

[20] T. Barker, H. Pun, J. Köhler, R. Warren, and S. Winne, "Decarbonizing the global economy with induced technological change: scenarios to 2100 using E3MG," *The Energy Journal*, vol. 27, pp. 241–258, 2006.

[21] T. Barker, T. Foxon, and S. S. Scrieciu, "Achieving the G8 50% target: modelling induced and accelerated technological change using the macro-econometric model E3MG," *Climate Policy*, vol. 8, pp. S30–S45, 2008.

[22] T. Barker and S. S. Scrieciu, "Unilateral climate change mitigation, carbon leakage and competitiveness: an application to the European Union," *International Journal of Global Warming*, vol. 1, no. 4, pp. 405–417, 2009.

[23] T. Barker, A. Dagoumas, and J. Rubin, "The macroeconomic rebound effect and the world economy," *Energy Efficiency*, vol. 2, no. 4, pp. 411–427, 2009.

[24] H. Jansen and G. Klaassen, "Economic impacts of the 1997 EU energy tax: simulations with three EU-wide models," *Environmental and Resource Economics*, vol. 15, no. 2, pp. 179–197, 2000.

[25] V. Bosetti, R. Gerlagh, and S. P. Schleicher, *Modelling Sustainable Development: Transitions to a Sustainable Future*, Edward Elgar, 2009.

[26] M. S. Andersen and P. Ekins, *Carbon Energy Taxation: Lessons from Europe*, Oxford University Press, 2010.

[27] P. Ekins and S. Speck, *Environmental Tax Reform (ETR)*, Oxford University Press, 2011.

[28] L. H. Goulder, "Environmental taxation and the double dividend: a reader's guide," *International Tax and Public Finance*, vol. 2, no. 2, pp. 157–183, 1995.

[29] S. J. Park, "Double dividend of an environmental tax reform in Japan: a CGE analysis based on the 1995 input-output table," *Journal of Kyoto Sangyo University*, vol. 21, pp. 137–157, 2004.

20

Case Study on Incentive Mechanism of Energy Efficiency Retrofit in Coal-Fueled Power Plant in China

Donghai Yuan,[1,2] Xujing Guo,[3] Yuan Cao,[2] Liansheng He,[4] Jinggang Wang,[5] Beidou Xi,[4] Junqi Li,[1] Wenlin Ma,[2] and Mingshun Zhang[2]

[1] *Key Laboratory of Urban Stormwater System and Water Environment, Ministry of Education, Beijing University of Civil Engineering and Architecture, Beijing 100044, China*
[2] *Beijing Climate Change Response Research and Education Center, Beijing University of Civil Engineering and Architecture, Beijing 100044, China*
[3] *Key Laboratory of Development and Application of Rural Renewable Energy, Biogas Scientific Research Institute of the Ministry of Agriculture, Chengdu 610041, China*
[4] *Water Environment System Project Laboratory, Chinese Research Academy of Environmental Sciences, Beijing 100012, China*
[5] *Department of Environmental Science and Engineering, Beijing University of Chemical Technology, Beijing 100029, China*

Correspondence should be addressed to Donghai Yuan, yuansy79@yahoo.com.cn

Academic Editors: B. Chen, Z.-M. Chen, and H.-S. Tang

An ordinary steam turbine retrofit project is selected as a case study; through the retrofit, the project activities will generate emission reductions within the power grid for about 92,463 tCO_2e per annum. The internal rate of return (IRR) of the project is only −0.41% without the revenue of carbon credits, for example, CERs, which is much lower than the benchmark value of 8%. Only when the unit price of carbon credit reaches 125 CNY/tCO_2, the IRR could reach the benchmark and an effective carbon tax needs to increase the price of carbon to 243 CNY/tce in order to make the project financially feasible. Design of incentive mechanism will help these low efficiency enterprises improve efficiency and reduce CO_2 emissions, which can provide the power plants sufficient incentive to implement energy efficiency retrofit project in existing coal-fuel power generation-units, and we hope it will make a good demonstration for the other low efficiency coal-fueled power generation units in China.

1. Introduction

With the rapid development of industrialization and urbanization, the global climate-change issue has become an important factor to affect the world economic order, political regime and international relations, as well as to determine the key of the world's energy future [1, 2]. Global climate change is closely related to energy, a variety of greenhouse gases (GHG) that cause climate change. Carbon dioxide (CO_2) contribution rate is more than 50%, and 70% of human activities' CO_2 emissions are from the burning of fossil fuels [3]. At present, China is one of the countries with the most CO_2 emissions in the word, and CO_2 emissions are still growing rapidly. The main reasons for CO_2 emissions of China are as follow: (1) The existing energy resource characteristics of primary energy consumption are dominated by coal, and thermal power accounts for about 70% of the total power generation; (2) the rate of energy-intensive industries and products is high [4].

In the 12th Five-Year Plan, China has shown its intention to shift from a policy of maximizing growth to balance growth with social harmony and environmental sustainability [5]. But the dominant position of coal in the energy consumption will continue in a long period; by 2020, the GDP goal is quadrupling, and then the total installed capacity will reach 900–950 million kilowatts, generating capacity will reach 4.2 trillion kWh, of which thermal power installed capacity still contains about 70%. How to coordinate the relationship between the rapid development with CO_2 emissions is a severe challenge for the power industry.

Today, there are lots of low efficiency power plants in China, which is urgent to implement the steam turbine retrofit, and these power plants have the great potential to reduce CO_2 emissions. Because the development among different regions and industries in China is very uneven, the enterprises cannot afford the high cost of the steam turbine retrofit by themselves [6]. The introduction of an incentive mechanism to help these low efficiency enterprises to improve efficiency and reduce CO_2 emissions has been practically significant [7]. In this study, a steam turbine retrofit project in China is selected for a case study to find an effective GHG emission reduction mechanism for this project type, which can provide the power plants sufficient incentive to implement energy efficiency retrofit project in existing coal fuel-power generation-units, and we want to make a good demonstration for the other low efficiency coal-fuel power generation-units in China.

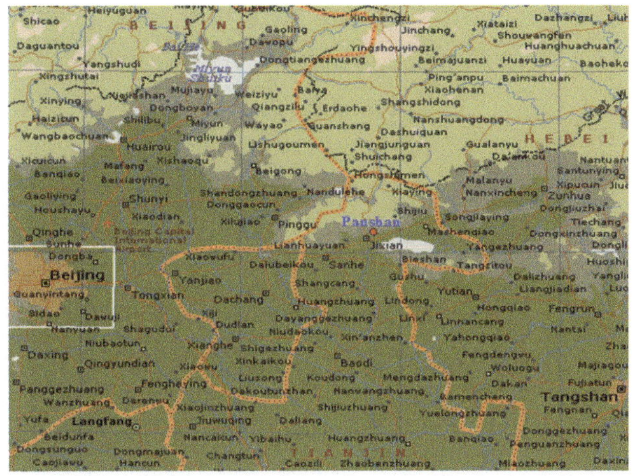

FIGURE 1: The map of the project location–Panshan Power Plant.

2. Description of the Project Activity

2.1. Site Description. Panshan Power Plant is located in the southeast of the Ji County, Tianjin City (117°16′58″E, 39°59′26″N). The mean annual temperature in Ji County is 10-11°C, and the mean annual precipitation is approximately 700 mm, of which three-quarters are distributed from July to September. Figure 1 is the location of Panshan Power Plant.

2.2. Description of the Project. Steam Turbine Retrofit Project of Tianjin Panshan Power Plant (hereafter refers to as the project) involves retrofitting supercritical steam turbine. The steam turbine with rated power of 500 MW (hereafter refers to as PAT, project activity turbine), which was designed in the early 1970s and introduced from Russia, has been put into commercial operation since April 16th 1996. The technical lifetime of the power unit is 24 years. Designed as a super critical turbine set, the technological level of the PAT is relatively higher than the subcritical turbine sets that are commonly used and regarded as a good practice currently in current China. So up to 2009 (the year of retrofit action), the remaining life is 11 years, more than 8 years. The principal specifications of the steam turbine are shown below (Table 1).

The main target of the project is retrofitting the low pressure cylinder to reduce the coal consumption of the power generation, in particular, by promoting the performance of the low pressure cylinder. The components, including rotor, blades, diaphragm and its set, inner cylinder, and shaft butt seal of the low pressure cylinder, are retrofitted. Steam seal installed in the surrounding bend of the first stage of high-pressure cylinder, steam seal of each turbine stage, and shaft butt seal of high-cylinder and medium-cylinder will also be altered (Figure 2).

The current practice with low efficiency would be continued in the absence of the proposed project. By adopting retrofit measures, the proposed project will not only reduce GHG emissions, but also contributes to sustainable development for local communities by the means of: (1) reducing the emissions of SO_2, NO_x, and coal ash due to the reduction of standard coal consumption; (2) improving the energy efficiency of power plant and promoting development of manufacturing industry. The project is expected to give a lead on turbine retrofit of supercritical power plants in China.

3. Performance Assessment

3.1. Emission Reductions. To evaluate the effect of GHG emission reduction by the project with a quantitative way, the approved CDM methodology AM0062 in Unit Nation Framework Convention on Climate Change (UNFCCC) is applied to this study on the project in the following steps: calculate the baseline emissions, the project emissions, and the emission reductions [8].

3.1.1. Calculate the Baseline Emissions. (1) Determine baseline emission for the scenario of project electricity generation. Electricity generation in the project power plant will displace in the baseline scenario less efficient electricity generation in the project plant and can, in addition, displace electricity to the grid, if the quantity of electricity generation is increased as a result of the project. The calculation of baseline emissions is therefore based on different emission factors for different quantities of electricity generated. In China the annual power generation is determined by the annual dispatch order from grid company where the PAT connects to, thus it is assumed that annual power generation dispatch will be according to the original capacity of PAT. In this case, the annual power generation after retrofit unlikely exceeds the historical average level. Therefore the emission reduction is only relevant to the efficiencies of PAT before and after the retrofit.

To evaluate the emission reduction due to the retrofit, the follow case from AM0062 [9] is selected in a future analysis: the quantity of electricity generated in the project turbine ($EG_{PJ,y}$) is lower or the same as the historic average annual generation level (EG_{AVR}). Baseline emissions are calculated as:

$$BE_y = EG_{PJ,y} * EF_{BL,y}, \tag{1}$$

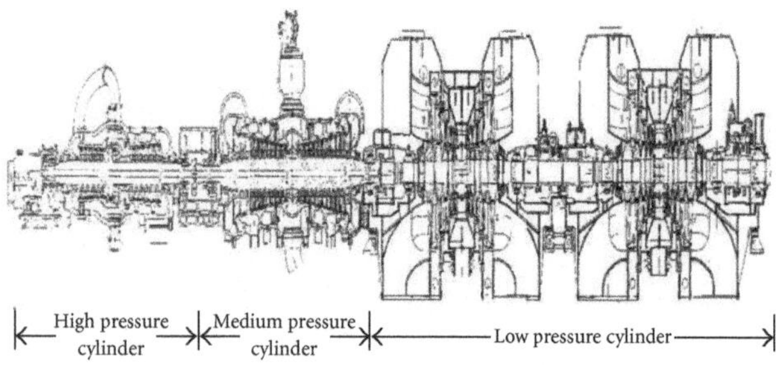

FIGURE 2: Profile of turbine.

TABLE 1: The principal specifications of the steam turbine.

Item	Designed value
Manufacturer	Leningrad metal factory of Russia (designed in the beginning of 1970s)
Type of turbine	Supercritical, once reheated, single shaft, 4 cylinder and 4 steam exhaust, condensing turbine
Rated power	500 MW
Rated main steam flow	1528.8 t/h
Main steam pressure	23.54 MPa
Main steam temperature	540°C
Reheated steam pressure	3.51 MPa
Reheated steam temperature	540°C
Exhausted steam pressure	4.27/5.44 kPa
Number of blade stage	54
Net heat rate	8146 kJ/kWh
Lifetime	24 years

where BE_y = baseline emissions in year "y" (tCO_2/yr), $EG_{PJ,y}$ = quantity of electricity supplied by the project turbine to the grid in year "y" (MWh/yr), adjusted for changes in efficiency, $EF_{BL,y}$ = baseline emission factor of the project turbine in year y (tCO_2/MWh).

(2) Determine Baseline Emission factor. The project turbine is steam turbine and the fuel is fired in a boiler, so its emission factor is calculated as follows [10]:

$$EF_{BL,y} = \frac{3.6}{1000} \times \frac{EF_{FF,BL} \times FC_{PJ,y} \times NCV_{FF,PJ}}{\eta_{BL,y} \times HI_{PJ,y}}, \quad (2)$$

where $EF_{BL,y}$ = baseline emission factor of the Project turbine in year "y" (tCO_2/MWh), $EF_{FF,BL}$ = CO_2 emission factor of the fossil fuel used in the Project turbine prior to the implementation of the Project (tCO_2/TJ), $NCV_{FF,PJ}$ = net calorific value (NCV) of fossil fuel used in the Project turbine during year y (TJ/tonne of fuel).

$\eta_{BL,y}$ = energy efficiency of the turbine without retrofitting estimated using the latest version of approved "tool to determine the baseline efficiency of thermal or electric energy generation systems," determined the efficiency based on measurements and used a conservative value, based on performance tests before the implementing the project following national/international standards, at discrete loads within the operating range or over the entire rated capacity, $FC_{PJ,y}$ = Actual fuel consumption by project in year "y" (tonne of fuel), $HI_{PJ,y}$ = Heat input to the steam turbine in year "y" (TJ). In case of multicylinder steam turbines, this is the sum of the heat input at the inlet of first stage and the heat inputs in the re-heaters of steam between various cylinders (e.g., high-pressure, medium pressure, and low-pressure cylinders).

3.1.2. Calculate the Project Emissions. The CO_2 emissions from fossil fuel consumption in the project (PE_y) should be calculated using the latest approved version of the "tool to calculate project or leakage CO_2 emissions from fossil fuel combustion," where the process j in the tool corresponds to the combustion of fossil fuels in the project for electricity generation in the project power plants [11]:

$$PE_{FC,j,y} = \sum_i FC_{i,j,y} \times COEF_{i,y}, \quad (3)$$

where $PE_{FC,j,y}$ = are the CO_2 emissions from fossil fuel combustion in process j during the year y, $FC_{i,j,y}$ = is the quantity of fuel type i combusted in process j during the year y (mass or volume unit/yr), $COEF_{i,y}$ = is the CO_2 emission coefficient of fuel type i in year y (tCO_2/mass or volume unit), i = are the fuel types combusted in process j during the year y.

The CO_2 emission coefficient $COEF_{i,y}$ can be calculated using one of two options proposed in the "tool to calculate project or leakage CO_2 emissions from fossil fuel combustion".Depending on the availability of data on the fossil fuel type i, the CO_2 emission coefficient $COEF_{i,y}$ is calculated based on net calorific value and CO_2 emission factor of the fuel type i:

$$COEF_{i,y} = NCV_{i,y} \times EF_{CO_2,i,y} \quad (4)$$

where $COEF_{i,y}$ = is the CO_2 emission coefficient of fuel type i in year y (tCO_2/mass or volume unit), $NCV_{i,y}$ = is the weighted average net calorific value of the fuel type i in year y (GJ/mass or volume unit). $EF_{CO_2,i,y}$ = is the weighted average CO_2 emission factor of fuel type i in year y (tCO_2/GJ), i = are the fuel types combusted in process j during the year y.

3.1.3. Calculate the Emission Reductions. (1) Emission reductions are calculated as follows:

$$ER_y = BE_y - PE_y, \quad (5)$$

where ER_y = emissions reductions in year y (tCO_2e/yr), BE_y = baseline emissions in year y (tCO_2e/yr), PE_y = project emissions in year y (tCO_2e/yr).

(2) Data and parameters which are available at validation. Because the regional specific value is not available, the Intergovernmental Panel on Climate Change (IPCC) default value of the CO_2 emission factor of coal was selected. Choose the CO_2 emission factor corresponding to the applicable fuel type. IPCC default values may be used. CO_2 emission factor of the fossil fuel used in the Project turbine prior to the

TABLE 2: The key parameters of the project.

Item	Value	Unit	Data source
Electricity supplied to the grid in year y ($EG_{PJ,y}$)	2,632,000	MWh	Feasibility study report of the project (FSR)
Standard coal consumption after retrofit	322.5956	kg/MWh	Efficiency test report
Net calorific value of standard coal	29,306	MJ/ton	Efficiency test report
Net calorific value of fossil coal	23,026	MJ/ton	Calculated
Annual consumption of fossil coal	1,080,643.24	ton	Calculated
Average net heat consumption of turbine after retrofit	8738.92	kJ/kWh	Efficiency test report

implementation the Project ($EF_{FF,BL}$) = 89.5 tCO_2/TJ of fuel [12].

Use the latest version of approved "tool to determine the baseline efficiency of thermal or electric energy generation systems." Depending upon the option selected from the latest version of approved "tool to determine the baseline efficiency of thermal or electric energy generation systems." Energy efficiency of the turbine without retrofitting in a year y($\eta_{BL,y}$) = 39.553% (from Efficiency Test Report of the PAT prior to the retrofit).

(3) Ex-ante calculation of emission reductions. In order to estimate the emission reductions generated by the project, the following assumptions are brought into account.

(4) Calculate the baseline emissions:

$$\begin{aligned} EF_{BL,y} &= \frac{3.6}{1000} \times \frac{EF_{FF,BL} \times FC_{PJ,y} \times NCV_{FF,PJ}}{\eta_{BL,y} \times HI_{PJ,y}} \\ &= \frac{3.6}{1000} \\ &\quad \times \frac{89.5 \times (322.5956 \times 2,632,000) \times 0.029306}{0.39553 \times 2,632,000 \times 8738.92} \\ &= \frac{0.88126\, tCO_2e}{MMh}, \\ BE_y &= EG_{PJ,y} * EF_{BL,y} = 2,632,000 * 0.88126 \\ &= 2,319,482\, tCO_2e. \end{aligned} \tag{6}$$

(5) Calculate the project emissions:

$$\begin{aligned} PE_y &= FC_{coal,y} \times NCV_{coal,y} \times EF_{CO2,coal,y} \\ &= 1,080,643.24\, \text{ton} \times 0.023026\, \text{TJ/ton} \\ &\quad \times 89.5\, tCO_2e/TJ \\ &= 2,227,019\, tCO_2e. \end{aligned} \tag{7}$$

(6) Calculate the emission reductions:

$$\begin{aligned} ER_y &= BE_y - PE_y = 2,319,482 - 2,227,019 \\ &= 92,463\, tCO_2e. \end{aligned} \tag{8}$$

3.1.4. Summary of the Ex-Ante Estimation of Emission Reductions. It is expected that the project activities will generate emission reductions within the power grid for about 92,463 tCO_2e per annum over an 8-year fixed crediting period from 01/01/2012 to 31/12/2019. And the total emission reductions will come to 739,704 tCO_2e.

3.2. Investment Analysis

3.2.1. Determine the Suitable Financial Indicator for the Project Type. Financial indicator of internal rate of return (IRR) is the most suitable for such power retrofit project and decision making context. According to Trial Implementation Methods for Economic Assessment of technology retrofit Project in Power Engineering, the benchmark of IRR of power retrofit project is set at 8% [13]. The financial attractiveness of this project will be determined by comparing the IRR after tax (without Certified Emission Reductions (CERs)) with its benchmark applied in power retrofit project of China power industry, which is a standard recommended by the industry experts and is widely used at present in China. If the IRR after tax (without CERs) is less than 8%, the project is considered not being financially attractive in the absence of CDM revenues, and is therefore considered to be additional.

3.2.2. Calculation and Comparison of Financial Indicators. The key parameters of the project are showed in Tables 2 and 3.

The financial analysis results are shown in Table 4. As shown in this table, without carbon credits, the IRR is −0.41%, which is much lower than the benchmark rate of 8%. This therefore indicates that in comparison to other alternative investments, the project without carbon credits is not financially attractive to a rational investor.

3.2.3. Sensitivity Analysis. The sensitivity analysis is conducted to check whether, under reasonable variations of the sensitive factors in the critical assumptions, the results from the analysis remain unaltered. After overall checking of the IRR calculation sheet, five factors have been selected for the sensitivity analysis, which are the total investment, annual operation hours, electricity tariff, standard coal price, and the decrease of standard coal consumption.

Assuming the five factors within a fluctuation range from −20% to 20%, the IRR (after tax) of the project (without income from selling CERs) varies to a different extent, as shown in Table 5.

As shown in the sensitivity analysis, even the varying range of the uncertain factors reaches ±20%, the IRR (after tax) could not reach the benchmark. The conclusion that the

TABLE 3: The key parameters of the proposed project.

Item	Value	Unit	Data source
Rated continuous power	500	MW	FSR
Total investment	100,000,000	RMB	FSR
Equity proportion	100	%	FSR
Incremental annual power generation	0	MWh	FSR
Profit loss due to the retrofitting	24,730,000	RMB	FSR
Annual operation hours	5600	h	FSR
Power consumption rate (for self-use)	6	%	FSR
Standard coal consumption for power generation before retrofit (designed in FSR)	316	g/KWh	FSR
Decrease of standard coal consumption after retrofit (designed in FSR)	13	g/KWh	FSR, P51
Standard coal price (without VAT)	389	RMB/ton	FSR
Electricity tariff (without VAT)	341.2	RMB/MWh	FSR
New added O and M cost	−11,675,000	RMB/year	FSR
Of which fuel cost saving due to retrofit	−14,175,000	RMB/year	FSR
New added repair cost	2,500,000	RMB/year	FSR
Income tax	25	%	FSR
Value added tax (VAT)	17	%	FSR
Town building maintenance tax	7 (of VAT)	%	FSR
Surcharge for education	3 (of VAT)	%	FSR
Project operating period	11	Years	FSR
Rate of residual value of the fixed assets	3 (out of total investment)	%	FSR
Depreciation period	11	Years	FSR
Amount of CERs	92,463	tCO_2e/year	ER calculation sheet

TABLE 4: The financial indicators of the project.

Financial indicators	Rate
IRR (after tax) without CERs	−0.41%
Benchmark	8%
IRR with CERs	8.27%

project is definitely not financially attractive would not be influenced.

As above, the internal rate of return (IRR) of the project is only −0.41% without the revenue of carbon credits for example, CERs, which is much lower than the benchmark value of 8%. In addition, only when the unit price of carbon credit comes to 125 CNY/tCO_2, the IRR could reach the benchmark and become financially feasible. Therefore the GHG emission reduction cost of such retrofit project is 125 CNY/tCO_2.

3.3. Design of Incentive Mechanism

3.3.1. Incentive from Carbon Tax. Carbon tax is a Pigovian tax levied on the carbon content of fuels [14], which is a form of carbon pricing. Carbon is present in every hydrocarbon fuel (coal, petroleum, and natural gas) and is released as CO_2, when they are burnt [15].

In this case study, it is assumed that government levies the carbon tax on coal consumption in the power sector. It means that the carbon tax will be a part of the fuel cost in operating the power plant, being of the same effect of raising the price of coal. According to the sensitive analysis, an effective carbon tax needs to increase the price of coal by 243 CNY/tce, so as to make the project financially feasible. Then the cost of carbon tax should be 90 CNY/tCO_2 (1tce leads to GHG emission of 2.77 tCO_2).

3.3.2. Incentive from Carbon Market (Credit Trading). Emissions trading or cap-and-trade is a market-based approach used to control pollution by providing economic incentives for achieving reductions in the emissions of pollutants. For GHG the largest is the European Union Emission Trading Scheme (ETS), whose purpose is to avoid dangerous climate change [16]. In the carbon market created by ETS, GHG emission reduction credits are a kind of eligible unit, generated from the project that is implemented by entity outside the ETS. As analyzed above, the price of carbon credits in the market for example, CERs must be not less than 125 CNY/tCO_2 to make the project financially attractive.

3.3.3. Incentives Combined with Encouraging Dispatch Plan. Since the electricity tariff is determined by the government in China, which is fixed unless the new tariff policy approved by the government, most of the coal fuel power plants resist

Table 5: The sensitivity analysis of the project within a fluctuation range from −20% to 20%.

Fluctuation range	−20%	−10%	0	10%	20%
Total investment	2.34%	0.87%	−0.41%	−1.55%	−2.57%
Annual operation hours	−3.64%	−1.98%	−0.41%	1.08%	2.50%
Electricity tariff	−0.41%	−0.41%	−0.41%	−0.41%	−0.41%
Standard coal price	−3.64%	−1.98%	−0.41%	1.08%	2.50%
Decrease of standard coal consumption	−3.64%	−1.98%	−0.41%	1.08%	2.50%

Table 6: The necessary cost of carbon tax and credit price with/without encouraging dispatch plan.

Dispatch plan	Necessary cost of Carbon tax	Necessary cost of credit price
The dispatch electricity generation does not change	90 CNY/tCO_2	125 CNY/tCO_2
Dispatched electricity generation increased by 20%	50 CNY/tCO_2	85 CNY/tCO_2

carbon tax in practice. A very high cost of carbon tax is not very likely to occur in China. On the other hand, given the experience of the EU ETS, the credit price is of high volatility in the market [17]. Thus, there is always uncertainty about the credit price to make the investment decisions of this kind of retrofit. To mitigate the shortage of carbon tax and carbon market, the encouraging dispatch plan is expected to enhance the effect from above mechanisms.

Given the sensitivity analysis, the annual operation hours of the PAT are a very crucial factor to the IRR. The annual operation hours of a power plant are normally determined by power dispatch arrangement of grid company. In this section, the effect of encouraging dispatch plan is an analysis together with carbon tax and credit price in an ETS. Clearly, the more annual operation hours of the PAT, the higher IRR appears in the Project (Table 6). Thus, the necessary cost of either carbon tax and credit cost in ETS will decrease by incorporating with such encourage to dispatch plan.

4. Conclusions

This study presented the emission reductions of an ordinary thermal power plant after a steam turbine retrofit project with an incentive mechanism of carbon tax and carbon market (ETS). In particular, the most cost-efficient method is the combination of this mechanism made by central authorities with encouraging dispatch plan by grid company. The results presented within this paper indicate that the project will make a good demonstration for the other low efficiency thermal power plants in China.

Acknowledgments

This work was financially supported by the National Natural Science Foundation of China (nos. 50908219 and 51209003), the Major Special Technological Program of Water Pollution Control and Management (no. 2010ZX07320-002), and Central Public-interest Scientific Institution Basal Research Fund (no. 0032012013).

References

[1] T. M. L. Wigley, R. Richels, and J. A. Edmonds, "Economic and environmental choices in the stabilization of atmospheric CO_2 concentrations," *Nature*, vol. 379, no. 6562, pp. 240–243, 1996.

[2] G. J. Boer, G. Flato, and D. Ramsden, "A transient climate change simulation with greenhouse gas and aerosol forcing: projected climate to the twenty-first century," *Climate Dynamics*, vol. 16, no. 6, pp. 427–450, 2000.

[3] Z. Z. Hu, S. Yang, and R. Wu, "Long-term climate variations in China and global warming signals," *Journal of Geophysical Research D*, vol. 108, no. 10, pp. 4614–4626, 2003.

[4] B. Bai, X. Li, Y. Liu, and Y. Zhang, "Preliminary study on CO_2 industrial point sources and their distribution in China," *Chinese Journal of Rock Mechanics and Engineering*, vol. 25, no. 2, pp. 2918–2923, 2006.

[5] The Climate Group, "Delivering low carbon growth: a guide to China's 12th Five Year Plan," 2011.

[6] M. I. Hoffert, K. Caldeira, G. Benford et al., "Engineering: advanced technology paths to global climate stability: energy for a greenhouse planet," *Science*, vol. 298, no. 5595, pp. 981–987, 2002.

[7] D. Mollicone, F. Achard, S. Federici et al., "An incentive mechanism for reducing emissions from conversion of intact and non-intact forests," *Climatic Change*, vol. 83, no. 4, pp. 477–493, 2007.

[8] UNFCCC/CCNUCC, "Approved baseline and monitoring methodology AM0062: energy efficiency improvements of a power plant through retrofitting turbines," AM0062/Version 02, Sectoral Scope: 01, EB 55.

[9] UNFCCC/CCNUCC, "Methodological tool "combined tool to identify the baseline scenario and demonstrate additionality" (Version 04.0.0)," EB 66, Report Annex 48.

[10] UNFCCC/CCNUCC, "Methodological tool "tool to calculate the emission factor for an electricity system" (Version 02.2.1)," EB 63, Report Annex 19.

[11] UNFCCC/CCNUCC, "Methodological tool "tool to calculate project or leakage CO_2 emissions from fossil fuel combustion" (Version 02)," EB 41, Report Annex 11.

[12] UNFCCC/CCNUCC, "Methodological tool "tool to calculate the emission factor for an electricity system" (Version 02.2.1)," EB 63, Report Annex 19.

[13] State Power Corporation, "Trial Implementation Methods for Economic Assessment of technology retrofit Project in Power Engineering," 2003.

[14] P. Hoeller and M. Wallin, "Energy prices, taxes and carbon dioxide emissions," OECD Economics and Statistics Department Working Papers 106, Organisation for Economic Co-operation and Development, Paris, Farnce, 1991.
[15] A. Staudt, *Understanding and Responding to Global Climate Change: Highlights of National Academies Reports*, National Academy of Sciences, 2008.
[16] EU Emissions Trading System (EU ETS), "UK Department of Energy and Climate Change," 2009.
[17] R. S. J. Tol, "The marginal damage costs of carbon dioxide emissions: an assessment of the uncertainties," *Energy Policy*, vol. 33, no. 16, pp. 2064–2074, 2005.

Permissions

The contributors of this book come from diverse backgrounds, making this book a truly international effort. This book will bring forth new frontiers with its revolutionizing research information and detailed analysis of the nascent developments around the world.

We would like to thank all the contributing authors for lending their expertise to make the book truly unique. They have played a crucial role in the development of this book. Without their invaluable contributions this book wouldn't have been possible. They have made vital efforts to compile up to date information on the varied aspects of this subject to make this book a valuable addition to the collection of many professionals and students.

This book was conceptualized with the vision of imparting up-to-date information and advanced data in this field. To ensure the same, a matchless editorial board was set up. Every individual on the board went through rigorous rounds of assessment to prove their worth. After which they invested a large part of their time researching and compiling the most relevant data for our readers. Conferences and sessions were held from time to time between the editorial board and the contributing authors to present the data in the most comprehensible form. The editorial team has worked tirelessly to provide valuable and valid information to help people across the globe.

Every chapter published in this book has been scrutinized by our experts. Their significance has been extensively debated. The topics covered herein carry significant findings which will fuel the growth of the discipline. They may even be implemented as practical applications or may be referred to as a beginning point for another development. Chapters in this book were first published by Hindawi Publishing Corporation; hereby published with permission under the Creative Commons Attribution License or equivalent.

The editorial board has been involved in producing this book since its inception. They have spent rigorous hours researching and exploring the diverse topics which have resulted in the successful publishing of this book. They have passed on their knowledge of decades through this book. To expedite this challenging task, the publisher supported the team at every step. A small team of assistant editors was also appointed to further simplify the editing procedure and attain best results for the readers.

Our editorial team has been hand-picked from every corner of the world. Their multi-ethnicity adds dynamic inputs to the discussions which result in innovative outcomes. These outcomes are then further discussed with the researchers and contributors who give their valuable feedback and opinion regarding the same. The feedback is then collaborated with the researches and they are edited in a comprehensive manner to aid the understanding of the subject.

Apart from the editorial board, the designing team has also invested a significant amount of their time in understanding the subject and creating the most relevant covers. They scrutinized every image to scout for the most suitable representation of the subject and create an appropriate cover for the book.

The publishing team has been involved in this book since its early stages. They were actively engaged in every process, be it collecting the data, connecting with the contributors or procuring relevant information. The team has been an ardent support to the editorial, designing and production team. Their endless efforts to recruit the best for this project, has resulted in the accomplishment of this book. They are a veteran in the field of academics and their pool of knowledge is as vast as their experience in printing. Their expertise and guidance has proved useful at every step. Their uncompromising quality standards have made this book an exceptional effort. Their encouragement from time to time has been an inspiration for everyone.

The publisher and the editorial board hope that this book will prove to be a valuable piece of knowledge for researchers, students, practitioners and scholars across the globe.

List of Contributors

Wen-Sheng Gao and Cui-fen Bai
Department of Electrical Engineering, Tsinghua University, Beijing 100084, China

Tong Liu
Electric Power Research Institute, CSG, Guangzhou, Guangdong 510080, China

Rong-Gang Cong
School of Economics and Management, North China Electric Power University, Beijing 102206, China
Centre for Environmental and Climate Research (CEC), Lund University, 223 62 Lund, Sweden

Shaochuan Shen
State Key Laboratory Breeding Base of Green Chemistry Synthesis Technology, College of Chemical Engineering and Materials Science, Zhejiang University of Technology, Hangzhou 310014, China

Wenqiang Sun, Jiuju Cai and Zhu Ye
State Environmental Protection Key Laboratory of Eco-Industry, Institute of Thermal and Environmental Engineering, Northeastern University, Shenyang, Liaoning 118019, China

Yi Lu, Meirong Su and Gengyuan Liu
State Key Joint Laboratory of Environmental Simulation and Pollution Control, School of Environment, Beijing Normal University, Beijing 100875, China

Bin Chen
State Key Joint Laboratory of Environmental Simulation and Pollution Control, School of Environment, Beijing Normal University, Beijing 100875, China
Beijing Development Area Co. Ltd, Beijing 100176, China

Shiyi Zhou and Meiming Jiang
Beijing Development Area Co. Ltd, Beijing 100176, China

Yi Huang
Sustainable Development Research Center, Guangzhou Institute of Geochemistry, Chinese Academy of Sciences, Guangzhou 510640, China
University of Chinese Academy of Sciences, Beijing 100049, China

Bin Xia
Sustainable Development Research Center, Guangzhou Institute of Geochemistry, Chinese Academy of Sciences, Guangzhou 510640, China
School of Marine Sciences, Sun Yat-Sen University, Guangzhou 510006, China

Lei Yang
College of Resources and Environment, Guangdong University of Business Studies, Guangzhou 510320, China

G. Shi
Department of Techno-Economic Research, Development Research Center of the State Council, Beijing 10010, China

X. Zheng and F. Song
School of Economics, Renmin University of China, Beijing 100872, China

Yang Wang, Cong Lin, Na Duan, Xue Li and Yanyan Fu
College of Water Resources and Civil Engineering, China Agricultural University, Beijing 100083, China

Jing Li
China Zhongyuan Engineering Corporation, Beijing 100191, China

Emily J. Shumchenia, Robert D. Kenney, John W. King and Malia Schwartz
Graduate School of Oceanography, University of Rhode Island, South Ferry Road, Narragansett, RI 02882, USA

Sarah L. Smith, Jennifer McCann and Michelle Carnevale
Coastal Resources Center and Rhode Island Sea Grant, University of Rhode Island, Narragansett, RI 02882, USA

Grover Fugate
Rhode Island Coastal Resources Management Council, South Kingstown, RI 02879, USA

Peter Paton and Kristopher J. Winiarski
College of the Environmental and Life Sciences, University of Rhode Island, Kingston, RI 02881, USA

Malcolm Spaulding
College of Engineering, University of Rhode Island, Narragansett, RI 02882, USA

Sami D. Salman
Department of Chemical and Process Engineering, Faculty of Engineering and Built Environment, Universiti Kebangsaan Malaysia, 43600 Bangi, Selangor, Malaysia
Biochemical Engineering Department, Al-Khwarizmi College of Engineering, University of Baghdad, Baghdad 47024, Iraq

Abdul Amir H. Kadhum, Mohd S. Takriff and Abu Bakar Mohamad
Department of Chemical and Process Engineering, Faculty of Engineering and Built Environment, Universiti Kebangsaan Malaysia, 43600 Bangi, Selangor, Malaysia

X. H. Xia
Institute of China's Economic Reform and Development, Renmin University of China, Beijing 100872, China

Yi Hu
Academy of Mathematics and Systems Science, Chinese Academy of Sciences, Beijing 100190, China

Zhibin Wang, Yang Hao, Xian Liu, Qian Zhang and Dandan Zhu
College of Electrical Engineering, Yanshan University, Qinhuangdao 066004, China

Zhongdong Wang
Mechanical Science and Engineering Institute, Northeast Petroleum University, Daqing 163318, China

Yingying Wu, Peng Zhao, Hongwei Zhang, Yuan Wang and Guozhu Mao
School of Environmental Science and Engineering, Tianjin University, Tianjin 300072, China

S. Amos-Abanyie and V. Kutin-Sanwu
College of Architecture and Planning, Kwame Nkrumah University of Science and Technology, Kumasi, Ghana

F. O. Akuffo
College of Engineering, Kwame Nkrumah University of Science and Technology, Kumasi, Ghana

Sheng Wang
Chongqing Academy of Social Sciences, Chongqing 400020, China

Jing Dai and Meirong Su
State Key Joint Laboratory of Environmental Simulation and Pollution Control, School of Environment, Beijing Normal University, Beijing 100875, China

Pascale Massot
Department of Political Science,The University of British Columbia, Vancouver, BC, Canada V6T 1Z1

Zhan-Ming Chen
School of Economics, Renmin University of China, Beijing 100872, China

Bo Zhang
School of Management, China University of Mining and Technology (Beijing), Beijing 100083, China
State Key Laboratory of Coal Resources and Safe Mining, China University of Mining and Technology (Beijing), Beijing 100083, China

Beihua Peng
Shenhua International Trading Co. Ltd., Beijing 100011, China

Mingchu Liu
Pingxiang College, Jiangxi Province, Pingxiang 337000, China

Gokhan Omeroglu
Department of Mechanical Engineering, Faculty of Engineering, Bayburt University, 69000 Bayburt, Turkey

Omer Comakli and Sendogan Karagoz
Department of Mechanical Engineering, Faculty of Engineering, Atatürk University, 25240 Erzurum, Turkey

Bayram Sahin
Department of Mechanical Engineering, Faculty of Engineering and Architecture, Technical University of Erzurum, Erzurum, Turkey

Bin Chen, Guoxuan He, Jing Qi and Meirong Su
State Key Joint Laboratory of Environmental Simulation and Pollution Control, School of Environment, Beijing Normal University, Beijing 100875, China

Shiyi Zhou and Meiming Jiang
Beijing Development Area Co., Ltd., Beijing 100176, China

Soocheol Lee
Faculty of Economics, Meijo University, 1-501 Shiogamaguchi Tenpak-ku, Nagoya 468-8502, Japan

Hector Pollitt
International Modelling, Cambridge Econometrics, Covent Garden, Cambridge CB1 2HT, UK

Kazuhiro Ueta
Graduate School of Economics, Kyoto University, Yoshida-Honmachi Sakyo-ku, Kyoto 606-8501, Japan

Donghai Yuan
Key Laboratory of Urban Stormwater System and Water Environment, Ministry of Education, Beijing University of Civil Engineering and Architecture, Beijing 100044, China
Beijing Climate Change Response Research and Education Center, Beijing University of Civil Engineering and Architecture, Beijing 100044, China

Xujing Guo
Key Laboratory of Development and Application of Rural Renewable Energy, Biogas Scientific Research Institute of the Ministry of Agriculture, Chengdu 610041, China

Yuan Cao, Wenlin Ma and Mingshun Zhang
Beijing Climate Change Response Research and Education Center, Beijing University of Civil Engineering and Architecture, Beijing 100044, China

Liansheng He and Beidou Xi
Water Environment System Project Laboratory, Chinese Research Academy of Environmental Sciences, Beijing 100012, China

Jinggang Wang
Department of Environmental Science and Engineering, Beijing University of Chemical Technology, Beijing 100029, China

Junqi Li
Key Laboratory of Urban Stormwater System and Water Environment, Ministry of Education, Beijing University of Civil Engineering and Architecture, Beijing 100044, China